AF389361

Infection in Honey Bees: Host–Pathogen Interaction and Spillover

Infection in Honey Bees: Host–Pathogen Interaction and Spillover

Editor

Giovanni Cilia

MDPI • Basel • Beijing • Wuhan • Barcelona • Belgrade • Manchester • Tokyo • Cluj • Tianjin

Editor
Giovanni Cilia
CREA Research Centre for
Agriculture and Environment
Italy

Editorial Office
MDPI
St. Alban-Anlage 66
4052 Basel, Switzerland

This is a reprint of articles from the Special Issue published online in the open access journal *Pathogens* (ISSN 2076-0817) (available at: https://www.mdpi.com/journal/pathogens/special_issues/Infection_Honey_Bees).

For citation purposes, cite each article independently as indicated on the article page online and as indicated below:

LastName, A.A.; LastName, B.B.; LastName, C.C. Article Title. *Journal Name* **Year**, *Volume Number*, Page Range.

ISBN 978-3-0365-2962-2 (Hbk)
ISBN 978-3-0365-2963-9 (PDF)

Cover image courtesy of Giovanni Cilia

Contents

About the Editor

Giovanni Cilia graduated cum laude with a Veterinary Biotechnological Sciences master's degree at the University of Milan (Italy) in 2016. He obtained a Ph.D. in Veterinary Sciences from the Department of Veterinary Sciences at the University of Pisa. Since 2021, he has been working at CREA Research Centre for Agriculture and Environment in Bologna, working on honey bee pathogens and molecular biology. He performs support teaching and seminar activities for the "Infectious Bacterial Diseases", "Microbiology and Immunology", "Microbiology and Biotechnology Applied to Animal Production", and "Veterinary Hygiene" courses in the Department of Veterinary Sciences at the University of Pisa. During this time, he applied for an internship at a different research institutes, including the National Reference Center for Leptospirosis of Institute Pasteur (Paris, France). At present, his research topics aim toward the research of bees, with particular attention to honey bee infectious diseases (viruses, bacteria and fungi), insect molecular biology and wild bees. He is the author/co-author of several scientific publications in international journals, the Editor of some infectious disease Special Issues, and the Topic Editor of "Veterinary Sciences".

Editorial

Special Issue: "Infection in Honey Bees: Host–Pathogen Interaction and Spillover"

Giovanni Cilia

CREA Research Centre for Agriculture and Environment, Via di Saliceto 80, 40128 Bologna, Italy; giovanni.cilia@crea.gov.it

Citation: Cilia, G. Special Issue: "Infection in Honey Bees: Host–Pathogen Interaction and Spillover". *Pathogens* **2022**, *11*, 77. https://doi.org/10.3390/pathogens11010077

Received: 23 December 2021
Accepted: 6 January 2022
Published: 8 January 2022

Publisher's Note: MDPI stays neutral with regard to jurisdictional claims in published maps and institutional affiliations.

Honey bee health is a very important topic that has recently raised the interest of researchers. In addition, the health of honey bees is strictly related to ecosystem health; therefore, honey bee species act as reservoirs for several pathogens widely spread and able to infect wild pollinators, contributing to their decline. The basis of this Special Issue is to contribute to the knowledge on host–pathogen interactions, in honey bees as well as wild bees.

To explore all possible features of the dynamics of honey bee pathogens, this Special Issue entitled "Infection in Honey Bees: Host–Pathogen Interaction and Spillover" aimed to explore a series of research articles focused on different aspects of honey bee pathogens and their interaction with the hosts. The published papers highlighted this theme at different levels—namely, considering any aspect of the host–pathogen interaction, focusing on different aspects of pathologies, or giving particular attention also to the spillover. All 13 Published articles explored this theme and emphasized the importance of this issue.

Chang et al. compared the genomic sequencing of Sacbrood virus (SBV) strains from the Asian honey bee, *Apis cerana*, and European honey bee, *Apis mellifera*, in Taiwan. Each viral genome encoded a polyprotein, which consisted of 2841 aa in *A. cerana* and 2859 aa in *A. mellifera*, and these sequences shared 95% identity. Compared with the other 54 SBV sequences, the structural protein and protease regions showed high variation, while the helicase region was the most highly conserved. Moreover, 17 amino acids resulted deleted in the viral protein 1 (VP1) region of *A. cerana*, compared with *A. mellifera*. The amino acid difference in the VP1 region might serve as a molecular marker for describing SBV cross-infection [1].

The effects of the application of the commercialized herbal supplements NOZEMAT HERB® and NOZEMAT HERB PLUS® for treating *Nosema ceranae* infection were investigated in 45 selected honey bee colonies. The obtained results reveal that both herbal supplements showed statistically significant activity against *N. ceranae* in infected apiaries. The results suggest a new approach as an alternative therapy to control nosemosis, even if the mechanism of their action is still not elucidated [2].

Nanetti et al. assessed the presence of *Lotmaria passim, Crithidia mellificae*, and replicative forms of deformed wing virus (DWV) and Kashmir bee virus (KBV) in *Aethina tumida*, using specimens collected from *A. mellifera* colonies in Gainesville (Florida, USA), in summer 2017. The replicative forms of KBV have not previously been reported. The results provide evidence of pathogen spillover between managed honey bees and small hive beetles, and these dynamics require further investigation [3].

Ptaszyńska et al. analyzed data collected from honey bees at various time points from anthropogenic landscapes in relation to amplicon sequencing of the *16S rRNA* from bacteria and ITS2 regions of fungi and plants. The differences found between samples were mainly influenced by the bacteria, plant pollen, and fungi. Additionally, honey bees fed with a sugar-based diet were more susceptible to *N. ceranae* and neogregarines, even in cases of co-infection. Healthy honey bees had a higher load of plant pollen and several bacterial groups. Finally, the period when honey bees switch to winter generation is the most sensitive to diet perturbations, and hence pathogen attack, for the entire beekeeping season. Evolutionary adaptation of bees may fail to benefit them in modern anthropomorphized environments [4].

Dechatre et al. proposed models for the prediction of *Varroa destructor* infestation in *Apis mellifera*. The models are based on easy and rapid use of measurable data—namely, phoretic *Varroa* load and capped brood cell numbers. Using these models, beekeepers will be able to either evaluate the risks and benefits of treating *Varroa* or anticipate the reduction in colony performance due to mites during the beekeeping season [5].

Naree et al. evaluated the effects of propolis extract of stingless bee *Tetrigona apicalis* and chitooligosaccharides (COS) on *N. ceranae* infection in giant honey bees *Apis dorsata*. In the infected bees, propolis extracts and COS caused a significant increase in trehalose levels in hemolymph, protein contents, survival rates, and acini diameters of the hypopharyngeal glands. All these changes suggest that both natural compounds could improve the health of infected honey bees [6]

Honey bee virus infections were studied in wild bumblebees, in Croatia. Acute bee paralysis virus (ABPV), black queen cell virus (BQCV), chronic bee paralysis virus (CBPV), and DWV were found in the investigated specimens. BQCV reported a higher prevalence, followed by DWV, ABPV, and CBPV, respectively. Moreover, BQCV and DWV strains showed a high similarity of 95.7 % and 98.09% nucleotide identity, respectively, with previously identified honey bees in Croatia and Slovenia, providing insights into highly diverse strains circulating in wild bees [7].

Power et al. investigated the histopathological features of 25 symptomatic and asymptomatic honey bees naturally infected with DWV. The results showed degenerative alterations of hypopharyngeal glands (in 19 specimens) and flight muscles (in 6 specimens) in symptomatic samples, while in 4 asymptomatic samples, evidence revealed an inflammatory response in the midgut and hemocele. All these findings suggested a possible pathogenic action of DWV in both symptomatic and asymptomatic honey bees, improving their immune response by keeping the virus under control in asymptomatic honey bees [8].

In 2017 and 2018, clinically healthy workers of bumblebees and honey bees were collected on flowers in four different areas of Slovenia to assess the spillover of honey bee pathogens. The results evinced a prevalence of 58.5% for BQCV, 24.5% for SBV, 17.0% for *Crithidia bombi*, 16.3% for *Nosema bombi*, 15.6% for Lake Sinai virus (LSV), 15.0% on *Apicystis bombi*, 8.8% for ABPV, 8.2% on *N. ceranae*, and 6.8% for DWV. The study confirmed that several pathogens are regularly detected in both bumblebees, suggesting important spillover events [9].

Cappa et al. highlighted the association between bee decline and the type of land surrounding the apiary. The authors developed a risk map to identify the areas with the highest risk of bee decline in Lombardy. The apiaries were considered "declined" if they reported at least one event of decline or tested positive for plant protection products, while the apiaries were "not declined" if they did not report any events of bee decline during the study period. Out of 14,188 apiaries analyzed, 80 were considered declined. Furthermore, the risk maps highlighted that the probability of apiary deterioration increases by 10% in orchards and 2% in arable lands. This information can be used by Italian Veterinary Services as a predictive measure for planning prevention and control activities [10].

Braglia et al. investigated the control of *N. ceranae* by using several compounds. The results showed that some of the ingredients administered, such as acetic acid at high concentration, p-coumaric acid, and *Saccharomyces* sp. strain KIA1, were effective in the control of nosemosis. On the other hand, wine acetic acid strongly increased the *N. ceranae* amount. The effects of all tested compounds can be investigated in more detail, especially to improve honey bee health [11].

Alonso-Prados et al. investigated the possible underlying causes behind the poor health of a professional *A. mellifera iberiensis* apiary located in Gajanejos (Guadalajara, Spain). The case report highlighted several factors that potentially favor colony collapse, including pathogen infections and accumulation in the beebread of coumaphos and tau-fluvalinate (acaricides commonly used to control *Varroa destructor*). The high level of acaricides and unusual climatic conditions of the year suggested a possible increase in vulnerability to infection by *N. ceranae* and the consequentially collapse events. This case report highlighted the importance of evaluating all possible factors in future monitoring programs, to adopt adequate preventive measures aimed to guarantee the health and fitness of bees [12].

Finally, in a systematic review, the pathogen spillover from honey bees to other arthropods was analyzed by Nanetti et al. The systematic review amassed and summarized spillover cases having in common *Apis mellifera* as the maintenance host and some of its pathogens. The collected data were grouped by final host species and condition, year, and geographic area of detection and the co-occurrence in the same host. In total, 81 articles in the time frame of 1960–2021 were analyzed. The reported spillover cases were evaluated in a wide range of hymenopteran species, generally sharing the same environment with the honey bees. Moreover, the honey bee pathogens are able to infect non-hymenopteran arthropods, such as spiders and roaches, which are either likely or unlikely to live near honey bees. The plasticity of bee pathogens and ecological consequences of spillover necessitate an approach that emphasizes bee health as well as the health of the ecosystem, fully implementing a One-Health outlook [13].

Funding: This research received no external funding.

Conflicts of Interest: The authors declare no conflict of interest.

References

1. Chang, J.-C.; Chang, Z.-T.; Ko, C.-Y.; Chen, Y.-W.; Nai, Y.-S. Genomic Sequencing and Comparison of Sacbrood Viruses from Apis cerana and Apis mellifera in Taiwan. *Pathogens* **2020**, *10*, 14. [CrossRef] [PubMed]
2. Shumkova, R.; Balkanska, R.; Hristov, P. The Herbal Supplements NOZEMAT HERB® and NOZEMAT HERB PLUS®: An Alternative Therapy for N. ceranae Infection and Its Effects on Honey Bee Strength and Production Traits. *Pathogens* **2021**, *10*, 234. [CrossRef] [PubMed]
3. Nanetti, A.; Ellis, J.D.; Cardaio, I.; Cilia, G. Detection of Lotmaria passim, Crithidia mellificae and Replicative Forms of Deformed Wing Virus and Kashmir Bee Virus in the Small Hive Beetle (*Aethina tumida*). *Pathogens* **2021**, *10*, 372. [CrossRef] [PubMed]
4. Ptaszyńska, A.A.; Latoch, P.; Hurd, P.J.; Polaszek, A.; Michalska-Madej, J.; Grochowalski, Ł.; Strapagiel, D.; Gnat, S.; Załuski, D.; Gancarz, M.; et al. Amplicon Sequencing of Variable 16S rRNA from Bacteria and ITS2 Regions from Fungi and Plants, Reveals Honeybee Susceptibility to Diseases Results from Their Forage Availability under Anthropogenic Landscapes. *Pathogens* **2021**, *10*, 381. [CrossRef] [PubMed]
5. Dechatre, H.; Michel, L.; Soubeyrand, S.; Maisonnasse, A.; Moreau, P.; Poquet, Y.; Pioz, M.; Vidau, C.; Basso, B.; Mondet, F.; et al. To Treat or Not to Treat Bees? Handy VarLoad: A Predictive Model for Varroa destructor Load. *Pathogens* **2021**, *10*, 678. [CrossRef] [PubMed]
6. Naree, S.; Ponkit, R.; Chotiaroonrat, E.; Mayack, C.L.; Suwannapong, G. Propolis Extract and Chitosan Improve Health of Nosema ceranae Infected Giant Honey Bees, Apis dorsata Fabricius, 1793. *Pathogens* **2021**, *10*, 785. [CrossRef] [PubMed]
7. Gajger, I.T.; Šimenc, L.; Toplak, I. The First Detection and Genetic Characterization of Four Different Honeybee Viruses in Wild Bumblebees from Croatia. *Pathogens* **2021**, *10*, 808. [CrossRef] [PubMed]
8. Power, K.; Martano, M.; Altamura, G.; Piscopo, N.; Maiolino, P. Histopathological Features of Symptomatic and Asymptomatic Honeybees Naturally Infected by Deformed Wing Virus. *Pathogens* **2021**, *10*, 874. [CrossRef] [PubMed]
9. Ocepek, M.P.; Toplak, I.; Zajc, U.; Bevk, D. The Pathogens Spillover and Incidence Correlation in Bumblebees and Honeybees in Slovenia. *Pathogens* **2021**, *10*, 884. [CrossRef] [PubMed]
10. Cappa, V.; Cerioli, M.P.; Scaburri, A.; Tironi, M.; Farioli, M.; Nassuato, C.; Bellini, S. Analysis of Bee Population Decline in Lombardy during the Period 2014–2016 and Identification of High-Risk Areas. *Pathogens* **2021**, *10*, 1004. [CrossRef] [PubMed]
11. Braglia, C.; Alberoni, D.; Porrini, M.P.; Garrido, M.P.; Baffoni, L.; Di Gioia, D. Screening of Dietary Ingredients against the Honey Bee Parasite Nosema ceranae. *Pathogens* **2021**, *10*, 1117. [CrossRef] [PubMed]
12. Alonso-Prados, E.; González-Porto, A.V.; Bernal, J.L.; Bernal, J.; Martín-Hernández, R.; Higes, M. A Case Report of Chronic Stress in Honey Bee Colonies Induced by Pathogens and Acaricide Residues. *Pathogens* **2021**, *10*, 955. [CrossRef] [PubMed]
13. Nanetti, A.; Bortolotti, L.; Cilia, G. Pathogens Spillover from Honey Bees to Other Arthropods. *Pathogens* **2021**, *10*, 1044. [CrossRef] [PubMed]

Article

Analysis of Bee Population Decline in Lombardy during the Period 2014–2016 and Identification of High-Risk Areas

Veronica Cappa [1,*], Monica Pierangela Cerioli [1], Alessandra Scaburri [1], Marco Tironi [1], Marco Farioli [2], Claudia Nassuato [2] and Silvia Bellini [1]

1   Istituto Zooprofilattico Sperimentale della Lombardia ed Emilia-Romagna, Via Bianchi 9, 25124 Brescia, Italy; monicapierangela.cerioli@izsler.it (M.P.C.); alessandra.scaburri@izsler.it (A.S.); marco.tironi@izsler.it (M.T.); silvia.bellini@izsler.it (S.B.)
2   Direzione Generale Welfare di Regione Lombardia, Unità Organizzativa Veterinaria, Piazza Città di Lombardia, 20124 Milan, Italy; Marco_Farioli@regione.lombardia.it (M.F.); claudia_nassuato@regione.lombardia.it (C.N.)
*   Correspondence: veronica.cappa@izsler.it; Tel.: +39-030-2290635

Abstract: The first events of bee decline in Italy were reported during 1999. Since then, population decline has frequently been reported in Lombardy. In this study, the association between bee decline and the type of land surrounding the apiary was evaluated. A risk map was developed to identify areas with the highest risk of decline. Apiaries in Lombardy were selected from the national beekeeping database (BDA). The study period was from 2014 to 2016. Apiaries were deemed "declined" if they reported at least one event of decline or tested positive for plant protection products; apiaries were "not declined" if they did not report any events of bee decline during the study period. Out of 14,188 apiaries extracted from the BDA, 80 were considered declined. The probability of an apiary being declined increases by 10% in orchards and by 2% in arable land for each additional km^2 of land occupied by these crops. The study showed an association between bee decline and the type of territory surrounding the apiaries, and the areas at the greatest risk of decline in Lombardy were identified. This information can be used by Veterinary Services as a predictive parameter for planning prevention and control activities.

Keywords: bees; population decline; plant protection; soil; risk; GIS

Citation: Cappa, V.; Cerioli, M.P.; Scaburri, A.; Tironi, M.; Farioli, M.; Nassuato, C.; Bellini, S. Analysis of Bee Population Decline in Lombardy during the Period 2014–2016 and Identification of High-Risk Areas. *Pathogens* **2021**, *10*, 1004. https://doi.org/10.3390/pathogens10081004

Academic Editor: Giovanni Cilia

Received: 12 July 2021
Accepted: 5 August 2021
Published: 9 August 2021

Publisher's Note: MDPI stays neutral with regard to jurisdictional claims in published maps and institutional affiliations.

1. Introduction

According to the International Union for Conservation of Nature (IUCN), more than 40% of the invertebrate species that are responsible for pollination, especially bees and butterflies, are at risk of disappearing. In particular, in Europe, 9.2% of bee species are currently threatened with extinction [1]. The Food and Agriculture Organization of the United Nations (FAO) estimates that of the 100 crops that provide 90% of the world's food, 71 are pollinated by bees. Pollinators, therefore, play a key role in regulating the processes that sustain food production, habitat conservation, and natural resources, and are thus also fundamental to the conservation of biological diversity, the basis of economies, and very existence [2].

In Italy, the first reports from beekeepers on bee deaths and declining populations date back to 1999 and increased sharply in 2008 [3,4]. The causes of decline in population are manifold and involve a number of factors including climate change, the presence of pathogens, air pollution, habitat changes with a decrease in melliferous plants, intensive agriculture, and the use of plant protection products [5–7]. Furthermore, according to Porrini et al. [8] and Martinello et al. [9], the cause of decline in population can be defined based on the period in which it occurs. If the die-off occurs in the spring–summer, in intensively farmed areas, and during the period of sowing coated maize seed, weeding wheat, and treating fruit trees, the cause is often attributable to the use of plant protection products. On the other hand, if damage is detected between late summer and the end of the following winter, population decline is mainly due to diseases affecting the bees and the hive.

A study carried out in Italy in the period 2011–2014 to establish the state of health of bees and identify the possible causes of population decline collected data on the health of bees from 63 apiaries located throughout the country [6]. The study found that, during the period under study, 241 samples of bee matrices collected following reports of mortality in Italy tested positive for pesticides. Among the highly toxic insecticides, tests were positive for neonicotinoids (19%), pyrethroids (18%), and organophosphorus (16%).

In Lombardy, a worrying decrease in the number of hives and apiaries had already been observed in the period 2008–2009, and, in 2008, population decline was reported in connection with the maize sowing period. Following these reports, the Veterinary Services of the Lombardy Region coordinated the monitoring of reports by taking samples of bees and sending them to the laboratory for further analysis on the basis of the suspicions. The purpose of the above-mentioned monitoring was to acquire data and information both on the state of health of the regional bee population, and on the possible causes of population decline.

This is the context for this study, which seeks to determine whether there is an association between the decline in bee population and the types of crops surrounding the apiary. In addition, risk maps were created and areas at high risk of depopulation were identified. The data analysed refer to occurrence of population decline recorded in Lombardy from 2014 to 2016.

2. Results

For this study, a total of 14,188 apiaries from 5646 farms in Lombardy were analysed in the period 2014 to 2016.

The provinces with the highest number of apiaries are Bergamo (2125) and Brescia (2482); those with the highest percentage of population decline are Mantua (1.07%), Sondrio (0.94%), Brescia (0.85%), Lodi (0.83%), Milan (0.83%), Cremona (0.63%), and Pavia (0.51%) (Table 1).

Table 1. Provincial distribution of apiary bee population decline.

	No Bee Decline	With Bee Decline	%
Province			
Bergamo	2119	6	0.28%
Brescia	2461	21	0.85%
Como	1133	4	0.35%
Cremona	631	4	0.63%
Lecco	993	3	0.30%
Lodi	238	2	0.83%
Monza Brianza	544	0	0.00%
Milan	1077	9	0.83%
Mantua	742	8	1.07%
Pavia	1369	7	0.51%
Sondrio	1264	12	0.94%
Varese	1537	4	0.26%
Classification *			
Sedentary	11,273	68	
Nomadic	2703	5	
missing data	132	7	
TOTAL	*14,108*	*80*	

* Last update of the Beekeeping Database (BDA) 31 December 2020.

Of the 14,188 apiaries in Lombardy in the period 2014 to 2016, 80 apiaries (0.56%) reported at least one event of population decline or positive results from tests for plant protection products; of these, 75 apiaries experienced population decline only once in the three-year period, and five apiaries reported it twice (Table 2).

Table 2. Apiaries and bee decline in the region in the period 2014–2016.

	Year			BDA-Registered Apiaries in the Three-Year Period **
	2014	**2015**	**2016**	
Bee Decline *	33	25	27	80
No Bee Decline	7810	8970	11,014	14,108
Total	7843	8995	11,041	**14,188**

* among the apiaries recording population decline, 13 also tested positive for infectious diseases or infestations (American Plague, *Nosema* spp., DWV and *Varroa* spp.). ** apiaries that were recorded at least once in the three-year period.

Table 3 shows the month in which the decline in population was reported for each year considered. Population decline occurred in the spring (April–May) in 64% of cases, which is the period in which maize and orchards are treated.

Table 3. Month of reported decline in population.

	Year			Total
	2014	**2015**	**2016**	
January	0	0	1	1
February	1	0	0	1
March	2	3	0	5
April	16	6	7	29
May	6	8	11	25
June	1	2	5	8
July	1	3	2	6
August	4	0	0	4
September	1	0	0	1
October	0	0	1	1
November	1	0	0	1
December	0	1	0	1
Total	33	23	27	83 *

* for 2 apiaries, the month in which population decline occurred is not available.

Table 4 shows the type of land surrounding the apiaries (within 1.5 km). We can see that 93.4% of the apiaries without population decline have at least one wooded area next to them, compared with 88.8% of the apiaries that were declined. There are bushes close to 97.9% of the apiaries that were not declined and 98.8% of the apiaries that did see decline in population. Apiaries with population decline were more frequently located near arable land (93.8% compared to 89.6% of apiaries with no decline), orchards (61.3% compared to 54.8% of apiaries with no decline) and vineyards (52.5% compared to 45.8% of apiaries with no decline).

Table 5 shows the percentage of land occupied by the different types of land surrounding the apiaries by province. Woodland is prevalent in the province of Como (64.3%), Lecco (62.8%), Sondrio (62.8%), and Varese (69.7%), while apiaries in the provinces of Cremona (0.7%), Lodi (2.4%), and Mantua (2.0%) are located far from these types of land. Sondrio is the province with the highest percentage of orchards (3.1%), followed by Mantua (1.6%). Apiaries located in the provinces of Cremona and Mantua are mostly located close to arable land (82.8% and 82.0%, respectively). Finally, 9.9% of Pavia's apiaries are located near vineyards, followed by the provinces of Brescia (4.2%) and Sondrio (3.3%).

Table 4. Number and proportion of apiaries occupying different types of land.

	No Bee Decline n (%)	Bee Decline n (%)
Total number of apiaries	14,108	80
Number (%) of locations near *:		
Woodland	13,171 (93.4%)	71 (88.8%)
Bushes	13,815 (97.9%)	79 (98.8%)
Orchards	7734 (54.8%)	49 (61.3%)
Arable land	12,641 (89.6%)	75 (93.8%)
Vineyards	6459 (45.8%)	42 (52.5%)
Other	14,068 (99.7%)	79 (98.8%)

* the percentages were calculated considering the denominator of 14,108 apiaries with no population decline and 80 with population decline.

Table 5. Provincial distribution of apiaries by type of land occupied.

Province	Woodland	Bushes	Orchards	Arable Land	Vineyards	Other	Total
Bergamo	50.4%	3.1%	0.2%	25.1%	2.0%	19.2%	100%
Brescia	41.8%	3.7%	0.4%	34.6%	4.2%	15.3%	100%
Como	64.3%	2.5%	0.1%	14.2%	0.2%	18.9%	100%
Cremona	0.7%	1.3%	0.1%	82.8%	0.03%	15.0%	100%
Lecco	62.8%	3.0%	0.1%	14.1%	0.5%	19.4%	100%
Lodi	2.4%	1.7%	0.1%	75.4%	0.2%	20.2%	100%
Monza Brianza	24.2%	3.0%	0.2%	63.3%	0.3%	9.2%	100%
Milan	10.4%	2.7%	0.2%	66.6%	1.2%	19.0%	100%
Mantua	2.0%	1.3%	1.6%	82.0%	1.8%	11.3%	100%
Pavia	17.2%	5.1%	0.9%	39.7%	9.9%	27.1%	100%
Sondrio	62.8%	5.5%	3.1%	2.3%	3.3%	23.0%	100%
Varese	69.7%	2.0%	0.1%	19.6%	0.8%	8.5%	100%

Table 6 shows that the risk of population decline is associated with the size of the type of land surrounding the apiary (within 1.5 km). In other words, the larger the area of a given type of land, the greater the probability of a decline in population. In particular, the probability of an apiary declining in population statistically increases by 10% (OR: 1.10 95%CI (1.05;1.14), $p < 0.0001$) in the presence of orchards, and 2% of 1.02 (95%CI (1.01;1.04), $p = 0.003$) in the presence of arable land, for every additional km^2 of land. This means that the risk of population decline increases with the unit increase (1 km^2) of the surrounding territory. Furthermore, the average area occupied by these types of land is 0.20 km^2 and 2.67 km^2 for apiaries with population decline, and 0.06 km^2 and 1.98 km^2 for those without population decline, respectively.

Figure 1 shows the apiaries with population decline in Lombardy in relation to the type of land, while Figure 2—via the density estimated using the Kernel algorithm—shows the regional areas most at risk of population decline (in red), where the color intensity denotes the degree of risk; the more intense the color, the more the area is at risk of population decline. The map shows that declined apiaries are often located near areas rich in orchards (as in the case of the province of Sondrio), areas rich in orchards and arable land (as in the lower Garda area), and areas rich in arable land (as in the west, in the province of Milan, and on the border between the provinces of Brescia and Cremona).

Table 6. Association between population decline and the area occupied by types of land near the apiaries.

Type of Land	OR *	(95%CI)	*p*-Value	Average Area Occupied, with Population Decline (km^2)	Average Area Occupied, No Population Decline (km^2)
Forests	1.01	(0.99; 1.02)	0.32	1.84	2.24
Shrubland	1.05	(0.98; 1.10)	0.12	0.19	0.17
Orchards	1.10	(1.05; 1.14)	<0.0001	0.20	0.06
Arable land	1.02	(1.01; 1.04)	0.003	2.67	1.98
Vineyards	1.01	(0.98; 1.04)	0.40	0.33	0.31
Other	1.01	(0.99; 1.03)	0.41	0.84	0.92

* OR = odds ratio estimated via the logistic regression model.

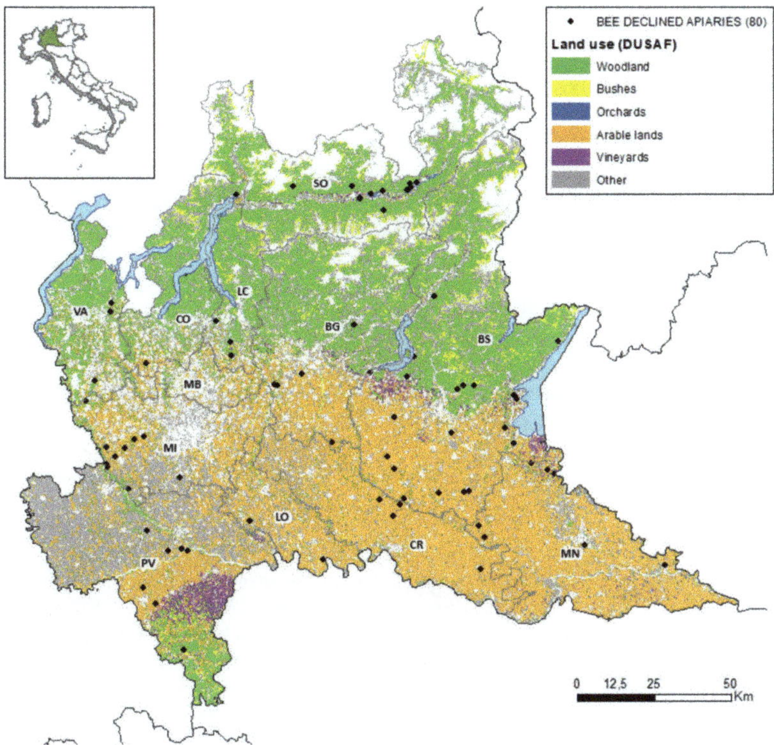

Figure 1. Map of land use (DUSAF 2015) and apiaries where population decline events have been recorded. Provinces: BG = Bergamo, BS = Brescia, CO = Como, CR = Cremona, LC = Lecco, LO = Lodi, MB = Monza Brianza, MI = Milan, MN = Mantua, PV = Pavia, SO = Sondrio, VA = Varese.

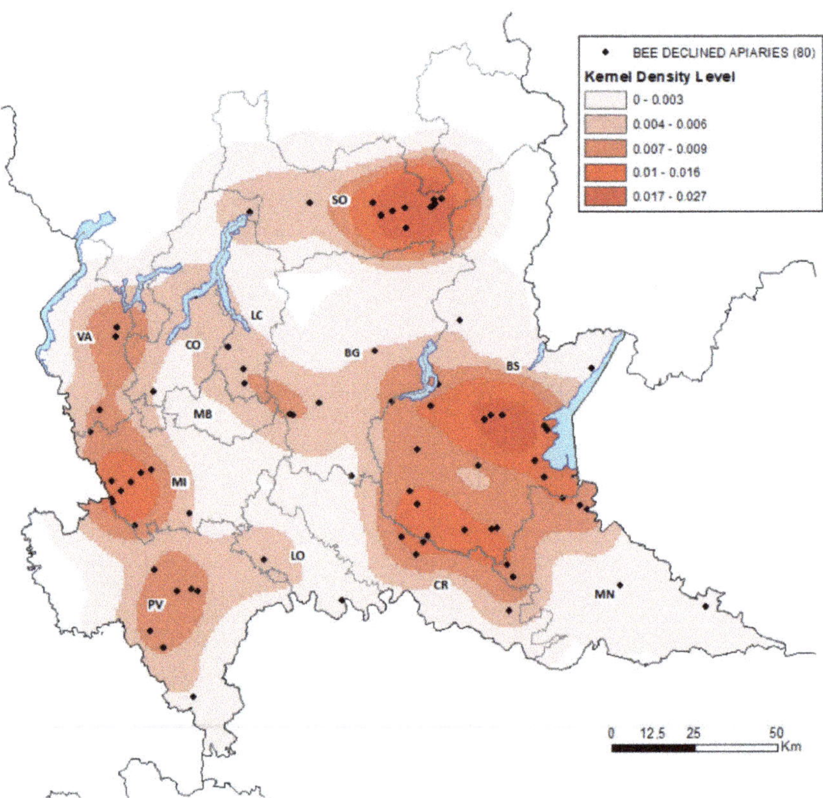

Figure 2. Lombardy with areas at risk of population decline. The density was estimated by means of the Kernel algorithm using the Epanechinikov distribution with bandwidth of 10 km; Provinces: BG = Bergamo, BS = Brescia, CO = Como, CR = Cremona, LC = Lecco, LO = Lodi, MB = Monza Brianza, MI = Milan, MN = Mantua, PV = Pavia, SO = Sondrio, VA = Varese.

3. Discussion

It has been amply demonstrated that the treatment of arable land, which leads to the dispersion of dust containing insecticide, is among the major contributors to the decline in bee population, supporting data found in other studies [7,9]. However, the quantification of the risk associated with the type of crop remains unknown.

This study aimed to quantify the risk of population decline by identifying the type of land surrounding the apiary according to the bee's flight radius and a statistically significant association emerged between arable land and bee deaths. The risk was quantified considering the area occupied by type of land near the apiary, and the study shows a statistically significant association between orchards and bee deaths. The mortality recorded in apiaries near fruit orchards was traced to treatments carried out on fruit trees, as they are susceptible to insect and aphid attacks and plant protection products are widely used to combat these infestations.

The results of this study are in line with those obtained in the study conducted by Porrini in 2016, where a statistically significant linear correlation was demonstrated between the percentage of land area used for agriculture and the mortality rate of the colonies [8,10]. However, in Porrini's study the authors did not distinguish between different type of crops.

Moreover, the timing of the population declines recorded in the Lombardy Region in the three-year period under consideration confirms the seasonality already reported in the literature [7,9]. Indeed, in this study it was observed that 64% of the population declines occurred in the spring (April–May), which is the preferred period for treating maize crops and orchards.

4. Methods

4.1. Data

The study area is the region of Lombardy in northern Italy, which includes three distinct natural zones: mountains, hills, and plains. Data processing was carried out using the following databases: (i) the national Beekeeping Database (BDA) [11]; (ii) reports of population decline from 2014 to 2016; (iii) the Agricultural and Forest Land Use (DUSAF) land use database; (iv) the lab results database from the Istituto Zooprofilattico Sperimentale della Lombardia e dell'Emilia Romagna (IZSLER).

4.1.1. Information on Apiaries

The data on population decline were collected from the report form that beekeepers filled in during the period from 2014 to 2016 for the Veterinary Services, from the IZSLER information system that collects the data and the results of the sample surveys carried out.

The following information was collected through these forms: details of the beekeeper and the apiary affected by population decline, information on health, territory, and date of observation of the population decline, information on the bees with an indication of the number of hives affected, indicative number of dead bees and behavior of surviving bees, and information provided by the beekeeper on the possible cause of the damage linked to treatments (type of treatment, crop treated, and type of product used).

For this study, we only considered apiaries declined following suspected use of plant protection products or in which, in the absence of a specific report, the active ingredients used for crop treatments were identified. Apiaries in which an infectious disease or infestation was suspected, or where only bacterial/viral/parasitic agents were detected, were excluded.

4.1.2. Agricultural and Forest Land Use Data

The Lombardy Region publishes the DUSAF land use data. This database collects information on the main types of land throughout the region and georeferences them. The last update was performed in 2018 using AGEA aerial photographs and satellite imagery. Thanks to this data source, it was possible to identify the agricultural use of the land surrounding the apiaries as well as the relative area of land occupied (in km^2) out of the total area covered by the bees.

The types of land included in the analyses are described in Table 7, while areas of human activity, wetlands, and water bodies were excluded. Rice fields, arboriculture for timber production (poplar groves and other agricultural wood species), permanent grassland, natural high-altitude grassland, and olive groves were grouped in the category "Other" because they are poorly represented if considered separately.

Table 7. DUSAF Legend.

Macro-Category	DUSAF Type	Description
Woodland	Coniferous forests	Low, medium, and high-density coniferous forests
	Broadleaf forests	Low, medium, and high-density broadleaf forests, underbrush formations, chestnut groves
	Mixed broadleaf and coniferous forests	Low, medium, and high-density mixed coniferous and broadleaf forests
	Recent reforestation	Artificial forest systems not yet established and under treatment or to be treated. There are typically young trees with limited development of plants; generally, a regular planting pattern is recognizable. Trees are typically under 15 years of age. Plantations of poplars or other timber producing species included in another class are excluded

Table 7. *Cont.*

Macro-Category	DUSAF Type	Description
Bushes	Bushes and shrubs	Bushes, vegetation of riverbanks, vegetation of raised banks
	Developing areas	Bushes with significant presence of tall shrub and tree species, bushes in abandoned agricultural areas
Orchards	Orchards and berry farms	Plantations of non-rotational fruit trees that occupy the soil even for long periods and can be used for many years before being renewed.
Arable land	Simple arable land	Arable land, arboretums, horticultural crops, floricultural crops, family vegetable gardens, and arable land in irrigated areas
	Irrigated agricultural land	Irrigated agricultural land
Vineyards	Vineyards	Plantations of vines intended forthe production of both table and wine grapes
Other	Olive groves	Plantations of olive trees forproducing olives
	Arboriculture for timber	Poplar groves, other agricultural wood species
	Rice fields	Areas used forrice cultivation
	Permanent grassland	Permanent grassland without tree and shrub species

4.2. Statistical Analysis

The Lombardy apiaries were divided into two groups: (i) apiaries were "bee decline" if in the three-year period 2014–2016 they reported at least one event of decline in bee population or tested positive for plant protection products, (ii) apiaries were "no bee decline" if in the three-year period 2014–2016 they did not report any decline in bee population. In the study, all apiaries for which a report of population decline was received from the beekeeper were considered declined, regardless of the laboratory test results, as chemical analyses are often unable to detect the toxic substance responsible for population decline. It is also difficult to detect chemical contamination because most bees affected by plant protection products do not return to the hive [3].

The characteristics of the farms and apiaries by population decline group (yes/no) were described using absolute frequencies and percentages.

The association between cases of population decline and the area occupied (in km^2) by each DUSAF macro-category was estimated through the logistic regression model, considering the population decline group as the dichotomous dependent variable and the area occupied by each type of land as the independent variable (woodland, bushes, orchards, arable land, vineyards, and others). Considering that a study conducted in 1984 by Crane [12] estimated the flight radius of a forager bee to be about 1500 metres (an estimated area of 7.065 km^2 around the apiary), we were able to use the georeferencing of the regional territory to identify for each apiary the types of crops located within this area.

All statistical analyses considered a threshold of statistical significance of 5%.

Finally, we mapped the density of population decline, calculated using Kernel's methodology [13], a spatial interpolation technique that identifies the areas at greatest risk of population decline. The density estimate was calculated using the Epanechinikov distribution and a bandwidth (h) of 10 km, with the red scale identifying areas at risk of population decline: the more intense the color, the greater the risk.

4.3. Software Used

The statistical analysis was performed using the software R version 3.6.1 [14]. The maps were created using the software QGIS Development Team [15].

5. Conclusions

Bees are considered indicators of the health of the environment in which they live and exposure to certain insecticides or other toxic substances used during the flowering period can lead to major population decline. In order to standardise monitoring activities following reports of a population decline, the national *Guidelines for the management of reports of hive deaths and population declines related to the use of crop protection products* were issued [16]. The purpose of these guidelines is to ensure coordinated, rapid, and effective interventions in response to reports. Protecting our bees is crucial for the protection of the heritage of beekeeping in terms of the role of bees as pollinating insects, but also for the food safety of beekeeping products. In fact, in addition to protecting beekeeping heritage, reports and the resulting control measures are also important for the protection of the end consumer. This study revealed a statistically significant association between bee population decline and arable fields and orchards, and we were able to quantify the risk according to the type of land surrounding the apiary. The analysis shows that the risk of bee decline increases statistically when it is located close to orchards and arable land. In the event of use of plant protection products, the active ingredients must be used in full compliance with the treatment instructions provided and, above all, users should be aware of the risks regarding toxicity to animals, specifically bees. Using the product on plants not in flower, as is sometimes recommended correctly, is not always enough to prevent the risks associated with using the product.

The timing of the population declines recorded in the Lombardy Region in the three-year period considered suggests that plant protection products are the cause; a result that would confirm the data in the literature. This means that the results of the territorial analysis, understood as the type of crops found around the apiaries and the correlated risk in terms of population decline, could be used as predictive parameters to be applied when planning prevention and control activities.

Author Contributions: Conceptualization, S.B., V.C., M.P.C. and C.N.; methodology, S.B., V.C., M.P.C. and A.S.; software, V.C., M.T. and A.S.; formal analysis, V.C. and A.S.; data curation, V.C. and A.S.; writing—original draft preparation, S.B., V.C. and M.P.C.; writing—review and editing, V.C., M.P.C., A.S., M.T., M.F., C.N. and S.B.; visualization, all authors; supervision, all authors; project administration, S.B. All authors have read and agreed to the published version of the manuscript.

Funding: This research received no external funding.

Institutional Review Board Statement: Not applicable.

Informed Consent Statement: Not applicable.

Data Availability Statement: Data is available upon request from the authors; the data that support the findings of this study are available from the corresponding author upon reasonable request.

Conflicts of Interest: The authors declare no conflict of interest.

References

1. Nieto, A.; Roberts, M.S.P.; Kemp, J.; Rasmont, P.; Kuhlmann, M.; Criado, G.M.; Biesmeijer, C.J.; Bogusch, P.; Dathe, H.H.; de la Rúa, P.; et al. European Red List of Bees. Luxembourg: Publication Office of the European Union. 2014. Available online: https://ec.europa.eu/environment/nature/conservation/species/redlist/downloads/European_bees.pdf/ (accessed on 9 August 2021).
2. ISPRA. IL Declino Delle Api e Degli Impollinatori. Le Risposte Alle Domande Più Frequenti. Quaderni Natura e Biodiversità N.12/2020. Available online: https://www.isprambiente.gov.it/files2020/pubblicazioni/quaderni/declino-impollinatori_quaderno-ispra_20maggio2.pdf/ (accessed on 9 August 2021).
3. Bortolotti, L.; Porrini, C.; Mutinelli, F.; Pochi, D.; Marinelli, E.; Balconi, C.; Nazzi, F.; Lodesani, M.; Sabatini, A.G. Salute delle api: Analisi dei fattori di rischio. *Il Progett. Apenet. APOidea* **2009**, *6*, 3–22.
4. Lodesani, M. Beenet, un monitoraggio dai risultati inequivocabili. *Ecoscienza* **2014**, *4*, 46–49.
5. Bellucci, V.; Bianco, P.; Formato, G.; Mutinelli, F.; Porrini, C.; Lodesani, M. Morìe di api e prodotti fitosanitari. *Apitalia* **2016**, *12*, 46–52.
6. Bellucci, V.; Bianco, P.; Silli, V. Le Api, Sentinelle Dell'inquinamento Ambientale. Available online: https://www.isprambiente.gov.it/it/attivita/biodiversita/lispra-e-la-biodiversita/articoli/le-api-sentinelle-dell2019inquinamento-ambientale/ (accessed on 5 August 2021).
7. Porrini, C.; Sabatini, A.G.; Mutinelli, F.; Astuti, M.; Lavazza, A.; Piro, R.; Tesoriero, D.; Medrzycki, P.; Sgolastra, F.; Bortolotti, L. Le segnalazioni degli spopolamenti e delle mortalità degli alveari in Italia: Resoconto 2008. *L'Apis* **2009**, *1*, 15–18.

8. Porrini, C.; Mutinelli, F.; Bortolotti, L.; Granato, A.; Laurenson, L.; Roberts, K.; Gallina, A.; Silvester, N.; Medrzycki, P.; Renzi, T.; et al. The status of honey bee health in Italy: Results from nationwide bee monitoring network. *PLoS ONE* **2016**, *11*, e0155411. [CrossRef] [PubMed]
9. Martinello, M.; Baratto, C.; Manzinello, C.; Piva, E.; Borin, A.; Toson, M.; Granato, A.; Boniotti, M.B.; Gallina, A.; Mutinelli, F. Spring mortality in honey bees in northeastern Italy: Detection of pesticides and viruses in dead honey bees and other matrices. *J. Apic. Res.* **2017**, *56*, 1–16. [CrossRef]
10. Martinello, M.; Manzinello, C.; Borin, A.; Avram, L.E.; Dainese, N.; Giuliato, I.; Gallina, A.; Mutinelli, F. A Survey from 2015 to 2019 to Investigate the Occurrence of Pesticide Residues in Dead Honeybees and Other Matrices Related to Honeybee Mortality Incidents in Italy. *Diversity* **2020**, *12*, 15. [CrossRef]
11. Italian Ministerial Decree (4/12/2009). Disposizioni per L'Anagrafe Apistica Nazionale. Available online: https://www.izsvenezie.it/documenti/temi/api/normativa/ministero-salute/2009-12-04-decreto.pdf/ (accessed on 9 August 2021).
12. Crane, E. Bees, honey and pullen as indicators of metals in environment. *Bee World* **1984**, *65*, 47–49. [CrossRef]
13. Silverman, B.W. *Density Estimation for Statistics and Data Analysis*; Chapman and Hall/CRC: New York, NY, USA, 1986.
14. R Core Team. *R: A Language and Environment for Statistical Computing*; R Foundation for Statistical Computing: Vienna, Austria, 2018. Available online: https://www.R-project.org/ (accessed on 18 November 2020).
15. QGIS Development Geographic Information System, Open Source Geospatial Foundation Project Guide. Available online: https://docs.qgis.org/3.16/en/docs/user_manual/ (accessed on 9 August 2021).
16. ILD GENERALE. Linee Guida per la Gestione Delle Segnalazioni DI Moria O Spopolamento Degli Alveari Connesse All'Utilizzo DI Agrofarmaci. Nota Del Ministero Della Saluten. Available online: https://www.izsvenezie.it/documenti/temi/api/normativa/ministero-salute/2014-07-31-nota-16168.pdf/ (accessed on 18 November 2020).

pathogens

Article

Screening of Dietary Ingredients against the Honey Bee Parasite *Nosema ceranae*

Chiara Braglia [1]**, Daniele Alberoni** [1,*]**, Martin Pablo Porrini** [2,3]**, Paula Melisa Garrido** [2,3]**,
Loredana Baffoni** [1] **and Diana Di Gioia** [1]

[1]   Department of Agricultural and Food Sciences (DISTAL), University of Bologna, Viale Fanin 44,
      40127 Bologna, Italy; chiara.braglia4@unibo.it (C.B.); loredana.baffoni@unibo.it (L.B.);
      diana.digioia@unibo.it (D.D.G.)
[2]   Instituto de Investigaciones en Producción Sanidad y Ambiente (IIPROSAM), Centro Científico Tecnológico
      Mar del Plata-CONICET-UNMdP-CIC-PBA, Funes 3350, Mar del Plata Zc 7600, Argentina;
      mporrini@mdp.edu.ar (M.P.P.); pmgarrid@mdp.edu.ar (P.M.G.)
[3]   Centro de Investigación en Abejas Sociales (CIAS), FCEyN, UNMdP, Funes 3350,
      Mar del Plata Zc 7600, Argentina
*   Correspondence: daniele.alberoni@unibo.it; Tel.: +39-051-2096269

Abstract: *Nosema ceranae* is a major pathogen in the beekeeping sector, responsible for nosemosis. This disease is hard to manage since its symptomatology is masked until a strong collapse of the colony population occurs. Conversely, no medicaments are available in the market to counteract nosemosis, and only a few feed additives, with claimed antifungal action, are available. New solutions are strongly required, especially based on natural methods alternative to veterinary drugs that might develop resistance or strongly pollute honey bees and the environment. This study aims at investigating the nosemosis antiparasitic potential of some plant extracts, microbial fermentation products, organic acids, food chain waste products, bacteriocins, and fungi. Honey bees were singularly infected with 5×10^4 freshly prepared *N. ceranae* spores, reared in cages and fed ad libitum with sugar syrup solution containing the active ingredient. *N. ceranae* in the gut of honey bees was estimated using qPCR. The results showed that some of the ingredients administered, such as acetic acid at high concentration, p-coumaric acid, and *Saccharomyces* sp. strain KIA1, were effective in the control of nosemosis. On the other hand, wine acetic acid strongly increased the *N. ceranae* amount. This study investigates the possibility of using compounds such as organic acids or biological agents including those at the base of the circular economy, i.e., wine waste production, in order to improve honeybee health.

Keywords: nosemosis; *Vairimorpha ceranae*; nisin; *Saccharomyces* sp.; acetic acid; para-coumaric acid; gut microbiota

Citation: Braglia, C.; Alberoni, D.; Porrini, M.P.; Garrido, P.M.; Baffoni, L.; Di Gioia, D. Screening of Dietary Ingredients against the Honey Bee Parasite *Nosema ceranae*. *Pathogens* **2021**, *10*, 1117. https://doi.org/10.3390/pathogens10091117

Academic Editor: Giovanni Cilia

Received: 14 July 2021
Accepted: 27 August 2021
Published: 1 September 2021

1. Introduction

Nosema ceranae is a unicellular sporogenous fungus belonging to the phylum microsporidia, which gives rise to a chronic debilitating infection in honey bees named nosemosis [1]. This pathogen co-evolved with *Apis cerana*, whose parasitism became endemic in Asia. Nevertheless, *Apis mellifera* colonies infected by *N. ceranae* were found for the first time in 2005 in Taiwan [2] and in 2006 in most European countries [3]. When this host species shift occurred is unknown, although if it is reasonable to believe that it happened at the time of *A. mellifera* introduction in Asia, in the 1880s [4]. The exact *N. ceranae* arrival period in Europe is not clear but evidence suggests that it has been present in Europe since 1998 [5], thanks to an active international trading of *A. mellifera* from Asia to the rest of the world. Recently, a revision and redefinition of the genera Nosema and Vairimorpha proposes to rename *N. ceranae* and *N. apis* as *Vairimorpha ceranae* and *Vairimorpha apis* [6] which, more than taxonomic consequences, could become relevant for future research in

15

the topic. Adult honey bees easily become infected by ingesting spores from stored honey and pollen and subsequent transmission through trophallaxis [7,8], also after the exposure to surfaces contaminated by spores, following the colony cleaning and visits to contaminated flowers and pollen in the foraging activity [7,9]. After the ingestion, microsporidia spores germinate extracellularly in the midgut lumen and then inject the sporoplasm in an epithelial cell through the polar tube [1]. Infected colonies at the beginning show no visible symptoms and when environmental conditions are favorable for the parasite, they may rapidly collapse [10], making nosemosis a disease that sometimes is hard to control and difficult to diagnose and cure. The mycotoxin fumagillin (dicyclohexylamine salt), produced by *Aspergillus fumigatus*, is the first and the only successful antibiotic for the treatment of nosemosis [11–13] since 1953 [14]. Currently it is available on the market in many American countries and Korea, but it is forbidden in the European Union because of the absence of a detailed threshold residue regulation in honey and hive products [15]. Fumagillin use is nowadays controversial: targeting the methionine aminopeptidase-2 (MetAP2) [16], an enzyme present in many eukaryotes, can cause metabolic imbalances also in non-target organisms, i.e., it has been classified as mutagenic and cytotoxic for mammals after a short-term exposure [17]. Fumagillin residues persist in hives and its degradation products pose a potential risk for human health [18,19]. For instance, fumagillin toxicity was assessed in honey bees, causing a reduction in their lifespans [20], an alteration of structural and metabolic proteins in midgut [21], a reduction in sperm quality [22] and health [13]. Moreover, its efficacy depends on several factors such as seasonality and plantations [23]. Furthermore, in laboratory conditions, Huang et al., 2013 [21] have demonstrated that mature *N. ceranae* spore proliferation was similar in treated and untreated bees at the recommended fumagillin concentration (250 µg/L) and induced *N. ceranae* hyper proliferation when fumagillin concentration was 10 folds lower. Therefore, the need to find new strategies, which combine honey bee health protection and governmental standards in terms of food safety, gave a new pulse to the research of alternative solutions. Studies aimed at verifying the effect of veterinary drugs or commercial dietary supplements on the honey bee gut microbiota composition have been published recently. In recent years [24,25], researchers have begun to evaluate the use of plant extracts as *Nosema* control agents, such as thymol, oregano oil, carvacrol, cinnamaldehyde, resveratrol, and garlic based products, and treated bees showed lower *N. ceranae* infection rates compared to control [26–28]. Moreover, *Andrographis paniculate*, *Cryptocarya alba*, *Gevuina avellane*, *Artemisia dubia*, and *Laurus nobilis* extracts were tested with positive results [29–33], supporting a plant extract based strategy as a promising tool in the control of bee diseases. Brassicaceae seeds also showed promising results for their protective effects against N. ceranae spores at the laboratory level [34]. Furthermore, the use of beneficial microorganisms like *Bifidobacterium* and *Lactobacillus* strains [35,36], and *Bacillus subtilis* metabolites [37,38] showed encouraging results. For example, Baffoni et al. (2016) [35] demonstrated that probiotic treatment with *Lactobacillus* and *Bifidobacterium* strains reduced the presence of *Nosema* spores in naturally infected bees, thus proving the efficacy of a preventive microorganism-based strategy. Commercial probiotic preparation also based on lactic acid bacteria were found to be effective against the same parasite [27,36]. Similarly, De Piano et al. [39] demonstrated a strong relation between bacterial metabolites and the count of *N. ceranae* spores, showing a significant decrease after *Lactobacillus johnsonii* AJ5 administration. Additionally, organic acids produced by lactic acid bacteria present in the honey bee's environment (flowers, beebread, and gut), such as lactic acid, acetic acid, and phenyl-lactic acid, were tested, through feeding, against these microsporidia and showed a strong reduction in spore load in bees [38]. Therefore, these compounds have particular interest for the beekeeping sector. The aim of the present study was to test feed ingredients belonging to 4 different groups (organic acids, *Saccharomyces* and antibiotics, wine derivatives, and plants extracts) for their antimicrobial activity against *N. ceranae* in *A. mellifera* workers, but also to evaluate their toxicity on individual honey bees.

2. Materials and Methods

2.1. Experimental Set up

Cage experiments were carried out in the microbiology laboratory of the Department of Agricultural and Food Sciences, University of Bologna, in the period between October 2019 and June 2021. Newly emerged honey bees (*Apis mellifera ligustica* were obtained from multiple brood frames of emerging honey bees within the University of Bologna experimental apiary (San Lazzaro, Bologna, Italy, 210 m a.s.l.) in a continental climate. At least two independent laboratory cage tests were performed for each feed ingredient in order to validate results. The first set of assays, referred to as "First screening", was organized in order to study the effects of the feed ingredients on honey bee survival and parasite development. After this first screening, a second one, referred to as "Derived tests", was performed in order to confirm the obtained results or to test new hypotheses derived from the first one. Moreover, if the results showed a promising trend, even if not significant, the dosage of the active ingredient was increased, determining two different dosages for some of the ingredients [AA, NisA and GRA] marked as "low" (_L) and "high" (_H). Only para-coumaric acid and acetic acid were tested a third time using summer honey bees due to the contrasting results obtained in the two tests. All the tests were performed with the same protocol, having only differences in the season at which the newly emerged honey bees were obtaine. The ingredients assayed were chosen considering their known antimicrobial properties in both food and feed safety (organic acids and antibiotics), their immune stimulation properties (mainly plant extracts and microorganisms), but also basing the choice on traditional homemade remedies used by beekeepers (e.g., wine derivatives) that do not have a proper scientific basis. The selected compounds reported in Table 1 and are divided in four main groups: i. organic acids (acetic acid [AA], abscisic acid [ABA], p-coumaric acid [pCA]); ii. wine derivatives (ethanol [EtOH], sulphites [SUL], wine vinegar [WA]); iii. *Saccharomyces*and antibiotics (*Saccharomyces* sp. strain KIA1 [SC], a mixture of gramicidin A, B, C, and D [GRA], Nisine A [NisA]); iv. plant extracts (extract of *Opuntia ficus-indica* [OPT], extract of brown alga *Padina pavonica* [PP], a mixture of manuka and tea tree oil [MT]). In all the tests, fumagillin [DCH] was used as positive control. Each assay had a dedicated control test, i.e., a group of infected bees not supplemented with any compounds in their diet [CTR]. Experimental cages had a dimension of $11 \times 7 \times 4$ cm and were made using plastic, including a ventilation mesh. Three replicate cages, containing 50 newly emerged honey bees, were prepared for every dietary treatment and controls. Mortality in every replicate cage was registered on a daily basis, extracting the dead individuals after counting. Ten worker bees for each experimental condition were sacrificed at day 9 (experiment end) and the guts (midgut and rectum) were collected individually and stored at $-20\ °C$ until tissue analysis.

Table 1. Description of the different active ingredients tested in this work and its relative dose or concentration. [a] Dose recommended by the manufacturer. Final concentration of the ingredients are expressed as the amount of ingredient per mL of sugar syrup (1:1 *w:v*).

Ingredient	Treatment Code	Source or Producers	Concentration of Ingredient Per Treatment	Reference (When Available)
		Test on Organic Acids		
Acetic acid lower concentration	AA_L	Acetic Acid;	84 mM	[40,41]
Acetic acid higher concentration	AA_H	Merck	0.35 M	[40,41]
Abscisic acid	ABA	S-(+)-Abscisic Acid Fanda Chem	50 µM	[42]
p-Coumaric acid	pCA	trans-4-Hydroxycinnamic acid; Merck	31.4 µM	[43]
		Test on *Saccharomyces* and Antibiotics		
Saccharomyces sp. strain KIA1	SC	Isolated by authors from soil	10^{11} CFU/mL	-
Gramicidin D lower concentration	GRA_L	Gramicidin from *Bacillus aneurinolyticus*; Merck	7.7 mM	-
Gramicidin D higher concentration	GRA_H	Gramicidin from *Bacillus aneurinolyticus*; Merck	15.4 mM	-
Nisin lower concentration	NisA_L	Nisin from *Lactococcus lactis* Merck	7.45 mM	-
Nisin higher concentration	NisA_H	Nisin from *Lactococcus lactis* Merck	74.5 mM	-
		Test on Wine Derivatives		
Ethanol	EtOH	Ethanol; Carlo Erba Reagents	0.69 M	-
Wine Sulphites (precipitates of potassium pyrosulfite)	SPH	Produced from red wine by a local winemaker and gifted	4 mM	-
Wine vinegar	WA	Produced from red wine by a local winemaker and gifted	0.3 M	-
		Test on Plant Extracts		
Extract of *Opuntia ficus-indica*	OPT	Produced by authors	0.005 µL/mL	-
Extract of *Padina pavonica*	PP	Produced by authors	0.005 µL/mL	-
Steam distilled Manuka and Tea tree essential oil	MT	Optima Naturalis and ESI s.r.l., respectively	0.75 µL/mL + 0.1 µL/mL	-
		Positive and Negative Controls (included in all tests)		
Fumagillin	DCH	Fumagilin-B; Medivet Ltd.	2.59 mM [a]	Medivet Ltd. guidelines
Untreated control	CTR	-	-	-

2.2. Production of N. ceranae Spores

Honey bee colonies infected with *N. ceranae* were identified in a apiary nearby Modena (Italy), in the city of Savignano sul Panaro (44°29′03.4″ N 11°03′28.1″ E). Spores were collected from diseased hives by capturing flying foragers honey bees over the colony entrance. Obtained foragers were sacrificed and the midgut and rectum extracted and broken up in distilled water. Spores were divided in multiple stocks and conserved in a 10% glycerol PBS solution at −80 °C to guarantee the same *N. ceranae* strain availability for all the assays. When an assay was established, spores were retrieved from cryostat and about thirty-five newly emerged bees were caged and infected by feeding bees with the prepared solution to allow *N. ceranae* proliferation and sporulation, in order to obtain fresh and highly infective spores. A sample of the spores obtained was characterized and confirmed as *Nosema ceranae* according to [44] and the same stock was used for all infections. Fresh spores were extracted from infected and caged honey bees, counted with Neubauer chamber to estimate the load per ml, diluted in a sugar syrup solution to the final concentration of 10^4 spore/mL and used as a fresh and standardized *N. ceranae* propagules for the ongoing assay.

2.3. Oral Infection with N. ceranae Spores

Newly emerged honey bees (*Apis mellifera ligustica*) were obtained from brood frames with worker bees ready to emerge. Brood frames were picked from three different colonies for every assay, in order to homogenize the genetic variability [45]. Frames were were maintained under controlled conditions (32 °C; 60% RH) until honey bees emerged. Then, newly emerged bees were kept in ventilated cages for 2 days until individual inoculation with control treatment (syrup) or 5×10^4 *N. ceranae* freshly prepared spores in syrup according to Porrini et al. [46]. The temperature and relative humidity were maintained during the experiments at 29 °C and 60% RH, respectively. Before starting the administration of the ingredients, a sample of newly emerged individuals was taken to test the potential basal infection with *N. ceranae*.

2.4. Cultivation of Saccharomyces sp.

Saccharomyces sp. was isolated from forest soil collected at the Bologna Apennines (Italy) 250 m a.s.l. (data not shown) and it was grown on PDB culture broth and incubated at 120 rpm for 5 days at 35 °C. Then, the obtained culture was mixed in sugar syrup 1:1 (*w:v*) in equal amounts in order to obtain 10^{11} CFUmL.

2.5. Production of Plant Extracts

Padina pavonica seaweeds were collected from the Mediterranean sea during summer 2019, by hand picking method from the submerged marine rocks. The seaweeds were cleaned from impurities, dried, and powdered with a mixer. The algal flour thus obtained was extracted sequentially with hot sodium oxalate solution, hot water, 1 M and 4 M KOH solution at 20 °C according to [47] and resuspended in glycerol. The *Opuntia ficus-indica* hydroglycolic extract was obtained from fresh cladodes collected in the Malta Island during summer 2019, which were washed with distilled water and cut into small pieces before extraction. A 20:80 (*w/w*) water/propylene glycol mixture was used with a 1:3 (*w/w*) plant-solvent ratio. The cladodes were macerated for 12 h and the resultant mixture was percolated to obtain the extract [48].

2.6. Treatment Administration

Ingredients were administered in sucrose syrup (1 kg sucrose in 1 L of water) using gravity feeders starting from a day after the artificial infection with *N. ceranae* spores. The active ingredients and concentrations for each dietary treatment administered in a matrix of sucrose syrup are reported in Table 1. The tested ingredients (treatments) were supplied to honeybees a day after the artificial infection with *N. ceranae*. The treatments were available *ad libitum* during 9 days. Honey bees in every experimental cage also received tap water *ad libitum* in gravity feeders.

2.7. DNA Extraction and qPCR of N. ceranae

Extracted gut were manually macerated with plastic micro pestles in 200 µL of buffer. DNA extraction of single honey bee guts was performed with PureLink^{TM} Genomic DNA Mini Kit (Invitrogen, Milan, Italy) following the manufacturer protocol with some modifications: guts were further smashed with glass beads (0.01–0.1 µm) at 50 Hz in Rotovortex. Moreover, samples were incubated at 55 °C in a water bath with 180 µL of lysis buffer and 20 µL of proteinase K per sample [49]. Fluorometric quantification of every sample was performed with Qubit Flex Fluorometer (Thermo Fisher Scientific). Extracted DNA was stored at −20 °C until further analysis. The 16S-like rRNA gene (SSU rRNA) was selected to perform *N. ceranae* specific qPCR relaying on specific primer Nc841f 5′-GAGAGAACGGTTTTTTGTTTGAGA-3′ and Nc980r 5′-ATCCTTTCCTTCCTACACTGA TTG-3′ [50]. The reactions were carried out on Step One thermal cycler (Applied Biosystems) with standard two-step PCR method using Fast SYBR Green PCR Master Mix (Life Technologies, Milan, Italy), according to [51,51]. The standard PCR fragment was diluted 1:10 to obtain the reference standards. The melting curve was performed in each real-time reaction to assess amplicons melting temperature (73.77 ± 0.23 °C St. Dev) according to the genetic variability of *N. ceranae*. According to Cilia et al. [52,53], the copy number of 16S-like rRNA gene in *N. ceranae* ranges from 5.7 to 11.5 per genome, therefore the obtained quantification was normalized according with the total amount of extracted DNA and divided by the average copy number of 16S-like rRNA (i.e., 8.6).

2.8. Statistical Analysis

Obtained data were divided according to groups of ingredients, as described above; each assay had its own control, therefore obtaining different datasets. Datasets were analyzed with the R software [54], tested for normality and homoscedasticity with Shapiro and Levene's tests. The dataset was corrected with Cooks Distance multivariate method, to identify outliers based on regression analysis comparison [55]. Datasets were analyzed with ANOVA when data were normal and homoscedastic, while a generalized linear model was applied for non-normal homoscedastic data. Bonferroni *p*-value correction for multiple comparisons was applied for every assay. Boxplots were generated with ggpubr and ggplot2 packages. *Nosema ceranae* infection load was expressed as Log *N. ceranae* units, considering the absolute quantification corrected for the average copy number as described above. To calculate, plot and compare the survival rates for every assay, daily mortality on each replicate cage was recorded and analyzed by means of Kaplan–Meier survival analysis and log-rank tests (SigmaStat Software, San Jose, CA, USA), which estimates also the median survival time.

3. Results

3.1. Survival Tests

Survival results of the test using organic acids, detailed in Figure 3.1A, showed a clear toxic effect of p-coumaric acid [pCA], abscisic acid [ABA] and the highest concentration of acetic acid [AA_H] compared with control treatment ($p < 0.001$). On the other hand, acetic acid [AA_L] at the lowest concentration (four times lower than [AA_H]) kept the survival response at the same level of the control treatment [CTR] and fumagillin [DCH]. The administration of *S. cerevisiae* live cells caused a progressive toxic effect to *N. ceranae* infected bees, reaching less than 50% of live bees at day 9 post-infection (Figure 3.1B). When bacteriocins were administered, a toxic effect was only detected for the highest dose of nisin [NIS_H]. The remaining treatments did not show significant changes in honey bees' survival. Toxicity curves shown in (Figure 3.1C), related to wine-derived products, demonstrated no toxicity of wine vinegar [WV] with control treatment and fumagillin [DCH]. On the contrary, sulphites caused significant toxicity in bees ($p < 0.001$). The two plant extracts assayed [OPT and PP] caused a toxicity level higher than CTR and DCH, as detailed in (Figure 3.1D).

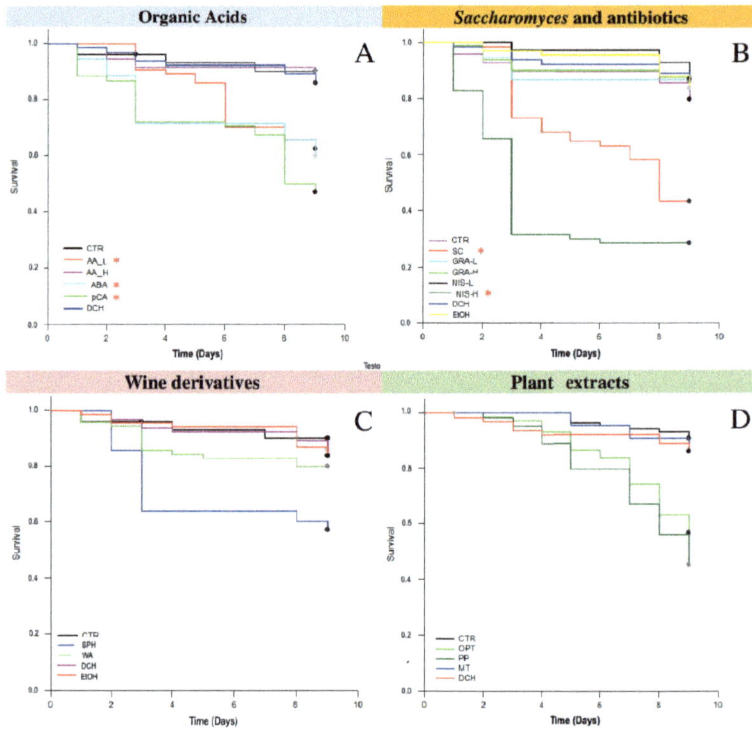

Figure 1. Survival of honey bees. (**A**) Survival curves for treatments in the test on "organic acids"; (**B**) Survival curves for treatments in the "test on microbial origin compounds"; (**C**) Survival curves for treatments in the test on "wine derivatives"; (**D**) Survival curves for treatments in the "plant extracts"; Asterisk (*) indicate significant differences with control treatment (Kaplan–Meier survival analysis, log-rank test; Statistical details for each test: 1A ($p < 0.001$, 52.256, df = 5); 1B ($p < 0.001$, 25.167, df = 7); 1C ($p < 0.001$, 176.870, df = 3) and 1D ($p < 0.001$, 51.352, df = 3). [AA_L] Acetic Acid

lower concentration; Acetic Acid higher concentration [AA_H]; *p*-Coumaric Acid [pCA]; Abscisic Acid [ABA] ; *Saccharomyces* sp. [SC]; a mixture of gramicidin A, B, C, and D [GRA]; Nisin higher concentration [NisA_H]; Nisin lower concentration [NisA_L]; Fumagillin [DCH]; infected control without treatments[CTR]. [*] $p < 0.001$).

3.2. N. ceranae Quantification

In collected honey bees, natural infection of *N. ceranae* was predominantly absent, indeed the highest detection did not exceed 2.0 Log NcU.

3.2.1. Acetic Acid and p-Coumaric Acid Decrease *N. ceranae* Units Only in Winter Honeybees

In the first screening, after 9 days, CTR samples reached an infection rate of 7.69 ± 0.24 Log of *N. ceranae* units (NcU). On the other hand, the positive control fumagillin [DCH], significantly decreased the NcU reaching 4.11 ± 0.32 and 4.86 ± 0.30 Log NcU ($p < 0.01$), respectively. Treatments based on "organic acids" [AA, ABA, pCA] did not show any significant variation with respect to the CTR (Figure 2A). In the derived test, pCA showed a significant reduction in NcU, to [DCH] (Figure 2B, $p < 0.01$), but again this was not confirmed in a third test performed ad hoc (Figure 2D). Acetic acid at the lowest dose confirmed its inefficacy in the derived test, whereas at the highest dose [AA_H] it showed contrasting results in performed tests (Figure 2C,D).

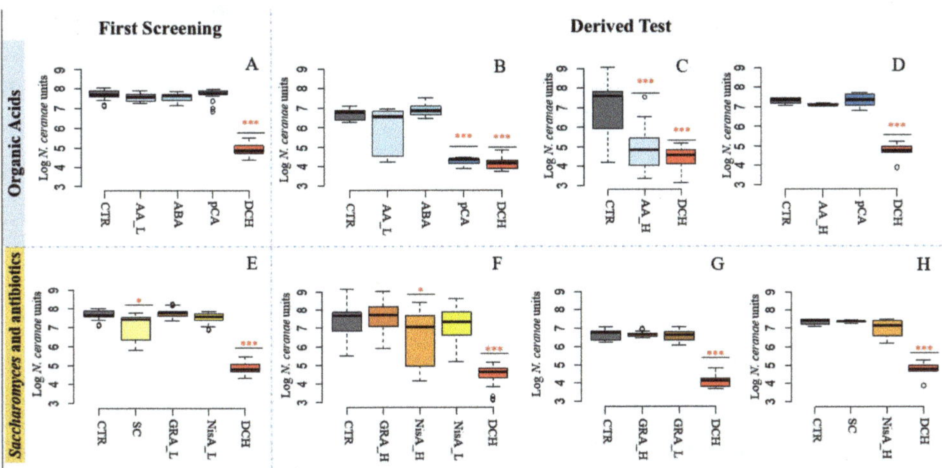

Figure 2. *N. ceranae* inhibition assays. Box plots from the first test (exploratory) and derived tests (confirmatory) are reporting the Log of *N. ceranae* units (NcU) per honey bee gut obtained at 9 days post inoculation with spores for every treatment with dietary ingredient. (I) results obtained from organic acids; (II) results obtained from *Saccharomyces* and antibiotics ; Acetic Acid lower concentration [AA_L];Acetic Acid higher concentration [AA_H]; p-Coumaric Acid [pCA]; Abscisic Acid [ABA]; *Saccharomyces* sp. [SC]; a mixture of (**A,B,C,D**) [GRA]; Nisin higher concentration [NisA_H]; Nisin lower concentration [NisA_L]; Fumagillin [DCH]; infected control without treatments [CTR]. [*] $p < 0.05$; [***] $p < 0.01$.; boxplots (**B,C,F**) shows results obtained with winter honey bees; boxplots (**A,D,E,G,H**) shows results obtained with summer honey bees. Organic Acids: First Screening (**A**); Derived Test (**B–D**). *Saccharomyces* and antibiotics: First Screening (**E**); Derived Test (**F–H**).

3.2.2. Nisin Has a Potential Effect on *N. ceranae*

In the test of "*Saccharomyces* and antibiotics" after 9 days of ingredients consumption CTR samples reached an infection rate of 6.64 ± 0.30 Log of *N. ceranae* units (NcU), respectively. A significant reduction was found for the treatments including *S. cerevisiae* [SC] (Figure 2E, $p < 0.05$) while nisin [NisA] showed a non-significant reduction in NcU with respect to the CTR. NisA at high dose showed a significant decrease ($p < 0.05$) which was not confirmed after repeating the test (Figure 2F,H). Finally, GRA confirmed the inefficacy at both high and low doses (Figure 2E–G).

3.2.3. Wine Derivatives and Plant Extracts Do Not Reduce *N. ceranae* Parasite Development Treatments

In the test including "wine derivatives", a significant increase was found for the treatment with wine vinegar [WV] in the first test up to 7.99 ± 0.24 Log of NcU ($p < 0.05$) with respect to the control [CTR] (Figure 3A,B), whereas the other substances tested [SPH and EtOH] did not cause significant differences, also in the derived test. None of the "plant extracts" tested [OPT, PP, and MT] showed significant changes, even if the mixture of Manuka and Tea oils [MT] showed a decreasing trend in the first test (7.12 ± 1.06 Log of NcU in [MT] vs. 7.69 ± 0.24 Log of NcU in [CTR]—Figure 3C,D). In the derived test OPT and PP caused the death of all honey bees until day 9 due to toxicity of the compounds used. All obtained data are reported in Table 2.

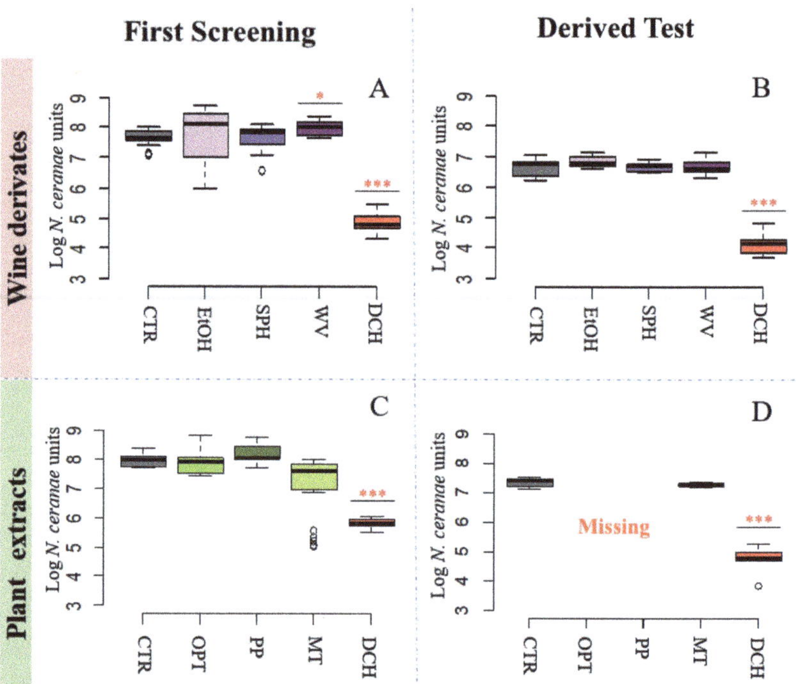

Figure 3. *N. ceranae* inhibition assays. Box plots from the first test (exploratory) and derived tests (confirmatory) are reporting the Log of *N. ceranae* units (NcU) per honey bee gut obtained at 9 days post inoculation with spores for every treatment with dietary ingredients. (I) results obtained from wine derivatives; (II) results obtained from plants extracts. Treatments included: Etanol [EtOH]; sulphites[SPH]; wine vinegar [WV]; *Opuntia ficus-indica*[OPT]; Padina pavonica [PP]; Manuka and Tea oil mixture [MT]; Fumagillin [DCH] ; infected control without treatments [CTR]. [*] $p < 0.05$; [***] $p < 0.01$. Wine derivates: First Screening (**A**); Derived Test (**B**). Plant Extracts: First Screening (**C**); Derived Test (**D**).

Table 2. The table from the first screening (exploratory) and derived tests (confirmatory) are reporting the Log of *N. ceranae* units (NcU) per honey bee gut obtained at 9 days post inoculation with spores for every treatment with dietary ingredients and the relative Standard Deviation. [CTR] = control [***] $p < 0.01$.

Reference Figure	Log NcU ± St.Dev in [CTR]	Experimental Conditions	Log NcU ± St.Dev	*p*-Value
		ORGANIC ACIDS		
		First Test		
		Acetic Acid_Low [AA_L]	7.57 ± 0.19	
2A	7.69 ± 0.24	Abscissic acid [ABA]	7.55 ± 0.21	
		Para-coumaric acid [pCA]	7.69 ± 0.29	
		Fumagillin [DCH]	4.86 ± 0.29	***
		Derived Test		
		Acetic Acid_Low [AA_L]	5.90 ± 1.20	
2B	6.64 ± 0.30	Abscissic acid [ABA]	6.86 ± 0.28	
		Para-coumaric acid [pCA]	4.23 ± 0.26	***
		Fumagillin [DCH]	4.11 ± 0.32	***
2C	7.08 ± 1.40	Acetic Acid_High [AA_H]	4.39 ± 1.20	***
		Fumagillin [DCH]	4.27 ± 0.91	***
		Acetic Acid_High [AA_H]	7.15 ± 0.07	
2D	7.36 ± 0.18	Para-coumaric acid [pCA]	7.36 ± 0.36	
		Fumagillin [DCH]	4.72 ± 0.47	***
		Saccharomyces **AND ANTIBIOTICS**		
		First Test		
		Saccharomyces sp. strain KIA1 [SC]	7.13 ± 0.65	***
2E	7.42 ± 0.24	mix, Low concentration [GRA_L]	7.78 ± 0.25	
		Nisin A, Low concentration [NisA_L]	7.53 ± 0.27	
		Fumagillin [DCH]	4.86 ± 0.29	
		Derived Test		
		Gramicidin mix, High concentration [GRA_H]	7.65 ± 0.85	
2F	7.42 ± 0.24	Nisin A, High concentration [NisA_H]	6.45 ± 1.48	***
		Nisin A, Low concentration [NisA_L]	7.18 ± 0.99	
		Fumagillin [DCH]	4.51 ± 0.58	***
		mix, High concentration [GRA_H]	6.67 ± 0.14	
2G	6.64 ± 0.28	mix, Low concentration [GRA_L]	6.57 ± 0.30	
		Fumagillin [DCH_A]	4.11 ± 0.32	***
		Saccharomyces sp. strain KIA1 [SC]	7.35 ± 0.09	
2H	7.36 ± 0.18	Nisin A, High concentration [NisA_H]	6.99 ± 0.59	
		Fumagillin [DCH]	4.72 ± 0.47	***
		WINE DERIVATES		
		First Test		
		Ethanol [EtOH]	7.78 ± 0.99	
3A	7.69 ± 0.24	Sulphites [SPH]	7.68 ± 0.40	
		Wine vinegar [WA]	7.99 ± 0.24	***
		Fumagillin [DCH]	4.86 ± 0.29	***
		Derived Test		
		Ethanol [EtOH]	6.83 ± 0.18	
3B	6.64 ± 0.31	Sulphites [SPH]	6.69 ± 0.16	
		Wine vinegar [WA]	6.67 ± 0.24	
		Fumagillin [DCH]	4.11 ± 0.32	***
		PLANT EXTRACTS		
		First Test		
		Opuntia ficus-indica extract [OPT]	7.92 ± 0.43	
3C	7.97 ± 0.24	Padina pavonica extract [PP]	8.20 ± 0.32	
		Manuka and tea oil [MT]	7.12 ± 1.06	
		Fumagillin [DCH]	5.82 ± 0.20	***
		Derived Test		
		Opuntia ficus-indica extract [OPT]	-	
3D	7.36 ± 0.18	Padina pavonica extract [PP]	-	
		Manuka and tea oil [MT]	7.28 ± 0.11	
		Fumagillin [DCH]	4.72 ± 0.47	***

4. Discussion

This work investigated the efficacy of different organic and biological agents against *N. ceranae* with the aim of finding dietary ingredients or medicaments potentially usable to control nosemosis. In recent years, this topic has become a priority in honey bee research since, at present, there are no effective and safe treatments commercially available. It is worth noting that some of the ingredients were tested against *Nosema* for the first time, although some of them are commonly handled in beekeeping practice to fight pathogens.

4.1. Organic Acids and Wine Derivatives

Feeding honey bee colonies on sugar syrup supplemented with acidifying substances was associated with an improved development of the honey bee colonies during the active season [56,57]. Moreover, it is not unusual that beekeepers add acetic acid or other acidic products to the winter food [58–60]. The justification for this practice relies on the assumption that this additive aids digestion, reduces granulation in syrup, diminishes robbing [61], and prevents the formation of molds in feeders. Moreover, the acidity may also have effects on *Nosema* spore germination. Nevertheless, there are contradictory results on the impact of acidified food on *Nosema* infections, as some studies have indicated it has no impact on the development of nosemosis caused by *Nosema apis* [62]. In spite of this, in the present research, a *N. ceranae* reduction at the highest concentration of acetic acid was shown. However, in winter bees, the reduction was significant in comparison with summer honey bees, suggesting a seasonal dependence of the described effect. Indeed, the first test was carried out in autumn, whereas the derived tests were carried out using summer honey bees. A possible explanation of the different effect is that in temperate climates worker honey bees can either develop into short-lived summer bees or long-lived winter bees. The latter show effective overall immune response [63], besides having a different protein metabolism [64] and a different composition of the microbiota [65] in order to adapt to cold temperatures. All these factors may influence *N. ceranae* development in comparison with newly emerged honey bees obtained in the summer. Further investigations are needed to establish the efficiency of acetic acid on early summer honey bees, also in field conditions. Acetic acid has a low toxicity to insects based on available data with an LD50 > 50 µg/honey bee in an acute contact toxicity study [66]. It is worth noting that the lowest concentration of the acid caused a significant mortality, whereas the highest one was well tolerated by treated honey bees. Those findings are possibly explained by differences in the amount of food consumed, related with cumulative toxic effects derived from the chronic exposure or deterrent effects. Although this hypothesis should be tested also in a longer exposure condition, our findings might be promissory to propose this compound as a candidate for in-field applications. This is also consistent with a previous research in which the administration of *Lactobacillus johnsonii* CRL1647 produced lower spore counts possibly due to the production of acidic metabolites including lactic acid, phenyl-lactic acid, and acetic acid at a concentration of 38 mM [37]. Wine vinegar, which is mainly composed of acetic acid and traditionally used by beekeepers to prevent *N. ceranae*, in our research induced an unexpected increase in the spore load. The different results between the use of acetic acid and vinegar may be explained by the presence in wine vinegar of other secondary metabolites like hydroxymethylfurfural (HMF) that is toxic for honey bees [58,67–69], although, in this study, the mortality was only 20% at the end point. In the study conducted by Ptaszyńska and collaborators [70], a strong correlation between the concentration of ethanol to feed honey bees and the proliferation of *Nosema* spp. infection was observed but, also, a higher toxic effect when 5% and 10% ethanol under a chronic administration design was given. In our study, we supplied 0.69 M (4%) of ethanol in the diet, a concentration sufficiently high to generate cellular stress in honey bees according to [71], but not enough to cause differences on mortality rates or on *N. ceranae* development, in agreement with [70]. Furthermore, Ptaszyńska and collaborators [70] suggested a correlation between acidification derived from ethanol metabolism and *Nosema* spp. development based on a facilitation of the spore germination, which could also explain some of our results.

In our research, *p*-coumaric acid was used as it was reported to possess antimicrobial potential against a bacterial infection affecting honey bee larvae, *Paenibacillus larvae* [72]. At the dose of 31.4 µM (31 ppm), the phytochemical induced only a slight reduction in the microsporidia amount in the first test, whereas in the derivative test the reduction was highly significant when compared to the respective control, but, again, this result was not confirmed when the acid was tested a third time. As mentioned above, these contrasting results might be explained by a different honey bee physiology correlated with the season and foraging resources. Summer and winter bees show distinct physiology: long-lived winter bees (diutinus stage) are characterized by oxidative stress tolerance and longevity [73,74] and senescence is almost negligible [75–77], whereas in summer worker bees, there is rapid senescence after the nurse-forager transition. In this study, *p*-coumaric acid reduced *Nosema* development only in winter honey bees. This is only partially in agreement with the results obtained by Bernklau et al. [78] that showed a significant reduction in *N. ceranae* spores with a concentration of 25 ppm in summer honey bees. Although abscisic acid dietary supplementation has been reported to stimulate the immune response and host health in honey bees [79], as well as to possibly influence the nosemosis prevalence under field condition [41], the administration of the phytohormone did not affect the development of the infection and, in addition, it was found to be toxic for honey bees. Therefore, further studies are necessary to elucidate the effect of this molecule on the bee physiology in relation to the environment.

Sulphite compounds are often used in agriculture to counteract fungal diseases, such as powdery mildew (*Uncinola necator*) in grapevine cultivation. They are also widely used in the wine industry for wine clearance and the obtained organic sulphites are a waste with potential antimicrobial activity. The honey bee food supplementation with sulphites resulted inactive against the microsporidia *N. ceranae*. This is in contrast with [80] that showed that sulphated-polysaccharides can counteract *N. ceranae*. The contrasting result is probably due to the different origin of the sulphated organic compounds used.

4.2. Saccharomyces and Antibiotics

The use of antimicrobial compounds to treat nosemosis was largely studied, being fumagillin the most effective against *N. ceranae* [81]. Its effectiveness was also deeply confirmed in our study, in contrast to Huang et al. [21], also causing low mortality rates. However, its use is controversial, due to its toxicity both to bees and humans, which limits its use in beekeeping because of the presence of possible residues in hive products. Therefore, searching for new antibiotic substances alternative to fumagillin, such as gramicidin and nisin, was one of the aims of this work. The cyclic peptide gramicidin S is produced by *Aneurinibacillus migulanus* DSM2895^T, formerly *Bacillus brevis*, that was reported to be highly effective against *N. apis* development [81,82] and active against fungi [83]. Since this substance is no longer available on the market, we opted for gramicidin D (a mixture of gramicidin A, B, and C, produced by *Bacillus brevis* ATCC 8185 [84], a linear channel forming peptides). To the best of our knowledge, gramicidin D has never been tested against microsporidia. The results obtained showed no effects but, on the contrary, an increase in the *N. ceranae* count. A possible explanation may be that gramicidin S compromises the integrity of lipid layer of the cytoplasmic membrane of gram-positive, gram-negative bacteria, and fungi [83], whereas the types A, B, and C are linear peptides with low solubility in water with an antimicrobial action described only against gram-positive bacteria [85]. Nisin, a lantibiotic produced by *Lactococcus lactis*, largely used in food packaging as a preservative and until now never tested as a potential nosemosis treatment, gave different results with respect to gramicidin. This polycyclic polypeptide caused a lower development of *N. ceranae* in honey bee gut, showing a significant effect when the concentration was increased in the derived test and administered to winter honey bees. Based on our results, nisin represents a potential anti-microsporidian for further lab and field tests. Considering the different results obtained to counteract *Nosema* proliferation of the different anti-microbial agents, a higher efficacy of the cyclic molecules with respect to the linear ones may be postulated. This is also confirmed by the work of [36] in which a cyclic surfactin produced by *Bacillus subtilis* was effective against the viability of *Nosema* spores. In the light of these considerations, other experiments are envisaged to validate the hypothesis but also to validate the efficacy of bacteriocin administration in the field. The use of *Saccharomyces* was determined by its presence in nectar and pollen [86,87] and it was also shown to be beneficial to honey bees [88]. The tested concentration (10^{11} CFU/ml) was found to be toxic to honey bees, whereas the effects on *N. ceranae* were not reproducible between the two performed test. Considering that a significant *N. ceranae* reduction was obtained in the first screening, we believe that the dose of the administered yeast should be adjusted.

4.3. Plant Extracts

The administration of plant extracts to honey bees, including extracts from plant material with different extraction methods, essential oils and single main components were widely tested for possible antiparasitic activity against nosemosis with different results [28,30,31,89]. In the case of the natural oils and plant extracts tested in our work, no significant results were obtained against *N. ceranae* development. A possible exception was found when manuka and tea oils in combinations were tested, obtaining a reduction of *Nosema* spores (not statistically significant), thus indicating that a combination of these oils may be a strategy to pursue. Conversely, *Opuntia* extract, notably rich in polyphenols, vitamins, polyunsaturated fatty acids, and amino acids, and whose antimicrobial activity has been studied against *Campylobacter* spp. in poultry [90], did not exhibit any positive effects but led to an increase in the microsporidia development.

Finally, Roussel et al. [80] found positive effects of some sulphated polysaccharides from different marine algae against *N. ceranae* infection. In this study, the Mediterranean seaweed *Padina pavonica* pure extract, possessing antibacterial and anti-*Candida* activities [91], caused a spore load increase in treated honey bees and high toxicity, differently from what expected.

5. Conclusions

In this work, we proposed a screening of innovative ingredients that were never tested against the development of *N. ceranae*. Many of these ingredients were selected based on treatments used by beekeepers that often apply some practices basing on empiric knowledge. In the present study it was pointed out that the cyclic antibiotic nisin is efficient in the control of *N. ceranae* even if it shows high mortality rates in cage texts. Moreover, organic acids, such as acetic acid, might be a valid alternative to control the disease avoiding contaminant residuals. It is important to highlight that the efficacy of some compounds seems to be strongly correlated with the seasonal physiology of honey bees, and this factor should be better considered in future studies. Therefore, the promising results from nisin, acetic acid, *p*-coumaric acid, and *Saccharomyces* sp. against the development of nosemosis point out that the use of these ingredients needs to be further explored both in laboratory and field conditions.

Author Contributions: Conceptualization, C.B. and D.A.; methodology, M.P.P.; validation C.B. and D.A.; formal analysis and investigation, C.B., D.A., M.P.P. and P.M.G.; data curation and statistics, L.B., D.A.; writing original draft preparation, C.B., D.A. and L.B.; writing review and editing, L.B., P.M.G., M.P.P. and D.D.G.; visualization C.B.; supervision, L.B. and D.D.G.; project administration D.D.G. All authors have read and agreed to the published version of the manuscript.

Funding: The research was partially funded by the EU project "NOurishingPROBiotics to bees to Mitigate Stressors" (NO PROBleMS), H2020-MSCA-RISE 2017, GA 777760, 2018-2022. The founder had no role in the study design, data collection and interpretation, or the decision to submit the work for publication.

Institutional Review Board Statement: Ethical review and approval were waived for this study, because the Italian law does not require and ethical approval for tests performed on arthropods with exceptions of cephalopods according to the D.L. 4 March 2014 n. 26, and national implementing decree following the European regulation 2010/63/UE.

Informed Consent Statement: Not applicable.

Data Availability Statement: Elaborated data presented in this study are available on reasonable request from the corresponding author.

Conflicts of Interest: The authors declare no conflict of interest.

Abbreviations

The following abbreviations are used in this manuscript:

qPCR	Quantitative-polymerase chain reaction
MetAP2	Methionine aminopeptidase-2
AA_L	Acetic acid low concentration
AA_H	Acetic acid high concentration
ABA	Abscissic acid
pCA	Para-coumaric acid
SC	*Saccharomyces* sp. strain KIA1
GRA_L	Mixture of gramicidin A, B, and C at low concentration
GRA_H	Mixture of gramicidin A, B, and C at high concentration
NisA_L	Nisin A at low concentration
NisA_H	Nisin A at high concentration
EtOH	Ethanol
SPH	Sulphites
WA	Wine vinegar
OPT	*Opuntia ficus-indica* extract
PP	*Padina pavonica* extract
MT	Manuka and tea oil mixture
DCH	Fumagillin
CTR	Control
PDB	Potatoes dextrose broth
CFU	Colony forming unit
RH	Relative humidity
PCR	Polymerase chain reaction
ANOVA	Analysis of variance
NcU	*Nosema ceranae* units

References

1. Higes, M.; Martín-Hernández, R.; Botías, C.; Garrido-Bailón, E.; González-Porto, A.V.; Barrios L.; del Nozal, M.; Bernal, J.M.; Jiménez, J.J.; García-Palencia, P.; et al. How natural infection by Nosema ceranae causes honey bee colony collapse. *Environ. Microbiol.* **2008** *10*, 2659–2669. [CrossRef]
2. Huang, W.F.; Jiang, J.H.; Chen, Y.W.; Wang, C.H. A Nosema ceranae isolate from the honeybee *Apis mellifera*. *Apidologie* **2007**, *38*, 30–37. [CrossRef]
3. Higes, M.; Martin-Hernández, R.; Meana, A. Nosema ceranae, a new microsporidian parasite in honey bees in Europe. *J. Invertebr. Pathol.* **2006**, *92*, 93–95. [CrossRef] [PubMed]
4. Rao, G.M.; Rao, K.S.; Chaudhary, O.P. Introduction of Apis mellifera in India. *Khadi Gramodyog* **1993**, *34*, 815–819.
5. Paxton, R.J.; Klee, J.; Korpela, S.; Fries, I. Nosema ceranae has infected Apis mellifera in Europe since at least 1998 and may be more virulent than Nosema apis. *Apidologie* **2007**, *38*, 558–565. [CrossRef]
6. Tokarev, Y.S.; Huang, W.F.; Solter, L.F.; Malysh, J.M.; Becnel, J.J.; Vossbrinck, C.R. A formal redefinition of the genera Nosema and Vairimorpha (Microsporidia: Nosematidae) and reassignment of species based on molecular phylogenetics. *J. Invertebr. Pathol.* **2020**, *169*, 107279. [CrossRef] [PubMed]
7. Higes, M.; Martín-Hernández, R.; Garrido-Bailón, E.; García-Palencia, P.; Meana, A. Detection of infective Nosema ceranae (Microsporidia) spores in corbicular pollen of foranger honey bees. *J. Invert. Pathol.* **2008**, *97*, 76–78. [CrossRef] [PubMed]
8. Higes, M.; Martín-Hernández, R.; Meana, A. Nosema ceranae in Europe: An emergent type C nosemosis. *Apidologie* **2010**, *41*, 375–392. [CrossRef]

9.   Graystock, P.; Goulson, D.; Hughes, W.O.H. Parasites in bloom: Flowers aid dispersal and transmission of pollinator parasites within and between bee species. *Proc. R. Soc.* **2015**, *282*, 20151371. [CrossRef]

10.   Genersch, E. Honey bee pathology: Current threats to honey bees and beekeeping. *Appl. Microbiol. Biotechnol.* **2010**, *87*, 87–107. [CrossRef]

11.   Higes, M.; Martín-Hernández, R.; Garrido-Bailón, E.; González-Porto, A.V.; García-Palencia, P.; Meana, A.; Del Nozal, M.J.; Mayo, R.; Bernal, J.L. honey bee colony collapse due to Nosema ceranae in professional apiaries. *Environ. Microbiol. Rep.* **2009**, *1*, 110–113. [CrossRef] [PubMed]

12.   Higes M.; Nozal M.J.; Alvaro A.; Barrios L.; Meana A.; Martín-Hernández R.; Bernal J.L.; Bernal J. The stability and effectiveness of fumagillin in controlling Nosema ceranae (Microsporidia) infection in honey bees (*Apis mellifera*) under laboratory and field conditions. *Apidologie* **2011**, *42*, 364–377. [CrossRef]

13.   Van den Heever, J.P.; Thompson T.S.; Otto S.J.G.; Curtis, J.M.; Ibrahim A.; Pernal, S.F. Evaluation of Fumagilin-B® and other potential alternative chemotherapies against Nosema ceranae -infected honey bees (*Apis mellifera*) in cage trial assays. *Apidologie* **2016**, *47*, 617–630. [CrossRef]

14.   Bailey L. Effect of fumagillin upon Nosema apis (*Zander*). *Nature* **1953**, *171*, 212–213. [CrossRef] [PubMed]

15.   Mutinelli, F. European legislation governing the authorization of veterinary medicinal products with particular reference to the use of drugs for the control of honey bee diseases. *Apiacta* **2003**, *38*, 156–168.

16.   Sin, N.; Meng, L.; Wang, M.Q.; Wen, J.J.; Bornmann, W.G.; Crews, C.M. The anti-angiogenic agent fumagillin covalently binds and inhibits the methionine aminopeptidase, MetAP-2. *Proc. Natl. Acad. Sci. USA* **1997**, *94*, 6099–6103. [CrossRef]

17.   Stanimirović, Z.; Jevrosima, S.; Kulić, M.; Stojic, V. Frequency of chromosomal aberrations in the evaluation of genotoxic potential of dicyclohexylamine (fumagillin) in vivo. *Acta Vet.* **2006**, *56*. [CrossRef]

18.   Lopez, M.I.; Pettis, J.S.; Smith, I.B.; Chu, P.S. Multiclass determination and confirmation of antibiot- ic residues in honey using LC–MS/MS. *J. Agric. Food Chem.* **2008**, *56*, 1553–1559. [CrossRef] [PubMed]

19.   Van den Heever, J.P.; Thompson, T.S.; Curtis, J.M.; Pernal, S.F. Stability of dicyclohexylamine and fumagillin in honey. *Food Chem.* **2015**, *179*, 152–158. [CrossRef] [PubMed]

20.   Webster, T.C. Fumagillin Affects *Nosema apis* and honey bees (Hymenoptera-Apidae). *J. Econ. Entomol.* **1994**, *87*, 601–604. [CrossRef]

21.   Huang, W.F.; Solter, L.F.; Yau, P.M.; Imai, B.S. Nosema ceranae escapes fumagillin control in honey bees. *PLoS Pathog.* **2013**, *9*, e1003185. [CrossRef]

22.   Abdelkader, F.B.; Çakmak, İ.; Çakmak, S.S.; Nur, Z.; İncebıyık, E.; Aktar, A.; Erdost, H. Toxicity assessment of chronic exposure to common insecticides and bee medications on colony development and drones sperm parameters. *Ecotoxicology* **2021**, *30*, 806–817. [CrossRef] [PubMed]

23.   Mendoza, Y.; Diaz-Cetti, S.; Ramallo, G.; Santos, E.; Porrini, M.; Invernizzi, C. Nosema ceranae winter control: Study of the effectiveness of different fumagillin treatments and consequences on the strength of honey bee (Hymenoptera: Apidae) colonies. *J. Econ. Entomol.* **2017**, *110*, 1–5. [CrossRef] [PubMed]

24.   Cilia, G.; Fratini, F.; Tafi, E.; Turchi, B.; Mancini, S.; Sagona, S.; Nanetti, A.; Cerri, D.; Felicioli, A. Microbial profile of the *ventriculum* of honey bee (*Apis mellifera ligustica* Spinola, 1806) fed with veterinary drugs, dietary supplements and non-protein amino acids. *Vet. Sci.* **2020**, *7*, 76. [CrossRef]

25.   Alberoni, D.; Baffoni, L.; Braglia, C.; Gaggìa, F.; Di Gioia, D. Honeybees Exposure to Natural Feed Additives: How Is the Gut Microbiota Affected?. *Microorganisms* **2021**, *9*, 1009. [CrossRef]

26.   Maistrello, L.; Lodesani, M.; Costa, C.; Leonardi, F.; Marani G.; Caldon, M.; Mutinelli, F.; Granato, A. Screening of natural compounds for the control of nosema disease in honey bees (*Apis mellifera*). *Apidologie* **2008**, *39*, 436–445. [CrossRef]

27.   Borges, D.; Guzman-Novoa, E.; Goodwin, P.H. Effects of prebiotics and probiotics on honey bees (*Apis mellifera*) infected with the microsporidian parasite Nosema ceranae. *Microorganism* **2021**, *9*, 481. [CrossRef]

28.   Cilia, G.; Garrido, C.; Bonetto, M.; Tesoriero, D.; Nanetti, A. Effect of Api-Bioxal® and ApiHerb® treatments against Nosema ceranae infection in Apis mellifera investigated by two qPCR methods. *Vet. Sci.* **2020**, *7*, 125. [CrossRef]

29.   Porrini, M.P.; Fernández, N.J.; Garrido, P.M.; Gende, L.B.; Medici, S.K.; Eguaras, M.J. In vivo evaluation of antiparasitic activity of plant extracts on *Nosema ceranae* (Microsporidia). *Apidologie* **2011**, *42*, 700–707. [CrossRef]

30.   Lee, J.K.; Kim, J.H.; Park, J.K. Evaluation of antimicrosporidian activity of plant extracts on Nosema ceranae. *J. Apic. Sci.* **2016**, *60*, 167–178. [CrossRef]

31.   Bravo, J.; Carbonell, V.; Sepúlveda, B.; Delporte, C.; Valdovinos, C.E.; Martín-Hernández, R.; Higes, M. Antifungal activity of the essential oil obtained from Cryptocarya alba against infection in honey bees by Nosema ceranae. *J. Invert Pathol.* **2017**, *149*, 141–147. [CrossRef]

32.   Arismendi, N.; Vargas, M.; López, M.D.; Barría, Y.; Zapata, N. Promising antimicrobial activity against the honey bee parasite Nosema ceranae by methanolic extracts from Chilean native plants and propolis. *J. Apicult Res.* **2018**, *57*, 522–535. [CrossRef]

33.   Chen, X.; Wang, S.; Xu, Y.; Gong, H.; Wu, Y.; Chen, Y.; Hu, F.; Zheng, H. Protective potential of Chinese herbal extracts against microsporidian Nosema ceranae, an emergent pathogen of western honey bees, *Apis mellifera* L. *J. Asia Pac. Entomol.* **2019**, *24*, 502–512. [CrossRef]

34.   Nanetti, A.; Ugolini, L.; Cilia, G.; Pagnotta, E.; Malaguti, L.; Cardaio, I.; Lazzeri, L. Seed meals from Brassica nigra and Eruca sativa control artificial Nosema ceranae infections in Apis mellifera. *Microorganisms* **2019**, *9*, 949. [CrossRef]

35. Baffoni, L.; Gaggìa, F.; Alberoni, D.; Cabbri, R.; Nanetti, A.; Biavati, B.; Di Gioia, D. Effect of dietary supplementation of Bifidobacterium and Lactobacillus strains in Apis mellifera L. against Nosema ceranae. *Benef. Microbes* **2016**, *7*, 45–51. [CrossRef] [PubMed]
36. El Khoury, S.; Rousseau, A.; Lecoeur, A.; Cheaib, B.; Bouslama, S.; Mercier, P.L.; Demey, V.; Castex, M.; Giovenazzo, P.; Derome, N. Deleterious interaction between honey bees (*Apis mellifera*) and its microsporidian intracellular parasite Nosema ceranae was mitigated by administrating either endogenous or allochthonous gut microbiota strains. *Front. Ecol. Evol.* **2018**, *6*, 58. [CrossRef]
37. Porrini, M.P.; Audisio, M.C.; Sabaté, D.C.; Ibarguren, C.; Medici, S.K.; Sarlo, E.G.; Garrido, M.P.; Eguaras, M.J. Effect of bacterial metabolites on microsporidian Nosema ceranae and on its host Apis mellifera. *Parasitol. Res.* **2010**, *107*, 381–388. [CrossRef]
38. Maggi, M.; Negri, P.; Plischuk, S.; Szawarski, N.; De Piano, F.; De Feudis, L.; Eguaras M.; Audisio, C. Effects of the organic acids produced by a lactic acid bacterium in Apis mellifera colony development, Nosema ceranae control and fumagillin efficiency. *Vet. Microbiol.* **2013**, *167*, 474–483. [CrossRef]
39. De Piano, F.G.; Maggi, M.; Pellegrini, M.C.; Cugnata, N.M.; Szawarski, N.; Buffa, F.; Negri, P.; Fuselli, S.R.; Audisio, C.M.; Ruffinengo, S.R. Effects of Lactobacillus johnsonii AJ5 metabolites on nutrition, Nosema ceranae development and performance of Apis mellifera. *J. Apic. Sci.* **2011**, *61*, 93–104. [CrossRef]
40. Vaillant, J. Nourrissement au sirop de sucre acidifié. *La Santé de L 'abeille* **1989**, *110*, 55–60.
41. Forsgren, E.; Fries, I. Acidic-benzoic feed and nosema disease. *J. Apia. Sci.* **2005** *49*, 81–88
42. Szawarski, N.; Saez, A.; Domínguez, E.; Dickson, R.; De Matteis, Á.; Eciolaza, C.; Justel, M.; Aliano, A.; Solar, P.; Bergara, I.; et al. Effect of abscisic acid (ABA) combined with two different beekeeping nutritional strategies to confront overwintering: Studies on honey bees' population dynamics and nosemosis. *Insects* **2019**, *10*, 329. [CrossRef]
43. Mao, W.; Schuler, M.A.; Berenbaum, M.R. Honey constituents up-regulate detoxification and immunity genes in the western honey bee Apis mellifera. *Proc. Natl. Acad. Sci. USA* **2013**, *110*, 8842–8846. [CrossRef] [PubMed]
44. Martín-Hernández, R.; Meana, A.; Prieto, L.; Martínez-Salvador, A.; Garrido-Bailon, E.; Higes, M. Outcome of colonization of Apis mellifera by Nosema ceranae. *Appl. Environ. Microbiol.* **2007**, *73*, 6331–6338. [CrossRef] [PubMed]
45. Williams, G.R.; Alaux, C.; Costa, C.; Csáki, T.; Doublet, V.; Eisenhardt, D.; Fries, I.; Kuhn, R.; McMahon, D.P.; Medrzycki, P.; et al. Standard methods for maintaining adult Apismellifera in cages under invitro laboratory conditions. *J. Apic. Res.* **2013**, *52*, 1–36. [CrossRef] ·
46. Porrini, M.P.; Garrido, P.M.; Eguaras, M.J. Individual feeding of honey bees: Modification of the Rinderer technique. *J. Apic. Res.* **2013**, *52*, 194–195. [CrossRef]
47. Robic, A.; Sassi, J.F.; Lahaye, M. Impact of stabilization treatments of the green seaweed Ulva rotundata (Chlorophyta) on the extraction yield, the physico-chemical and rheological properties of ulvan.*Carbohydr. Polym.* **2008**, *74*, 344–352. [CrossRef]
48. Ribeiro, R.C.D.A.; Barreto, S.M.A.G.; Ostrosky, E.A.; Rocha-Filho, P.A.D.; Veríssimo, L.M.; Ferrari, M. Production and characterization of cosmetic nanoemulsions containing Opuntia ficus-indica (L.) Mill extract as moisturizing agent. *Molecules* **2015**, *20*, 2492–2509. [CrossRef]
49. Baffoni, L.; Alberoni, D.; Gaggìa, F.; Braglia, C.; Stanton, C.; Ross, P.R.; Di Gioia, D. Honeybee exposure to veterinary drugs: How is the gut microbiota affected? *Microbiol. Spectr.* **2021**, *2021*, e0017621. [CrossRef]
50. Huang, W.F.; Solter, L.F. Comparative development and tissue tropism of Nosema apis and Nosema ceranae. *J. Invertebr. Pathol.* **2013**, *113*, 35–41. [CrossRef] [PubMed]
51. Alberoni, D.; Favaro, R.; Baffoni, L.; Angeli, S.; Di Gioia, D. Neonicotinoids in the agroecosystem: In-field long-term assessment on honey bee colony strength and microbiome. *Sci. Total Environ.* **2021**, *762*, 144116. [CrossRef]
52. Cilia, G.; Sagona, S.; Giusti, M.; Dos Santos, P.E.J.; Nanetti, A.; Felicioli, A. Nosema ceranae infection in honey bee samples from Tuscanian Archipelago (Central Italy) investigated by two qPCR methods. *Saudi J. Biol. Sci.* **2019**, *26*, 1553–1556. [CrossRef] [PubMed]
53. Cilia, G.; Cabbri, R.; Maiorana, G.; Cardaio, I.; Dall'Olio, R.; Nanetti, A. A Novel TaqMan® assay for Nosema ceranae quantification in honey bee, based on the protein coding gene Hsp70. *Eur. J. Protistol.* **2018**, *63*, 44–50. [CrossRef] [PubMed]
54. R Core Team. A Language and Environment for Statistical Computing. Available online: https://www.R-project.org/ (accessed on 23 March 2021).
55. Cook, R.D. Influential observations in linear regression. *J. Am. Stat. Assoc.* **1979**, *74*, 169–174. [CrossRef]
56. Pătruică, S.; Bogdan, A.T.; Bura, M.; Popovici D. Research on the effect of acidifying substances on bee colonies development and health in spring. *Agrobuletin AGIR* **2011**, *2*, 124–130.
57. Pătruică, S.; Dumitrescu, G.; Popescu, R.; Filimon, N.M. The effect of prebiotic and probiotic products used in feed to stimulate the bee colony (*Apis mellifera*) on intestines of working bees. *J. Food. Agricul. Environ.* **2013**, *11*, 2461–2464. [CrossRef]
58. Frizzera, D.; Fabbro, S.D.; Ortis, G.; Zanni, V.; Bortolomeazzi, R.; Nazzi, F.; Annoscia, D. Possible side effects of sugar supplementary nutrition on honey bee health. *Apidologie* **2020**, *51*, 594–608. [CrossRef]
59. Claing, G.; Dubreuil, P.; Ferland, J.; Bernier, M.; Arsenault, J. Beekeeping management practices in southwestern Quebec. *Can. J. Vet. Res.* **2019**, *1*, 103.
60. Mederle, N.; Kaya, A.; Morariu, S.; CrăCiun, C.; Balint, A.; Radulov, I.; Dărăbuș, G. Innovative therapeutic approaches for the control and prevention of honey bee nosemosis: Implications in the bee's life and the quality of their products. *Geogr. Timisiensis***2017**, *26* , 31–40.

61. Brighenti, D.M.; Brighenti, C.R.; Carvalho, C.F. Life spans of Africanized honey bees fed sucrose diets enhanced with citric acid or lemon juice. *J. Apicul. Res.* **2017**, *56*, 91–99. [CrossRef]

62. Forsgren, E.; Fries, I. Acidic Food and Nosema Disease. In Proceedings of the 38th International Apimondia Congress in Ljubljana, Ljubljana, Slovenia, 24–29 August 2003; Volume 488.

63. Gätschenberger, H.; Azzami, K.; Tautz, J.; Beier, H. Antibacterial immune competence of honey bees (*Apis mellifera*) is adapted to different life stages and environmental risks. *PLoS ONE* **2013**, *8*, e66415. [CrossRef]

64. Crailsheim K. Dependence of protein metabolism on age and season in the honey bee (*Apis mellifica carnica* Pollm). *J. Insect Physiol.* **1986**, *32*, 629–634. [CrossRef]

65. Kešnerová, L.; Emery, O.; Troilo, M.; Liberti, J.; Erkosar, B.; Engel, P. Gut microbiota structure differs between honey bees in winter and summer. *ISME J.* **2020**, *14*, 801–814. [CrossRef]

66. USEPA. Acetic Acid and Sodium Diacetate Interim Registration Review Decision, Docket Number EPA-HQ-OPP-2008-0016. Available online: https://www.regulations.gov/document?D=EPA-HQ-OPP-2008-0016-0017 (accessed on 7 July 2021).

67. Theobald, A.; Müller, A.; Anklam, E. Determination of 5- hydroxymethylfurfural in vinegar samples by HLPC. *J. Agric. Food Chem.* **1998**, *46*, 1850–1854. [CrossRef]

68. LeBlanc, B.W.; Eggleston, G.; Sammataro, D.; Cornett, C.; Dufault, R.; Deeby, T.; St. Cyr, E. Formation of hydroxymethylfurfural in domestic high-fructose corn syrup and its toxicity to the honey bee (*Apis mellifera*). *J. Agric. Food Chem.* **2009**, *57*, 7369–7376. [CrossRef] [PubMed]

69. Zirbes, L.; Nguyen, B.K.; de Graaf, D.C.; de Meulenaer, B.; Reybroeck, W.; Haubruge, E.; Saegerman, C. Hydroxymethylfurfural: A possible emergent cause of honey bee mortality? *J. Agric. Food Chem.* **2013**, *61*, 11865–11870. [CrossRef]

70. Ptaszyńska, A.A.; Borsuk, G.; Mułenko, W.; Olszewski, K. Impact of ethanol on Nosema spp. infected bees. *Med. Weter* **2013**, *69*, 736–740.

71. Hranitz, J.M.; Abramson, C.I.; Carter, R.P. Ethanol increases HSP70 concentrations in honey bee (*Apis mellifera* L.) brain tissue. *Alcohol* **2010**, *44*, 275–282. [CrossRef]

72. Szawarski, N.; Giménez-Martínez, P.; Mitton, G.; Negri, P.; Meroi-Arcerito, F.; Moliné, M.P.; Fuselli, S.; Eguaras, M.; Lamattina, L.; Maggi, M. Short communication: Antimicrobial activity of indoleacetic, gibberellic and coumaric acids against Paenibacillus larvae and its toxicity against Apis mellifera. *Span. J. Agric. Res.* **2020**, *18*, e05SC01. [CrossRef]

73. Amdam, G.V.; Hartfelder, K.; Norberg, K.; Hagen, A.; Omholt, S.W. Altered physiology in worker honey bees (Hymenoptera: Apidae) infested by the mite Varroa destructor (Acari: Varroidae): A factor in colony loss during over-wintering? *J. Econ. Entomol.* **2004**, *97*, 741–747. [CrossRef]

74. Amdam, G.V.; Norberg, K.; Omholt, S.W.; Kryger, P.; Lourenco, A.P.; Bitondi, M.M.G.; Simões, Z.L.P. Higher vitellogenin concentrations in honey bee workers may be an adaptation to life in temperate climates. *Insectes Soc.* **2005**, *52*, 316–319. [CrossRef]

75. Corona, M.; Velarde, R.A.; Remolina, S.; Moran-Lauter, A.; Wang, Y.; Hughes, K.A.; Robinson, G.E. Vitellogenin, juvenile hormone, insulin signaling, and queen honey bee longevity. *Proc. Natl. Acad. Sci. USA* **2007**, *104*, 7128–7133. [CrossRef] [PubMed]

76. Münch, D.; Amdam, G.V.; Wolschin, F. Ageing in a eusocial insect: Molecular and physiological characteristics of life span plasticity in the honey bee. *Funct. Ecol.* **2008**, *22*, 407–421. [CrossRef] [PubMed]

77. Johnson, B.R. Within-nest temporal polyethism in the honey bee. *Behav. Ecol. Sociobiol.* **2008**, *62*, 777–784. [CrossRef]

78. Bernklau, E.; Bjostad, L.; Hogeboom, A.; Carlisle, A.; HS, A. Dietary phytochemicals, honey bee longevity and pathogen tolerance. *Insects* **2019**, *10*, 14. [CrossRef]

79. Negri, P.; Maggi, M.; Ramirez, L.; De Feudis, L.; Szwarski, N.; Quintana, S.; Eguaras, M.; Lamattina, L. Abscisic acid enhances the immune response in Apis mellifera and contributes to the colony fitness. *Apidologie* **2015**, *46*, 542–557. [CrossRef]

80. Roussel, M.; Villay, A.; Delbac, F.; Michaud, P.; Laroche, C.; Roriz, D.; El Alaoui, H.; Diogon, M. Antimicrosporidian activity of sulphated polysaccharides from algae and their potential to control honey bee nosemosis. *Carbohydr. Polym.* **2015**, *133*, 213–220. [CrossRef]

81. Williams, G. R.; Sampson, M.A.; Shutler, D.; Rogers, R.E. Does fumagillin control the recently detected invasive parasite Nosema ceranae in western honey bees (Apis mellifera)? *J. Invertebr. Pathol.* **2008**, *99*, 342–344. [CrossRef]

82. Alpatov, V.V.; Kirikova, V.A. Apicultural Abstracts. *Bee World* **1954**, *35*, 181–192.

83. Prenner, E.J.; Lewis, R.N.; McElhaney, R.N. The interaction of the antimicrobial peptide gramicidin S with lipid bilayer model and biological membranes. *Biochim. Biophys. Acta-Biomembr.* **1999**, *1462*, 201–221. [CrossRef]

84. Kessler, N.; Schuhmann, H.; Morneweg, S.; Linne, U.; Marahiel, M.A. The linear pentadecapeptide gramicidin is assembled by four multimodular nonribosomal peptide synthetases that comprise 16 modules with 56 catalytic domains. *J. Biol. Chem.* **2004**, *279*, 7413–7419. [CrossRef]

85. Wang, F.; Qin, L.; Pace, C.J.; Wong, P.; Malonis, R.; Gao, J. Solubilized gramicidin A as potential systemic antibiotics. *Chem. BioChem.* **2012**, *3*, 51–55. [CrossRef] [PubMed]

86. Liu, H.; Macdonald, C.A.; Cook, J.; Anderson, I.C.; Singh, B.K. An ecological loop: Host microbiomes across multitrophic interactions. *Trends Ecol. Evol.* **2019** *34*, 1118–1130. [CrossRef]

87. Abe, T.; Toyokawa, Y.; Sugimoto, Y.; Azuma, H.; Tsukahara, K.; Nasuno, R.; Watanabe, D.; Tsukahara, M.; Takagi, H. Characterization of a new Saccharomyces cerevisiae isolated from hibiscus flower and its mutant with l-leucine accumulation for awamori brewing. *J. Front. Genet.* **2019**, *10*, 490. [CrossRef] [PubMed]

88. Tauber, J.P.; Nguyen, V.; Lopez, D.; Evans, J.D. Effects of a resident yeast from the honey bee gut on immunity, microbiota, and Nosema disease. *Insects* **2019**, *10*, 296. [CrossRef]
89. Damiani, N.; Fernández, N.J.; Porrini, M.P.; Gende, L.B.; Álvarez, E.; Buffa, F.; Brasesco, C.; Maggi, D.M.; Marcangeli A.J.; Eguaras, M.J. Laurel leaf extracts for honey bee pest and disease management: Antimicrobial, microsporicidal, and acaricidal activity. *Parasit. Res.* **2014**, *113*, 701–709. [CrossRef]
90. Castillo, S.L.; Heredia, N.; Contreras, J.F.; García, S. Extracts of edible and medicinal plants in inhibition of growth, adherence, and cytotoxin production of Campylobacter jejuni and Campylobacter coli. *J. Food Sci.* **2011**, *76*, M421–M426. [CrossRef]
91. Ismail, A.; Ktari, L.; Ahmed, M.; Bolhuis, H.; Boudabbous, A.; Stal, L.J.; Cretoiu, M.S.; El Bour, M. Antimicrobial activities of bacteria associated with the brown alga Padina pavonica. *Front. Microbiol.* **2016**, *7*, 1072. [CrossRef] [PubMed]

Article

The Herbal Supplements NOZEMAT HERB® and NOZEMAT HERB PLUS®: An Alternative Therapy for *N. ceranae* Infection and Its Effects on Honey Bee Strength and Production Traits

Rositsa Shumkova [1], Ralitsa Balkanska [2] and Peter Hristov [3,*]

[1] Research Centre of Stockbreeding and Agriculture, Agricultural Academy, Smolyan 4700, Bulgaria; rositsa6z@abv.bg
[2] Institute of Animal Science, Department Special Branches, Agricultural Academy, Kostinbrod 2230, Bulgaria; r.balkanska@gmail.com
[3] Institute of Biodiversity and Ecosystem Research, Department of Animal Diversity and Resources, Bulgarian Academy of Sciences, Sofia 1113, Bulgaria
* Correspondence: peter_hristoff@abv.bg; Tel.: +359-2-979-2327

Abstract: Honey bees (*Apis mellifera* L.) are the most effective pollinators for different crops and wild flowering plants, thus maintaining numerous ecosystems in the world. However, honey bee colonies often suffer from stress or even death due to various pests and diseases. Among the latter, nosemosis is considered to be one of the most common diseases, causing serious damage to beekeeping every year. Here, we present, for the first time, the effects from the application of the herbal supplements NOZEMAT HERB® (NH) and NOZEMAT HERB PLUS® (NHP) for treating *N. ceranae* infection and positively influencing the general development of honey bee colonies. To achieve this, in autumn 2019, 45 colonies were selected based on the presence of *N. ceranae* infections. The treatment was carried out for 11 months (August 2019–June 2020). All colonies were sampled pre- and post-treatment for the presence of *N. ceranae* by means of light microscopy and PCR analysis. The honey bee colonies' performance and health were evaluated pre- and post-treatment. The obtained results have shown that both supplements have exhibited statistically significant biological activity against *N. ceranae* in infected apiaries. Considerable enhancement in the strength of honey bee colonies and the amount of sealed workers was observed just one month after the application of NH and NHP. Although the mechanisms of action of NH and NHP against *N. ceranae* infection are yet to be completely elucidated, our results suggest a new holistic approach as an alternative therapy to control nosemosis and to improve honey bee colonies' performance and health.

Keywords: microsporidia; nosemosis control; phytotherapy; 16S rDNA gene

Citation: Shumkova, R.; Balkanska, R.; Hristov, P. The Herbal Supplements NOZEMAT HERB® and NOZEMAT HERB PLUS®: An Alternative Therapy for *N. ceranae* Infection and Its Effects on Honey Bee Strength and Production Traits. *Pathogens* **2021**, *10*, 234. https://doi.org/10.3390/pathogens10020234

Academic Editors: Giovanni Benelli and Giovanni Cilia

Received: 31 December 2020
Accepted: 18 February 2021
Published: 19 February 2021

Publisher's Note: MDPI stays neutral with regard to jurisdictional claims in published maps and institutional affiliations.

1. Introduction

The European honey bee, *Apis mellifera* (Linnaeus, 1758) is the most effective and globally distributed pollinator not only of a large number of important crops but also of wild flowering plants, some of which play an essential role in maintaining ecosystem services [1–3]. Over the past several decades, there has been a significant reduction in bee colonies, especially in some geographical regions (e.g., North America), which has raised great public and societal concern [4,5]. The decline of the honey bee population has the most tangible effect on food sources for human and livestock, disrupting wild plant pollination and diversity, altering ecological interactions and function, decreasing crop yields, reducing the yield of bee products, a large number of which have important medical value, etc. [6–9]. A honey bee colony may harbour a wide variety of disease agents and pests, bacteria, fungi, honey bee-associated viruses, parasitic mites and even other insects that try to take advantage of the rich resources contained within bee colonies [10,11]. Among them, *N. ceranae* (Fries et al. 1996) has been implicated in inflicting annually heavy losses on beekeeping [12,13]. Until 2017, two microsporidia were known—*N. apis* (Zander, 1909) (causing nosemosis Type A) and *N. ceranae* (causing nosemosis Type C) [14]. *N. ceranae* is specific for the Asiatic honey bee (*Apis cerana*, Fabricius, 1793); however, presumably after 2003, *N. ceranae* has switched its hosts and has begun to infect the European honey bee (*Apis mellifera*) [15,16]. Nowadays, *N. ceranae* has become widespread microsporidia in many regions in the world [17,18]. Based on ultra-structural and molecular investigations, a new *Nosema* species in *Apis mellifera*, namely *N. neumanni* (Chemurot et al. 2017), was described in Uganda in 2017 [19]. The importance of *N. ceranae* spillover is not limited to the honey bee but also to other insect species like the bee-eater *Merops apiaster* [20], the South American native bumblebee, *Bombus brasiliensis* (Lepeletier, 1836) [21,22], stingless bees and social wasps [23], solitary bees [24], the small hive beetle, Aethina tumida (Murray, 1867) [25], etc. Numerous studies have indicated that *N. ceranae* has become a worldwide distributed microsporidian pathogen, including in Central Italy [26,27], Croatia [28], Lithuania [29], etc. In an investigation on the prevalence of *Nosema* spp. in temperate and subtropical regions, pure *N. ceranae* infection and *N. ceranae*/*N. apis* co-infection were detected in apiaries from both regions, while pure *N. apis* infection was exclusively observed in the subtropical region [30].

It was found that *Apis cerana* showed a higher immune response and lower *N. apis* and *N. ceranae* spore loads than *A. mellifera*, suggesting that Asiatic honey bees may be better able to defend themselves against microsporidia infection [31].

The most prominent negative influences on honey bee colonies include: suppression of the honey bee immune system [32], shortening of worker bee lifespan [33], the decline in colony strength and productivity [34], queen supersedure [35], increased winter losses and colony collapse [36]. All these adverse effects of *Nosema* spp. on honey bee colonies require the search and development of effective strategies against these widespread parasites. For more than several decades, bicyclohexylammonium fumagillin (isolated from the fungus *Aspergillus fumigatus*) has been widely used as an anti-*N. ceranae* antibiotic [37]. Recent studies have shown that this antimicrobial agent is becoming less and less effective against *N. ceranae* infection [37,38]. Some researchers have further found that fumagilin is rather toxic and may provoke tumorigenic formations in humans. Moreover, it has negative effects on bee health and even leads to hyperproliferation of *N. ceranae* spores [39,40]. The observed toxic effects of fumagillin require strict measures regarding its use in many countries [41].

Considering the above, it seems essential to develop new, alternative approaches against nosemosis. Until now, there have been several new basic approaches to control nosemosis in honey bees.

1.1. Use of Small Molecules

The use of biologically active small molecules represents a promising approach against nosemosis [38]. A large number of organic compounds have been tested for control of nosemosis. These include: porphyrins (Porphyrin: PP(Asp)2 and Porphyrin: TMePyP), inhibitors of the enzyme methionine aminopeptidase type 2 (MetAP2), phenolic acids (formic acid, oxalic acid, etc.), polyphenol compounds (resveratrol and thymol), etc. [42–44]. Although they represent a reliable alternative therapy in the combat against nosemosis, a disadvantage of these compounds is that after their use viable spores remain in beehives, combs, and feces [38]. Thus, there is a real danger of infection or re-infection in the treated honey bee colonies.

1.2. RNA Interference as a Gene Regulating Expression Approach

Another approach for treating *N. ceranae* infection is associated with the use of RNA interference (RNAi). RNAi represents a biological process in which small RNA molecules (microRNA (miRNA) and small interfering RNA (siRNA)) inhibit gene expression or translation, by degrading targeted messenger RNA (mRNA) molecules via post-transcriptional gene silencing [45]. RNAi is widely used in human medicine and represents a promising new anticancer approach [46]. In beekeeping, RNAi technology has been used to protect honey bees from infection by various pathogens and parasites [47]. In vitro studies have shown that RNAi can be applied successfully against some honey bee-associated viruses and the ectoparasitic mite *Varroa destructor* (Anderson and Trueman, 2000) [48,49].

Using an RNAi strategy to reduce the expression of some honey bee genes (gene silencing) has been one of the key measures against nosemosis [50,51]. An example of this is the upregulation of the mRNA levels of the naked cuticle gene (nkd) in adult bees by means of *N. ceranae* infection provoking a suppressed host immune function [52]. It has been found that the oral application of nkd double-stranded RNA (dsRNA) in *N. ceranae*-infected bees, i.e., silencing the host nkd gene, can activate the immune response, suppress the reproduction of *N. ceranae*, and improve honey bees' health status [52]. Another similar strategy, but this time with the use of RNAi-based gene silencing on parasitic DNA, is the downregulation of the gene encoding *N. ceranae* polar tube protein 3 (ptp3) through the application of dsRNA that is homologous in gene sequence [50]. The ptp3 is the part of the polar tube structure relevant to host–parasite interaction, contributing to the parasitic invasion [53]. It has been demonstrated that the oral application of a dsRNA corresponding to the sequences of *N. ceranae* ptp3 gene silences the expression of the corresponding ptp3 in *N. ceranae*-infected bees. As a result, *N. ceranae* load reduction, improvement of host physiological status, and extension of lifespan in infected bees have been observed [49]. The application of this therapy has also its drawbacks. One of them is related to the degradation of the dsRNA molecules inside insects' guts, which is associated with additional costs regarding the protection of the dsRNA molecules from insect gut nucleases [54]. Other limiting factors include: the insects' gut pH and the related activity of the restricted enzyme (affecting the stability of dsRNA), the amount of the dsRNA molecules when administered orally in target insects, the length of the dsRNA molecules, and the life stage of the insects (larvae, pupae or adults) [54]. To overcome these obstacles, dsRNAs have been incorporated in liposomes or nanoparticles, and then these particles have been delivered to insects through feeding on an artificial diet [55]. Nanoparticles/liposomes stabilize the dsRNA molecules, thus ensuring the greater efficiency of the RNAi process.

The results obtained from the use of RNAi technology have clearly demonstrated the prospects of its applications in anti-nosemosis therapy, but more research is needed in order to be widely implemented in beekeeping practice.

1.3. Use of Organic Extracts and Natural Supplements as an Alternative Holistic Strategy

Another approach against nosemosis is the use of organic extracts and natural supplements. The major advantage is their lower toxicity for both bee colonies and the environment, compared to other chemical compounds [47]. Many investigations have shown that various organic and aqueous natural products do not show any toxicity to honey bees and lead to a decrease in both parasite load and mortality rate caused by *N. ceranae* infection [56–58]. Natural compounds, mostly flavonoids and polysaccharides contained in a number of medicinal plants, demonstrate anti-microsporidian activity in honey bees and are applied most often as alcoholic extracts, although some studies dispute the role of the biological activity of flavonoids against *N. ceranae* infection [58]. Propolis, a mixture of resins, wax, and pollen from buds and flowers of plants, enriched with enzymes and subjected to lactic acid fermentation in the digestive system of bees, has strong antimicrobial, antiviral, and antifungal properties [59]. These properties provoke the interest of the scientific community in propolis as an anti-microsporidian drug. In this relation, an ethanolic extract of propolis has been tested in different bee species experimentally infected with *N. ceranae* [59–61]. The obtained results from these investigations support the hypothesis that propolis represents an effective and safe product to control *N. ceranae*, while it is interesting to note that bees seem not to use it to self-medicate when infected with these microsporidia [61]. Some studies have indicated that honey and pollen from sunflowers (*Helianthus annuus* L.; Asteraceae) may also reduce the microsporidian infection and increase survival rate in honey bees [62,63].

A number of commercial supplements (HiveAlive[TM], Api-Bioxal[®] and ApiHerb[®], "BEEWELL AminoPlus", Nozevit[®], BeePro[®], MegaBee[®], etc.) have been tested for anti-*N. ceranae* activity as well [64–68]. For instance, administration of HiveAlive[TM] and ApiHerb[®] significantly reduces *N. ceranae* spores load [64,65]. Application of the dietary amino acid and vitamin complex called "BEEWELL AminoPlus" decreases *N. ceranae* spore and protects honey bees from immune suppression by upregulating the expression of genes for immune-related peptides (abaecin, apidaecin, hymenoptaecin, defensin and vitellogenin) [66]. The investigation of Nozevit[®] (a natural product from plant polyphenols) has shown that this commercial phytopharmacological supplement may improve bee health by decreasing colony spore loads [67]. However, DeGrandi-Hoffman et al. [68] have found that bee colonies fed with the commercial protein supplements BeePro[®] and MegaBee[®] exhibited higher levels of black queen cell virus and *N. ceranae* incidence and greater queen losses in comparison to bee colonies feeding on natural forage (*Brassica rapa*—rapini).

1.4. Probiotics and/or Prebiotics

The negative consequences associated with the use of antibiotics in the treatment against nosemosis are primarily related to the disruption of the host microbiota and, in some cases, the increased susceptibility to *N. ceranae* infection [69,70]. The use of microbial supplements (probiotics or prebiotics) represents another innovative approach not only for maintaining or restoring intestinal microbiota, but also in the combat against nosemosis. In this aspect, endogenous gut bacteria belonging to *Lactobacilliaceae*, *Bifidobacteriaceae*, and *Acetobacteraceae* families have been found to suppress the development of *N. ceranae*, by reducing the spore load [70,71]. Different commercial probiotics strains (Bactocell[®] and Levucell SB[®], Lallemand Inc., "Biogen-N", "Trilac", "Lakcid", etc.) have also been tested as an alternative therapy against *N. ceranae* infections in honey bees [72–76]. The obtained results from these investigations have shown endogenous bacterial strains to be as efficient as commercial strains in terms of survival of honey bees infected with *N. ceranae*, without a pronounced antagonistic effect on the parasite development.

However, some studies have indicated that uncontrolled and unbalanced administration of probiotics to honey bees may cause dysbacteriosis and increase pathogen susceptibility [76,77].

Some prebiotics are also used to control *N. ceranae* infection in honey bees. For example, prebiotic mannan-oligosaccharides (MOSs) are easily fermentable by gut microbiota and their accelerated reproduction allows them to compete with pathogenic bacteria for both nutrients and space in the gut. The β-glucans (glucose homopolymers) are known for their immune-modulating impact on different species, including honey bees [78]. The polysaccharide chitosan stimulates the bee immune system, leading to a decrease in the degree of infection with *N. apis* and to increased bee survival [79].

1.5. Other Approaches

The zeolite clinoptilolite is well-known for its powerful antioxidant activity as a dietary supplement in bees infected with *N. ceranae*. Using clinoptilolite has resulted in significant therapeutic effects on honey bees that are naturally infected with *N. ceranae*, without affecting bee physiology [80].

Many beekeepers prefer the use of a natural approach for the treatment of nosemosis. This is no accident, as prolonged use of antibiotics will inevitably lead to resistance, pollute the environment and honey bee products, which will consequently affect human health as well.

Inspired by the use of natural compounds for *N. ceranae* treatment, we decided to test two new herbal supplements, NOZEMAT HERB[®] (NH) and NOZEMAT HERB PLUS[®] (NHP), and their influence on *N. ceranae* infection and honey bee health and survival. We believe that this study will contribute to establishing the most effective approaches against nosemosis without a negative impact on the physiological state of the bee colonies and the resulting bee products.

2. Results

2.1. N. ceranae Infection and Spore Counts

The amplification of the part of the *16S rDNA* gene showed the presence of *N. ceranae* in both experimental groups (NH and NHP) and the control group (C) during the two pre-treatment sessions (August 2019 and April 2020) and about two months after the two post-treatment sessions (October 2019 and June 2020). There were no PCR products in the negative controls. The results from the obtained sequences have shown the presence of *N. ceranae* species after standard nucleotide BLAST (BLASTN programs) in the GenBank database.

The numbers of *N. ceranae* spores per bee for the pre- and post-treatment periods each year are shown in Figure 1. Even after the first treatment, a significant reduction in the number of *N. ceranae* spores per bee ($p < 0.01$) was observed (Figure 1a). Compared to the pre-treatment period (August 2019) (NH: $3.8 \pm 0.61 \times 10^6$; NHP: $3.5 \pm 0.78 \times 10^6$), in October 2019 the two supplement-treated groups (NH and NHP) showed a significant decrement in *N. ceranae* spore counts: about 68% in the NH group and about 60% in the NHP group (Figure 1a). In contrast to the two experimental groups, the control C group showed a reduction by only 19% of the number of *N. ceranae* spores (pre- treatment $2.7 \pm 0.68 \times 10^6$; post-treatment $2.2 \pm 0.59 \times 10^6$). During the next year, at the beginning of the second pre-treatment session (April 2020), in contrast to October 2019, a much more pronounced augmentation of *N. ceranae* spore counts was observed in the control C group—about 33% (ANOVA with Tukey's HSD post hoc test, $p = 0.6047$; $df = 8$) and less in the experimental groups NH and NHP—18% and 22% (ANOVA with Tukey's HSD post hoc test, $p = 0.6104$, $df = 8$; $p = 0.5406$, $df = 8$, respectively) (Figure 1b). At the end of the second experimental period in June 2020, compared to the pre-treatment period (April 2020), the NH and NHP groups treated with herbal supplements showed a significant reduction in *N. ceranae* spores ($p < 0.01$)—about 67% ($0.47 \pm 0.58 \times 10^6$) and 60% ($0.68 \pm 0.64 \times 10^6$), respectively (Figure 1b). The reduction in the number of *N. ceranae* spores in the control C group at the end of the second experimental period (June 2020) was only 11% ($2.61 \pm 0.72 \times 10^6$).

The treatment with the herbal supplements NH and NHP showed a significant reduction in the *N. ceranae* spore levels compared to the control C group (post-hoc Tukey HSD Test: $F = 24.199$, $p = 0.0004$; $F = 18.919$, $p = 0.0009$, respectively).

2.2. Effect of Herbal Supplements on Honey Bee Strength and Production Traits

2.2.1. Strength of the Honey Bee Colonies (Estimated Based on Mass)

The strength of the bee colonies was evaluated based on colony mass 12 times during the first pre-treatment (August 2019) and after that 4 times at 12-day intervals post-treatment until September 2020. After the winter period October 2019–February 2020, the same investigations were performed before the second pre-treatment (March 2020) and after that 4 times at 12-day intervals post-treatment until June 2020 (Figure 2). The first difference between the experimental groups (NH and NHP) and the control C group was observed about 25 days after the first treatment (at the end of August 2019). Then, the mass of the bee colonies in the NH and NHP groups was estimated to be 1.65 ± 0.05 kg and 1.63 ± 0.12 kg, respectively, while in the control C group it was 1.25 ± 0.10 kg ($F = 6.818$, $df = 6$, $p = 0.040$; $F = 6.943$, $df = 6$; $p = 0.038$, respectively). As the autumn period approached, there was a gradual decrease in the studied indicator of bee colony strength in both the control and the two experimental groups, reaching almost full equalization at the end of September(Figure 2). During the second pre-treatment (9 April 2020) the average mass values indicating the strength of the honey bee colonies were significantly higher in the experimental groups than in the control group (ANOVA with Tukey's HSD post hoc test, $F = 26.727$, $df = 8$, $p = 0.002$; $F = 30.857$, $df = 8$, $p = 0.001$, respectively) (Figure 2). The most significant difference was observed in the NHP group, where the mass exceeded 1.2 times that of the control group. This is an indicator of the successful wintering of honey bee colonies and a good start for spring development.

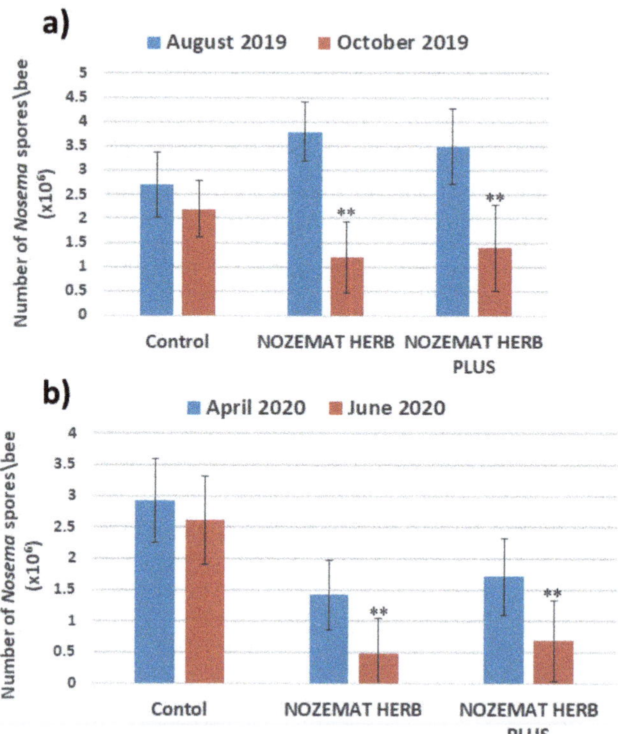

Figure 1. *N. ceranae* spore counts (±SD) pre- and post-treatment with NOZEMAT HERB® and NOZEMAT HERB PLUS® during 2019 (**a**) and 2020 (**b**). The number of *N. ceranae* spores was counted by microscopic examination. Asterisks indicated the level of significance as determined by an analysis of variance followed by Tukey's multiple comparison test. **: $p < 0.01$ compared to the Nosema spore during pre-treatment.

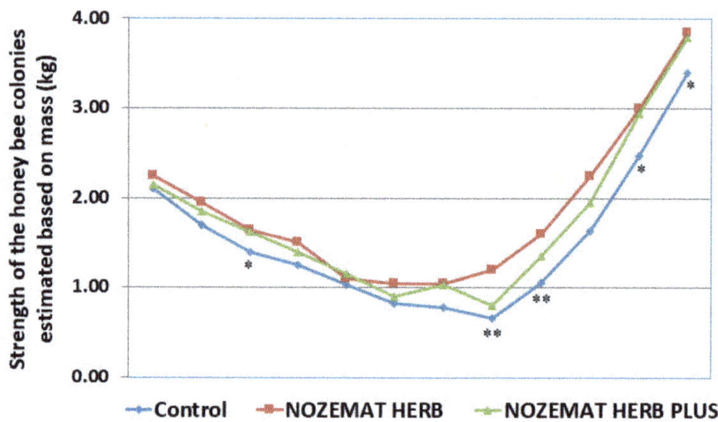

Figure 2. Average (±SD) values in colony strength in honey bee colonies treated with NOZEMAT HERB® and NOZEMAT HERB PLUS® and in an untreated (control) group during 2019 and 2020. Asterisks indicated the level of significance as determined by an analysis of variance followed by Tukey's multiple comparison test. *: $p < 0.05$, **: $p < 0.01$ compared to the untreated control group.

Similar to the first treatment in the previous year (August 2019), the first significant difference between the control and the experimental groups was observed around the 10th day after the second treatment (9 April 2020). Then, the mass (and, accordingly, the strength) of the colonies from the NH group was 1.8 times higher than that of the control C group (F = 31.176, df = 6, p = 0.001). The difference between the NHP and the control C was about 0.150 kg bees (F = 32.666, df = 6, p = 0.001). About two months after the second treatment, the difference between the two experimental groups and the control group was about 0.450 kg bees. The mass (and, accordingly, the strength) of the bee colonies of the NH group (1.60 ± 0.07 kg) and NHP group (1.35 ± 0.10 kg) was, respectively, about 1.5 and 1.2 times higher than that of the control group (1.06 ± 0.09 kg) (F = 19.091, df = 8, p = 0.002; F = 16.90, df = 8, p = 0.003, respectively). These significant differences persisted during the next measurement period (middle of May 2020), after which there was a less pronounced advantage of the experimental groups over the control group until the last reporting period (June 2020) (Figure 2).

2.2.2. Sealed Worker Brood Area within the Hives

A lot of factors determine the queen's egg-laying rate: age, genotypes, colony size, colony nutrition, etc. It is difficult to quantify the performance of queens relative to workers in the field, and there are not many laboratory assays on queen performance. For this reason, the obtained results on sealed worker brood areas within the hives may not be completely comprehensive, yet this is a parameter that needs to be considered. In contrast to the achieved results regarding the strength of the honey bee colonies, the parameter of sealed worker brood area showed fewer differences after the treatment of the honey bee colonies with herbal supplements (Figure 3). After the first application of NH and NHP (August 2019), the sealed worker brood area gradually decreased both in the experimental (NH and NHP) and in the control C group, which is a biologically determined process in the late autumn period (September 2019). After the winter period and the second pre-treatment (April 2020), the sealed worker brood area in the hives increased; however, only in the NH group (925 ± 29.3 cm^2) there were statistically significant differences, compared to the NHP (200 ± 31.7 cm^2) and the control C group (120 ± 26.9 cm^2) (F = 6.241, df = 6, p = 0.037; F = 0.303, df = 6, p = 0.610, respectively) (Figure 3). The second statistically significant difference between the experimental groups and the control group was observed after the fifth month following the second treatment (June 2020). Then, the NH and NHP groups had a significantly larger sealed worker brood area, compared to the control C group (ANOVA with Tukey's HSD post hoc test, df = 8, F = 9.424, p = 0.018; F = 13.256, df = 8, p = 0.008, respectively).

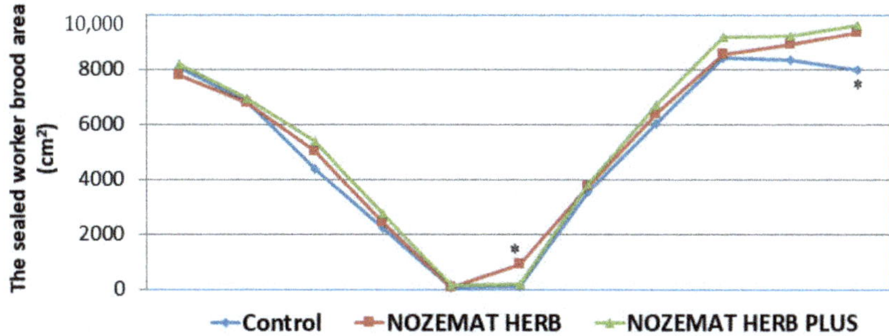

Figure 3. Average (±SD) worker sealed brood area (cm^2) in honey bee colonies treated with NOZEMAT HERB® and NOZEMAT HERB PLUS® and in an untreated (control) group during 2019 and 2020. Asterisks indicated the level of significance as determined by an analysis of variance followed by Tukey's multiple comparison test. *: p < 0.05 compared to the untreated control group.

2.2.3. Amount of Capped Honey in the Beehives

The results regarding the amount of capped honey in bee combs are presented in Figure 4. As this indicator is directly related to the strength of honey bee colonies (Figure 1), no significant difference was observed between the experimental groups and the control group in the period after the first treatment (August 2019) until the approach of the autumn (September 2019). During the first spring survey in 2020, there were more significant differences in the values of this indicator. About two months after the second treatment (May 2020), there were significant differences between the experimental groups and the control group regarding the amount of capped honey in the beehives. The experimental groups NH (1.18 ± 0.21 kg) and NHP (1.16 ± 0.10 kg) exceeded with about 490% the control C group (0.20 ± 0.13 kg) (F = 77.0395, df = 6, p = 0.001; F = 148.725, df = 6, p = 0.001, respectively). During the next two measurement periods (in the middle and at the end of May 2020) this trend persisted, as higher values were observed in the experimental NH group, compared to the control C group (F = 30.015, df = 8, p = 0.001; F = 62.157, df = 8, p = 0.001, respectively). During both these periods, the amount of capped honey in the experimental NH group exceeded nearly 4 times this in the control C group. A similar trend was observed between the NHP group and the control C group but with a less pronounced level of significance (Figure 4).

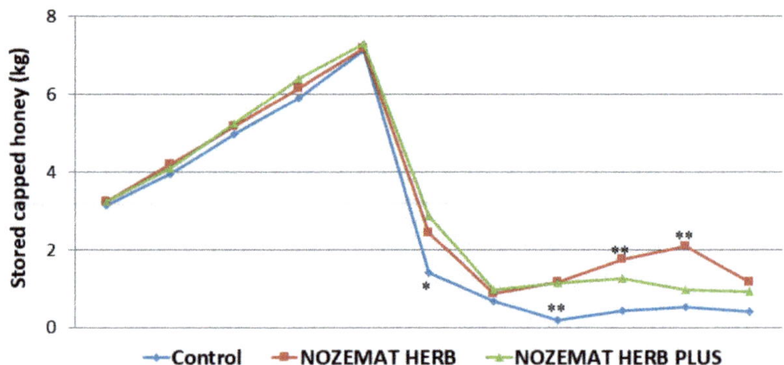

Figure 4. Average (±SD) amount of stored capped honey in honey bee colonies treated with NOZEMAT HERB® and NOZEMAT HERB PLUS® and in an untreated (control) group during 2019 and 2020. Asterisks indicated the level of significance as determined by an analysis of variance followed by Tukey's multiple comparison test. *: $p < 0.05$, **: $p < 0.01$ compared to the untreated control group.

2.2.4. Stored Pollen Area within the Hives

The stored pollen area is an indicator of the amount of collected pollen in beehives (a quantity that depends on environmental conditions, the needs of the colonies and the number of bees in the colonies). The stored pollen area was evaluated through direct surface measurements of the comb (cm^2), and the results are presented in Figure 5. Although the strength of honey bee colonies decreased in all groups, a significant difference was observed between the experimental groups and the control group about 25 days after the first treatment (9 August 2019). Then, the stored pollen area in the experimental groups NH (425 ± 30.8 cm^2) and NHP (313 ± 31.3 cm^2) exceeded approximately 7 and 5 times, respectively, that in the control C group (63 ± 32.3 cm^2) (F = 53.680, df = 8, p = 0.001, $p < 0.01$; F = 15.491, df = 8, p = 0.001, $p < 0.01$, respectively). Significantly higher values of this indicator in the two experimental groups compared to the control were observed during the next measurement (the beginning of September 2019) (Figure 5). Then, statistically significant differences were observed between the experimental groups NH and NHP and the control C group (F = 53.172, df = 8, p = 0.003, $p < 0.01$; F = 23.835, df = 8, p = 0.002, $p < 0.01$, respectively). During the next year (2020), the most significant difference between the experimental groups and the control group was observed about two months after the second treatment (15 May 2020). In terms of stored pollen area, the NH group surpassed by nearly 80% the control C group (ANOVA with Tukey's HSD post hoc test, $F = 5.631$, $df = 8$, $p = 0.049$, $p < 0.05$), while the NHP group exceeded by 170% the control C group (ANOVA with Tukey's HSD post hoc test, $F = 7.966$, $df = 8$, $p = 0.025$, $p < 0.05$) (Figure 5).

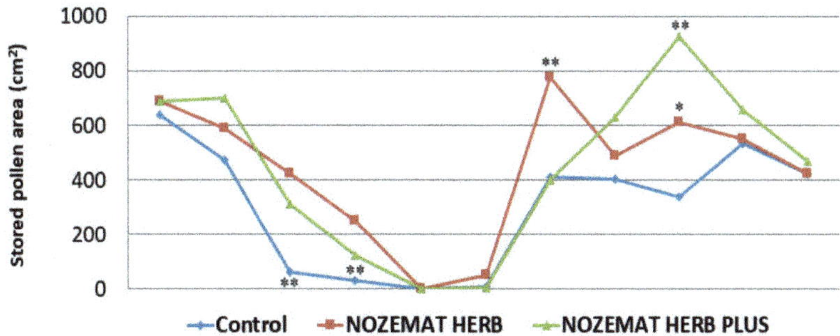

Figure 5. Average ($\pm$SD) stored pollen area (cm^2) in honey bee colonies treated with NOZEMAT HERB® and NOZEMAT HERB PLUS® and in an untreated (control) group during 2019 and 2020. Asterisks indicated the level of significance as determined by an analysis of variance followed by Tukey's multiple comparison test. *: $p < 0.05$, **: $p < 0.01$, compared to the untreated control group.

3. Discussion

About 15–20 years after the detection of *N. ceranae* in the Eastern honey bee (*Apis cerana*), these microsporidia became a globally distributed pathogen [81–84]. Currently, *N. ceranae* is considered one of the main causes of severe annual losses in beekeeping worldwide [85]. For this reason, various strategies have been developed for the control of nosemosis type C caused by *N. ceranae*. It is of crucial importance that the applied approaches are not toxic to honey bees and also to humans (through residues remaining in honey bee products), and are not polluting the environment.

One of the allopathic therapies related to the fight against nosemosis is the use of ecologic phytotherapy [86,87]. It is a suitable remedy against nosemosis due to the low toxicity to bees and the safety of the environment and human health.

A number of studies have focused on the impact of some herbal extracts on the regulation of the expression of certain genes in honey bees in order to reduce the damage caused by *N. ceranae*. For example, it has been shown that plant extracts or decoction from *Andrographis paniculate* promote Wnt and JNK pathways by upregulating the expression of certain genes (including armadillo, basket, frizzled and groucho) in intestinal cells [57]. These results have demonstrated that this Chinese herb can provide protection from *N. ceranae* infection under laboratory conditions, suggesting that it can be used in apiculture to control *N. ceranae*. Another study has revealed that *Eleutherococcus senticosus* extract contains eleutherosides (eleutheroside B + E), which have an impact on the honey bee immune system [88]. These eleutherosides increase phenoloxidase (PO, a major defense enzyme in many invertebrates) in hemolymph and inhibit the development of fungal spores. The piperine (an alkaloid in the roots of the Piperaceae family) and curcumin (a natural phenol produced by *Curcuma longa*) are known as natural supplements which increase the activity of the antioxidant system in honey bees [89,90]. They promote the activities of antioxidant enzymes such as superoxide dismutase, peroxidase, catalase and glutathione S-transferase, which reduces oxidative stress [91]. Moreover, it is observed that these herbal supplements exhibit hydroxyl radical scavenger action in honey bees, suppressing the destructive effects of the free radicals and reactive oxygen species [92]. Because *N. ceranae* is able to induce oxidative stress in bees [93], piperine and curcumin are potential candidates regarding antinosemosis therapy.

The present study is the first report of the in vivo application of two plant extracts, NOZEMAT HERB® (NH) and NOZEMAT HERB PLUS® (NHP), against *N. ceranae* and their impact on honey bee colonies. These products are patented herbal supplements for honey bees. NH contains herbal extracts, vegetable glycerin, water, citric acid, and preservative—potassium sorbate. NHP contains additional herbal extract as well as vitamin C (ascorbic acid). The exact quantitative composition of these two supplements is patent-protected and thereby not disclosed in this paper.

After a sequence analysis of the part of the 16S rDNA gene, we obtained a fragment with 219 bp. Then, we performed a BLAST analysis in the GenBank genetic sequence database to identify regions of local similarity between the available sequences for the *N. ceranae* 16S rDNA gene. We found the highest homology of our sequences (Acc. no. MG657260) with *N. ceranae* isolates from Argentina (Acc. no. KX024757) and Lithuania (Acc. no. JQ639314). Besides the conventional PCR analysis for detection of *N. ceranae* based on sequence analysis of 16S rDNA gene, a recently developed new method not only provides rapid and reliable detection of the presence of these microsporidia but also allows the quantification of *N. ceranae* (qPCR assay) based on the highly-conserved protein coding gene Hsp70 [94].

In both years of our study (2019–2020), a significant reduction of *N. ceranae* spore loads was observed after NH and NHP administration, compared to the pre-treatment level. Among the benefits from these two supplements is their positive effect on the intestinal microflora of honey bees. Numerous studies have indicated that microbial communities have an essential role in the resistance to pests and pathogens, environmental toxins and pesticide exposure [77,95–97]. As for the *N. caranae* infection, there is evidence that these microsporidia disrupt the microbiota, causing dysbiosis, which may have consequences on bee development and immune suppression [97–99]. As an intestinal infection, *N. ceranae* causes significant changes in the composition of honey bee gut microfauna, and the plant extracts that we use contain biologically active substances that have a beneficial effect on gut bacterial communities. In fact, both products (NH and NHP) contain basic polyphenols—flavonoids and phenolic acids—as biologically active compounds. According to some authors, the antimicrobial activity of plant extracts is not due to a single biologically active substance (flavonoids vs phenolic acid), but rather to the totality of all, with potentially synergistic effects [59,100]. Phenolic compounds extracted from *Artemisia dubia* and *Aster scaber* have shown a clear anti-nosemosis effect, which is a promising strategy for controlling nosemosis [101,102]. As far as flavonoids are concerned, the studies carried out so far have not identified a single representative that can be effective on its own against nosemosis [59,65]. Unfortunately, the mechanism of action of both flavonoids and phenolic compounds against *N. ceranae* has not been elucidated yet. The latter would certainly help to speed up the process against these microsporidia.

A lot of research has been carried out on the application of plant extracts supplementation against various pathogens in bee colonies, e.g., *V. destructor* and honey bee-associated viruses, *N. ceranae*, American foulbrood (*Paenibacillus larvae*) [66,87,103–105], or pests, e.g., the greater wax moth *Galleria mellonella* [106]. In addition to the effect of plant extracts, the potential beneficial effects of herbal extracts on honey bee colonies have been tracked as well. Certain researchers purposefully reflect on the impact of various plant extracts on honey bee performance [107–111].

Our study aimed to evaluate the impact of the herbal supplements NOZEMAT HERB® and NOZEMAT HERB PLUS®—on the one hand, as an alternative therapy against nosemosis, and on the other hand, considering the influence of the two products on honey bee colony strength and honey production and pollen collection for two consecutive years (2019 and 2020).

We found a significant reduction of *N. ceranae* spores load after administration of NH and NHP, compared to the pre-treatment period both in 2019 and in 2020. In contrast to the experimental groups, the control C group exhibited a less clear decrease of *N. ceranae* spores levels (Figure 1). These data are consistent with other studies in which a significant decrease of *N. ceranae* spores load was established after the application of other commercial food supplements, such as HiveAlive™ [64] and ApiHerb [65]. A similar effect was found in

Chinese herbal extracts (in particular *Andrographis paniculata*), when applied as a decoction to treat *N. ceranae* infections in *A. mellifera* [55]. These results indicate that plant extracts represent a potent alternative therapy against nosemosis.

During the first year (2019) of our research, the administration of the two herbal supplements did not reveal significant differences between the experimental NH and NHP groups compared to the control C group with regard to colony strength (Figure 2). These results are an indicator of a decrease in the strength of bee colonies as autumn approaches. These data support the findings of Charistos et al. [64], who also did not observe differences between the groups in terms of colony strength after the administration of HiveAlive™ food supplement during the autumn. The first informative review in the spring of the following year (2020) showed a significant increase in colony strength in both experimental groups in contrast to the control group, which is an indicator of the more successful overwintering of the treated bee colonies. This trend continued after the second treatment (April 2019) until the beginning of the autumn period. These data support previous investigations about the positive influence of herbal supplements such as HiveAlive™, *Laurus nobilis* L. and *Agaricus brasiliensis* extracts on honey bee colony strength [64,104,107].

Similar to the indicator of bee colony strength, the sealed worker brood area did not show significant differences in all the groups after the first herbal supplement administration (Figure 3). The first informative review in the spring of the following year (2020) showed a larger sealed worker brood area in the NH group, but as a whole, this indicator was not affected by the applied plant extracts.

The amount of capped honey was affected most significantly after the second treatment of the bee colonies in April 2020 (Figure 4). After that, the amount of capped honey was significantly higher in the two experimental NH and NHP groups compared to the control C group. This indicator maintained significantly higher levels in the two experimental groups throughout the summer, which is a prerequisite for successful wintering of the bee colonies and reducing the costs associated with additional feeding during the winter.

It is interesting to note that the stored pollen area was significantly affected after the first treatment (August 2019) despite the presence of sparse flowering vegetation (Figure 5). After the second treatment the following year, a larger stored pollen area was observed in the experimental groups, and it seemed that the NHP group had an advantage over the NH and the control C group in terms of this indicator.

Although the findings of the present study clearly demonstrate the benefits from applying the two plant extracts for the reduction of *N. ceranae* spore counts as well as for honey bee performance in general, these encouraging results call for further research in order to clarify the impact of NOZEMAT HERB® and NOZEMAT HERB PLUS® on *N. ceranae* spore loads as well as on honey bees.

4. Materials and Methods

4.1. Ethics Statement

All experimental procedures were reviewed and approved by the Animal Research Ethics Committee of the Bulgarian Food Safety Agency (BFSA), (Ar. 154 from of the Law on Veterinary Activity) in accordance with the European Union Directive 86/609.

4.2. Experimental Design

The study was conducted during the period of 11 months (5 August 2019–8 June 2020) at the Experimental Apiary of the Research Center of Stockbreeding and Agriculture—Smolyan, Bulgaria (41°35′7.01″ N, 24°41′30.98″ E). The apiary is located in Smolyan municipality—the Perelik-Prespa part of the Western Rhodope Mountains. The apiary consists of 60 colonies of *Apis mellifera rodopica* (local ecotype of *A. m. macedonica*) housed in Langstroth Rut hives. All hives have exposure to the same environment and the same forage conditions.

For the purposes of our investigation, 45 of those colonies were selected based on the presence of *N. ceranae* infection, which was detected in the 25 forager bees sampled from each colony both microscopically and with PCR analysis. Forager bees can be distinguished from house bees based on appearance (with less, darker hairs in the chest area), presence of visible pollen load on their legs, and location in the hive (mainly in the part of the combs occupied by food supplies). Considering the presence of *N. ceranae* infection, the 45 honey bee colonies were divided into three groups—two experimental groups NH ($n = 15$) and NHP ($n = 15$), and a control group C ($n = 15$). The bee colonies were equalized in terms of bee colony strength, sealed worker brood area, amount of honey and stored pollen area.

The treatment of the bee colonies with herbal supplements was carried out once in the autumn (5 August 2019) and once in the spring (9 April 2020). The bee colonies in the experimental groups were treated 4 times at 7-day intervals with NOZEMAT HERB® and NOZEMAT HERB PLUS® (Extract Pharma, Sofia, Bulgaria) at a dose of 10 mL of the product dissolved in 100 mL of sugar syrup (1:1, w/w), according to the manufacturer's instructions (Extract Pharma Ltd., Sofia, Bulgaria). The solution was sprayed with a syringe onto the bee combs in each experimental hive. The bee colonies of the control C group were sprayed only with sugar syrup (1:1, w/w), at the same dose as the experimental groups.

Before each pre-treatment (August 2019 and April 2020) and about two months post-treatment (October 2019 and June 2020), 20 forager honey bees were sampled from each colony from the three investigated groups for microscopic (stored at 4 °C) and PCR analysis (stored at −20 °C) for *N. ceranae* examination.

4.3. Microscopic Detection of N. ceranae

The bees were sent to the National Reference Laboratory for Bee Diseases at the National Diagnostic Science and Research Veterinary Medical Institute (Sofia, Bulgaria) in a cooler bag. The abdomens of 20 forager honey bees from each colony of the experimental groups and the control group were macerated in 3 mL of distilled water, and the pellets were filtered and centrifuged for 10 min at 1000× *g*. A 100 µL aliquot was placed on a microscope slide and covered with a coverslip. *N. ceranae* spores were counted at ×400 magnification. Positive samples were recounted for an accurate spore count using a Neubauer hemocytometer on (0.1 µL volume). One *N. ceranae* spore observed in the entire hemocytometer's grid (25 × 16 = 400 small squares) was equal to an average of 1500 spores per bee. The reported information was the number of spores per bee [112].

4.4. DNA Extraction, PCR Amplification and Sequencing

After light microscopy, a total of 150 spore samples from each investigated colony were investigated by a PCR analysis in the experimental groups and the control group in the pre-treatment period. For each of the suspensions of isolated *N. ceranae* spores, an aliquot of 50 µL was transferred to a new tube and centrifuged at 15,000× *g* for 10 min. The total DNA was isolated from the obtained supernatant by using a GeneMATRIX Tissue DNA purification kit (Cat. no. E3550, EURx Ltd., Gdansk, Poland) as per the manufacturer's instruction. Briefly, the pellet was resuspended in a buffer Lyse T, and 20 µL of Proteinase K were added and incubated overnight at 56 °C with shaking. The quality and quantity of the isolated DNA were checked by 1% agarose gel electrophoresis and then visualized under UV trans-illuminator gel documentation systems after staining with SimpliSafe™ (cat. no. E4600; EURx Ltd., Gdansk, Poland). The isolated DNA was stored at −20 °C before analysis.

Considering that *N. ceranae* is becoming a globally distributed pathogen, we decided to perform molecular detection on these microsporidia.

The small-subunit (SSU) rRNA (*16S rDNA*) gene was chosen for molecular detection of *N. ceranae*, using the primers 218MITOC—FOR (5′-CGGCGACGATGTGATATGAAAATATTAA-3′) and 218MITOC—REV (5′-CCCGGTCATTCTCAAACAAAAAACCG-3′) designed by Martín-Hernández et al. [113]. Negative controls were included in all PCR experiments. As a positive control, cytochrome c oxidase subunit 1 (*coI*) gene fragment of *Apis mellifera* was used in all the studied samples. The sequence of primers used for positive control was CoI2-F (5′-CCTGATATAGCATTTCCTCG-3′) and CoI2-R (5′-TGTGAATGATCTAAAGGTGG-3′) designed on the basis of the complete mitochondrial genome of *A. m. ligustica* (Acc. no. L06178) [114].

All PCR reactions were performed with 10 ng template DNA in a final volume of 50 µL (NZYTaq II 2 × Colourless Master Mix, cat. no. MB354; NZYTech, Lda.—Genes and Enzymes, Lisbon, Portugal). The PCR conditions were as follows: initial denaturation at 94 °C for 5 min; 30 cycles of denaturation at 94 °C for 30 s, primers hybridization at 50 °C for 30 s, elongation at 72 °C for 1 min, and final elongation at 72 °C for 10 min. The successfully amplified products for *N. ceranae* were purified with a GeneMATRIX PCR/DNA Clean-Up Purification Kit (cat. no. E3520; EURx Ltd., Gdansk, Poland) and sequenced in both directions using a PlateSeq kit (Eurofins Genomics Ebersberg, Germany).

4.5. Evaluation of Honey Bee Strength, Sealed Worker Brood and Food Supplies

During the investigation period, 12 measurements were performed in all investigated groups (experimental and control) to determine the strength of the colonies (based on mass) and 11 measurements to determine the sealed worker brood area, the amount of honey and the stored pollen area. These measurements were performed once pre-treatment (August 2019) and four times post-treatment, at 12-day intervals, until the end of September (2019). After the winter period, the same measurements were performed once before the second pre-treatment (April 2020) and four times post-treatment, at 12-day intervals, until the beginning of June (2020).

The following parameters characterizing the development of the bee colonies were determined:

1. Strength of the bee colony based on its mass (kg)—the mass is calculated based on the number of frames occupied by bees, considering that one frame in a Langstroth–Rut hive contains approximately 200 g of bees [115].
2. Sealed worker brood area—a measuring frame with the size of the squares 5 × 5 cm was used. In 1 cm^2 there were 4 worker cells in the bee comb. The area of 25 cm^2 corresponded to 100 worker cells [115].
3. Amount of honey in the beehives—a measuring frame with 5 × 5 cm squares was used to measure the capped honey in the bee combs. Eight squares of the measuring frame corresponded to 0.350 kg honey [115].
4. Stored pollen area in the beehives—the amount of the collected pollen was evaluated through direct surface measurements of the comb (cm^2) [115].

4.6. Statistical Analysis

The dependent variables studied and normalized by linear transformation were: number of *N. ceranae* spores, strength (mass) of honey bee colonies, sealed worker brood area, amount of honey, and stored pollen area. These variables in the investigated groups were compared using F-statistic, a one-way ANOVA analysis of variance (IBM SPSS Statistics 23.0 for Windows) with a post hoc Tukey HSD test for multiple comparisons with Bonferroni and Holm correction.

The obtained sequences (219 bp the part of *16S rRNA* gene) were deposited in the GenBank database under accession numbers MG657260.

5. Conclusions

This is the first study to evaluate in vivo the effects of NOZEMAT HERB® and NOZE-MAT HERB PLUS® on *N. ceranae* spore loads, honey bee strength and production traits for two consecutive years. Both herbal supplements increase the strength of the bee colonies, with NOZEMAT HERB® showing a stronger impact. The sealed worker brood area does not seem to be significantly affected by the treatment. However, both supplements seriously increase the amount of capped honey, and this effect is maintained throughout the year. The supplements also increase the stored pollen area throughout the year, with NOZEMAT HERB PLUS® having a greater effect.

Both herbal supplements can be successfully used as an alternative therapy against nosemosis. Furthermore, they are not toxic to bees and are easily ingested by the latter. As plant extracts, they are completely safe for humans, animals, and the environment. Therefore, they can be fed as natural food supplements all year round in combination with sugar syrup or as honey bread for winter feeding.

Author Contributions: Conceptualization, R.S. and R.B.; methodology, R.S., R.B. and P.H.; investigation, R.S. and R.B.; data curation, R.S., R.B. and P.H.; writing—original draft preparation, R.S., R.B. and P.H.; writing—review and editing, R.S., R.B. and P.H.; All authors have read and agreed to the published version of the manuscript.

Funding: This work was funded by the National Scientific Fund of the Bulgarian Ministry of Education and Science, (grant numbers 06/10 17.12.2016). The funders had no role in study design, data collection and analysis, decision to publish, or preparation of the manuscript.

Institutional Review Board Statement: The study was conducted according to the guidelines of the Declaration of Helsinki, and approved by the Ethics Committee of the RESEARCH CENTRE OF STOCKBREEDING AND AGRICULTURE, Smolyan, Bulgaria (protocol code 627/30.03.2015).

Informed Consent Statement: Not applicable.

Data Availability Statement: The data presented in this study are available on request from the corresponding author.

Acknowledgments: We thank Ilian Gechev for kindly providing both herbal supplements. We also thank Maya Marinova (IBPhBME-BAS, Bulgaria) for the English editing.

Conflicts of Interest: The authors declare no conflict of interest.

References

1. Paudel, Y.P.; Mackereth, R.; Hanley, R.; Qin, W. Honey Bees (*Apis Mellifera* L.) and Pollination Issues: Current Status, Impacts and Potential Drivers of Decline. *J. Agric. Sci.* **2015**, *7*, 93. [CrossRef]
2. Sidhu, C.S.; Joshi, N.K. Establishing Wildflower Pollinator Habitats in Agricultural Farmland to Provide Multiple Ecosystem Services. *Front. Plant Sci.* **2016**, *7*, 363. [CrossRef] [PubMed]
3. Dufour, C.; Fournier, V.; Giovenazzo, P. The impact of lowbush blueberry (*Vaccinium angustifolium* Ait.) and cranberry (*Vaccinium macrocarpon* Ait.) pollination on honey bee (*Apis mellifera* L.) colony health status. *PLoS ONE* **2020**, *15*, e0227970.
4. Moritz, R.F.A.; Erler, S. Lost Colonies Found in a Data Mine: Global Honey Trade but Not Pests or Pesticides as a Major Cause of Regional Honeybee Colony Declines. *Agric. Ecosyst. Environ.* **2016**, *216*, 44–50. [CrossRef]
5. Lozier, J.D.; Zayed, A. Bee Conservation in the Age of Genomics. *Conserv. Genet.* **2017**, *18*, 713–729. [CrossRef]
6. Potts, S.G.; Imperatriz-Fonseca, V.; Ngo, H.T.; Aizen, M.A.; Biesmeijer, J.C.; Breeze, T.D.; Dicks, L.V.; Garibaldi, L.A.; Hill, R.; Settele, J.; et al. Safeguarding Pollinators and Their Values to Human Well-Being. *Nature* **2016**, *540*, 220–229. [CrossRef] [PubMed]
7. Stanley, D.A.; Msweli, S.M.; Johnson, S.D. Native Honeybees as Flower Visitors and Pollinators in Wild Plant Communities in a Biodiversity Hotspot. *Ecosphere* **2020**, *11*, e02957. [CrossRef]
8. Aslan, C.E.; Liang, C.T.; Galindo, B.; Kimberly, H.; Topete, W. The Role of Honey Bees as Pollinators in Natural Areas. *Nat. Areas J.* **2016**, *36*, 478–488. [CrossRef]
9. Yohannes, G. Review on Medicinal Value of Aloe Vera in Veterinary Practice. *Biomed. J. Sci. Tech. Res.* **2018**, *6*, 1–6. [CrossRef]
10. Fürst, M.A.; McMahon, D.P.; Osborne, J.L.; Paxton, R.J.; Brown, M.J.F. Disease Associations between Honeybees and Bumblebees as a Threat to Wild Pollinators. *Nature* **2014**, *506*, 364–366. [CrossRef]
11. Hristov, P.; Shumkova, R.; Palova, N.; Neov, B. Factors Associated with Honey Bee Colony Losses: A Mini-Review. *Vet. Sci.* **2020**, *7*, 166. [CrossRef] [PubMed]
12. Seitz, N.; Traynor, K.S.; Steinhauer, N.; Rennich, K.; Wilson, M.E.; Ellis, J.D. A national survey of managed honey bee 2014–2015 annual colony losses in the USA. *J. Apic. Res.* **2015**, *54*, 292–304. [CrossRef]
13. Chauzat, M.P.; Jacques, A.; Laurent, M.; Bougeard, S.; Hendrikx, P.; Ribière-Chabert, M.; EPILOBEE Consortium. Risk indicators affecting honey bee colony survival in Europe: One year of surveillance. *Apidologie* **2016**, *47*, 348–378. [CrossRef]
14. Paxton, R.J.; Klee, J.; Korpela, S.; Fries, I. *N. ceranae* has infected *Apis mellifera* in Europe since at least 1998 and may be more virulent than *N. ceranae apis*. *Apidologie* **2007**, *38*, 558–565. [CrossRef]
15. Fries, I. *N. ceranae* in European Honey Bees (*Apis mellifera*). *J. Invertebr. Pathol.* **2010**, *103*, S73–S79. [CrossRef]

16. Klee, J.; Besana, A.M.; Genersch, E.; Gisder, S.; Nanetti, A.; Tam, D.Q.; Chinh, T.X.; Puerta, F.; Ruz, J.M.; Kryger, P.; et al. Widespread dispersal of the microsporidian *Nosema ceranae*, an emergent pathogen of the western honey bee, *Apis mellifera*. *J. Invertebr. Pathol.* **2007**, *96*, 1–10. [CrossRef] [PubMed]

17. Higes, M.; Meana, A.; Bartolomé, C.; Botías, C.; Martín-Hernández, R. *N. ceranae* (Microsporidia), a Controversial 21st Century Honey Bee Pathogen. *Environ. Microbiol. Rep.* **2013**, *5*, 17–29. [CrossRef] [PubMed]

18. Emsen, B.; Guzman-Novoa, E.; Hamiduzzaman, M.M.; Eccles, L.; Lacey, B.; Ruiz-Pérez, R.A.; Nasr, M. Higher Prevalence and Levels of *N. ceranae* than *N. apis* Infections in Canadian Honey Bee Colonies. *Parasitol. Res.* **2016**, *115*, 175–181. [CrossRef] [PubMed]

19. Chemurot, M.; De Smet, L.; Brunain, M.; De Rycke, R.; de Graaf, D.C. *N. neumanni* n. sp. (Microsporidia, Nosematidae), a New Microsporidian Parasite of Honeybees, *Apis mellifera* in Uganda. *Eur. J. Protistol.* **2017**, *61*, 13–19. [CrossRef] [PubMed]

20. Higes, M.; Martín-Hernández, R.; Garrido-Bailón, E.; Botías, C.; García-Palencia, P.; Meana, A. Regurgitated pellets of *Merops apiaster* as fomites of infective *Nosema ceranae* (Microsporidia) spores. *Environ. Microbiol.* **2008**, *10*, 1374–1379. [CrossRef]

21. Plischuk, S.; Martín-Hernández, R.; Prieto, L.; Lucía, M.; Botías, C.; Meana, A.; Abrahamovich, A.H.; Lange, C.; Higes, M. South American native bumblebees (Hymenoptera: Apidae) infected by *Nosema ceranae* (Microsporidia), an emerging pathogen of honeybees (*Apis mellifera*). *Environ. Microbiol. Rep.* **2009**, *1*, 131–135. [CrossRef]

22. Plischuk, S.; Lange, C.E. *Bombus brasiliensis* Lepeletier (Hymenoptera, Apidae) infected with *Nosema ceranae* (Microsporidia). *Rev. Bras. Entomol.* **2016**, *60*, 347–351. [CrossRef]

23. Porrini, M.P.; Porrini, L.P.; Garrido, P.M.; Silva Neto, K.M.; Porrini, D.P.; Muller, F.; Nuñez, L.A.; Alvarez, L.; Iriarte, P.F.; Eguaras, M.J. *Nosema ceranae* in South American native stingless bees and social wasp. *Microb. Ecol.* **2017**, *74*, 761–764. [CrossRef] [PubMed]

24. Ravoet, J.; De Smet, L.; Meeus, I.; Smagghe, G.; Wenseleers, T.; de Graaf, D.C. Widespread occurrence of honey bee pathogens in solitary bees. *J. Invertebr. Pathol.* **2014**, *122*, 55–58. [CrossRef]

25. Cilia, G.; Cardaio, I.; dos Santos, P.E.J.; Ellis, J.D.; Nanetti, A. The first detection of *Nosema ceranae* (Microsporidia) in the small hive beetle, *Aethina tumida* Murray (Coleoptera: Nitidulidae). *Apidologie* **2018**, *49*, 619–624. [CrossRef]

26. Papini, R.; Mancianti, F.; Canovai, R.; Cosci, F.; Rocchigiani, G.; Benelli, G.; Canale, A. Prevalence of the microsporidian *Nosema ceranae* in honeybee (*Apis mellifera*) apiaries in Central Italy. *Saudi J. Biol. Sci.* **2017**, *24*, 979–982. [CrossRef] [PubMed]

27. Cilia, G.; Sagona, S.; Giusti, M.; Dos Santos, P.E.J.; Nanetti, A.; Felicioli, A. *Nosema ceranae* infection in honeybee samples from Tuscanian Archipelago (Central Italy) investigated by two qPCR methods. *Saudi J. Biol. Sci.* **2019**, *26*, 1553–1556. [CrossRef] [PubMed]

28. Gajger, I.T.; Vugrek, O.; Grilec, D.; Petrinec, Z. Prevalence and distribution of *Nosema ceranae* in Croatian honeybee colonies. *Vet. Med.* **2010**, *55*, 457–462. [CrossRef]

29. Blažytė-Čereškienė, L.; Skrodenytė-Arbačiauskienė, V.; Radžiutė, S.; Nedveckytė, I.; Būda, V. Honey bee infection caused by *Nosema* spp. in Lithuania. *J. Apic. Sci.* **2016**, *60*, 77–88. [CrossRef]

30. Pacini, A.; Mira, A.; Molineri, A.; Giacobino, A.; Cagnolo, N.B.; Aignasse, A.; Zago, L.; Izaguirre, M.; Merke, J.; Orellano, E.; et al. Distribution and prevalence of *Nosema apis* and *N. ceranae* in temperate and subtropical eco-regions of Argentina. *J. Invertebr. Pathol.* **2016**, *141*, 34–37. [CrossRef]

31. Sinpoo, C.; Paxton, R.J.; Disayathanoowat, T.; Krongdang, S.; Chantawannakul, P. Impact of *Nosema ceranae* and *Nosema apis* on individual worker bees of the two host species (*Apis cerana* and *Apis mellifera*) and regulation of host immune response. *J. Insect Physiol.* **2018**, *105*, 1–8. [CrossRef]

32. Paris, L.; El Alaoui, H.; Delbac, F.; Diogon, M. Effects of the Gut Parasite *N. ceranae ceranae* on Honey Bee Physiology and Behavior. *Curr. Opin. Insect Sci.* **2018**, *26*, 149–154. [CrossRef]

33. Arismendi, N.; Caro, S.; Castro, M.P.; Vargas, M.; Riveros, G.; Venegas, T. Impact of Mixed Infections of Gut Parasites *Lotmaria passim* and *N. ceranae ceranae* on the Lifespan and Immune-related Biomarkers in *Apis mellifera*. *Insects* **2020**, *11*, 420. [CrossRef] [PubMed]

34. Mortensen, A.N.; Jack, C.J.; Bustamante, T.A.; Schmehl, D.R.; Ellis, J.D. Effects of Supplemental Pollen Feeding on Honey Bee (Hymenoptera: Apidae) Colony Strength and *N. ceranae* spp. Infection. *J. Econ. Entomol.* **2019**, *112*, 60–66. [CrossRef]

35. Farrar, C.L. *N. ceranae* Losses in Package Bees as Related to Queen Supersedure and Honey Yields. *J. Econ. Entomol.* **1947**, *40*, 333–338. [CrossRef]

36. Botías, C.; Martín-Hernández, R.; Barrios, L.; Meana, A.; Higes, M. *N. ceranae* spp. infection and its negative effects on honey bees (*Apis mellifera iberiensis*) at the colony level. *Vet. Res.* **2013**, *44*, 25. [CrossRef]

37. Mendoza, Y.; Diaz-Cetti, S.; Ramallo, G.; Santos, E.; Porrini, M.; Invernizzi, C. *N. ceranae* Winter Control: Study of the Effectiveness of Different Fumagillin Treatments and Consequences on the Strength of Honey Bee (Hymenoptera: Apidae) Colonies. *J. Econ. Entomol.* **2017**, *110*, 1–5. [PubMed]

38. Burnham, A.J. Scientific Advances in Controlling *N. ceranae* (Microsporidia) Infections in Honey Bees (*Apis Mellifera*). *Front. Vet. Sci.* **2019**, *6*, 79. [CrossRef] [PubMed]

39. van den Heever, J.P.; Thompson, T.S.; Curtis, J.M.; Ibrahim, A.; Pernal, S.F. Fumagillin: An Overview of Recent Scientific Advances and Their Significance for Apiculture. *J. Agric. Food Chem.* **2014**, *62*, 2728–2737. [CrossRef]

40. Huang, W.-F.; Solter, L.F.; Yau, P.M.; Imai, B.S. *N. ceranae* Escapes Fumagillin Control in Honey Bees. *PLoS Pathog* **2013**, *9*, e1003185. [CrossRef] [PubMed]

41. Van den Heever, J.P.; Thompson, T.S.; Otto, S.J.; Curtis, J.M.; Ibrahim, A.; Pernal, S.F. The effect of dicyclohexylamine and fumagillin on *Nosema ceranae*-infected honey bee (*Apis mellifera*) mortality in cage trial assays. *Apidologie* **2016**, *47*, 663–670. [CrossRef]

42. Ptaszyńska, A.A.; Trytek, M.; Borsuk, G.; Buczek, K.; Rybicka-Jasińska, K.; Gryko, D. Porphyrins Inactivate *N. ceranae* spp. Microsporidia. *Sci. Rep.* **2018**, *8*, 5523. [CrossRef] [PubMed]

43. Nanetti, A.; Rodriguez-García, C.; Meana, A.; Martín-Hernández, R.; Higes, M. Effect of Oxalic Acid on *N. ceranae* Infection. *Res. Vet. Sci.* **2015**, *102*, 167–172. [CrossRef] [PubMed]

44. Costa, C.; Lodesani, M.; Maistrello, L. Effect of thymol and resveratrol administered with candy or syrup on the development of *N. ceranae* and on the longevity of honeybees (*Apis mellifera* L.) in laboratory conditions. *Apidologie* **2010**, *41*, 141–150. [CrossRef]

45. Jiang, Z.; Cui, W.; Prasad, P.; Touve, M.A.; Gianneschi, N.C.; Mager, J.; Thayumanavan, S. Bait-and-Switch Supramolecular Strategy To Generate Noncationic RNA–Polymer Complexes for RNA Delivery. *Biomacromolecules* **2019**, *20*, 435–442. [CrossRef]

46. Zuckerman, J.E.; Davis, M.E. Clinical Experiences with Systemically Administered SiRNA-Based Therapeutics in Cancer. *Nat. Rev. Drug Discov.* **2015**, *14*, 843–856. [CrossRef]

47. Grozinger, C.M.; Robinson, G.E. The Power and Promise of Applying Genomics to Honey Bee Health. *Curr. Opin. Insect. Sci.* **2015**, *10*, 124–132. [CrossRef]

48. Brutscher, L.M.; Daughenbaugh, K.F.; Flenniken, M.L. Virus and DsRNA-Triggered Transcriptional Responses Reveal Key Components of Honey Bee Antiviral Defense. *Sci. Rep.* **2017**, *7*, 6448. [CrossRef]

49. Evans, J.D.; Cook, S.C. Genetics and Physiology of Varroa Mites. *Curr. Opin. Insect Sci.* **2018**, *26*, 130–135. [CrossRef] [PubMed]

50. Rodríguez-García, C.; Evans, J.D.; Li, W.; Branchiccela, B.; Li, J.H.; Heerman, M.C.; Banmeke, O.; Zhao, Y.; Hamilton, M.; Higes, M.; et al. Nosemosis control in European honey bees, *Apis mellifera*, by silencing the gene encoding *N. ceranae* polar tube protein 3. *J. Exp. Biol.* **2018**, *221*, jeb184606. [CrossRef] [PubMed]

51. Kim, I.; Kim, D.; Gwak, W.; Woo, S. Increased Survival of the Honey Bee *Apis mellifera* Infected with the Microsporidian *N. ceranae* by Effective Gene Silencing. *Arch. Insect Biochem. Physiol.* **2020**, *105*, e21734. [CrossRef] [PubMed]

52. Li, W.; Evans, J.D.; Huang, Q.; Rodríguez-García, C.; Liu, J.; Hamilton, M.; Grozinger, C.M.; Webster, T.C.; Su, S.; Chen, Y.P. Silencing the Honey Bee (*Apis mellifera*) Naked Cuticle Gene (nkd) Improves Host Immune Function and Reduces *N. ceranae* Infections. *Appl. Environ. Microbiol.* **2016**, *82*, 6779–6787. [CrossRef] [PubMed]

53. Yang, D.; Pan, L.; Chen, Z.; Du, H.; Luo, B.; Luo, J.; Pan, G. The Roles of Microsporidia Spore Wall Proteins in the Spore Wall Formation and Polar Tube Anchorage to Spore Wall during Development and Infection Processes. *Exp. Parasitol.* **2018**, *187*, 93–100. [CrossRef] [PubMed]

54. Mamta, B.; Rajam, M.V. RNAi Technology: A New Platform for Crop Pest Control. *Physiol. Mol. Biol. Plants* **2017**, *23*, 487–501. [CrossRef]

55. Zhang, X.; Zhang, J.; Zhu, K.Y. Chitosan/Double-Stranded RNA Nanoparticle-Mediated RNA Interference to Silence Chitin Synthase Genes through Larval Feeding in the African Malaria Mosquito (Anopheles Gambiae): Nanoparticle-Mediated RNAi in Mosquito Larvae. *Insect Mol. Biol.* **2010**, *19*, 683–693. [CrossRef]

56. Roussel, M.; Villay, A.; Delbac, F.; Michaud, P.; Laroche, C.; Roriz, D.; El Alaoui, H.; Diogon, M. Antimicrosporidian Activity of Sulphated Polysaccharides from Algae and Their Potential to Control Honeybee Nosemosis. *Carbohydr. Polym.* **2015**, *133*, 213–220. [CrossRef] [PubMed]

57. Chen, X.; Wang, S.; Xu, Y.; Gong, H.; Wu, Y.; Chen, Y.; Hu, F.; Zheng, H. Protective potential of Chinese herbal extracts against microsporidian *N. ceranae*, an emergent pathogen of western honey bees, *Apis mellifera* L. *J. Asia Pac. Entomol.* **2019**, in press. [CrossRef]

58. Arismendi, N.; Vargas, M.; López, M.D.; Barría, Y.; Zapata, N. Promising Antimicrobial Activity against the Honey Bee Parasite *N. ceranae* by Methanolic Extracts from Chilean Native Plants and Propolis. *J. Apic. Res.* **2018**, *57*, 522–535. [CrossRef]

59. Fokt, H.; Pereira, A.; Ferreira, A.M.; Cunha, A.; Aguiar, C. How do bees prevent hive infections? The antimicrobial properties of propolis. *Biology* **2010**, *1*, 481–493.

60. Suwannapong, G.; Maksong, S.; Phainchajoen, M.; Benbow, M.E.; Mayack, C. Survival and Health Improvement of *N. ceranae* Infected *Apis florea* (Hymenoptera: Apidae) Bees after Treatment with Propolis Extract. *J. Asia Pac. Entomol.* **2018**, *21*, 437–444. [CrossRef]

61. Mura, A.; Pusceddu, M.; Theodorou, P.; Angioni, A.; Floris, I.; Paxton, R.J.; Satta, A. Propolis Consumption Reduces *N. ceranae* Infection of European Honey Bees (*Apis mellifera*). *Insects* **2020**, *11*, 124. [CrossRef]

62. Gherman, B.I.; Denner, A.; Bobiş, O.; Dezmirean, D.S.; Mărghitaş, L.A.; Schlüns, H.; Moritz, R.F.A.; Erler, S. Pathogen-Associated Self-Medication Behavior in the Honeybee *Apis mellifera*. *Behav. Ecol. Sociobiol.* **2014**, *68*, 1777–1784. [CrossRef]

63. LoCascio, G.M.; Aguirre, L.; Irwin, R.E.; Adler, L.S. Pollen from Multiple Sunflower Cultivars and Species Reduces a Common Bumblebee Gut Pathogen. *R. Soc. Open Sci.* **2019**, *6*, 190279. [CrossRef]

64. Charistos, L.; Parashos, N.; Hatjina, F. Long Term Effects of a Food Supplement HiveAlive™ on Honey Bee Colony Strength and *N. ceranae* Spore Counts. *J. Apic. Res.* **2015**, *54*, 420–426. [CrossRef]

65. Cilia, G.; Garrido, C.; Bonetto, M.; Tesoriero, D.; Nanetti, A. Effect of Api-Bioxal® and ApiHerb® Treatments against *N. ceranae* Infection in *Apis mellifera* Investigated by Two QPCR Methods. *Vet. Sci.* **2020**, *7*, 125. [CrossRef]

66. Glavinic, U.; Stankovic, B.; Draskovic, V.; Stevanovic, J.; Petrovic, T.; Lakic, N.; Stanimirovic, Z. Dietary Amino Acid and Vitamin Complex Protects Honey Bee from Immunosuppression Caused by *N. ceranae*. *PLoS ONE* **2017**, *12*, e0187726. [CrossRef]

67. Van den Heever, J.P.; Thompson, T.S.; Otto, S.J.G.; Curtis, J.M.; Ibrahim, A.; Pernal, S.F. Evaluation of Fumagilin-B® and other potential alternative chemotherapies against *N. ceranae* -infected honeybees (*Apis mellifera*) in cage trial assays. *Apidologie* **2016**, *47*, 617–630. [CrossRef]

68. DeGrandi-Hoffman, G.; Chen, Y.; Rivera, R.; Carroll, M.; Chambers, M.; Hidalgo, G.; de Jong, E.W. Honey Bee Colonies Provided with Natural Forage Have Lower Pathogen Loads and Higher Overwinter Survival than Those Fed Protein Supplements. *Apidologie* **2016**, *47*, 186–196. [CrossRef]

69. Alberoni, D.; Baffoni, L.; Gaggìa, F.; Ryan, P.M.; Murphy, K.; Ross, P.R.; Stanton, C.; Di Gioia, D. Impact of Beneficial Bacteria Supplementation on the Gut Microbiota, Colony Development and Productivity of *Apis mellifera* L. *Benef. Microbes* **2018**, *9*, 269–278. [CrossRef]

70. Li, J.H.; Evans, J.D.; Li, W.F.; Zhao, Y.Z.; DeGrandi-Hoffman, G.; Huang, S.K.; Li, Z.G.; Hamilton, M.; Chen, Y.P. New evidence showing that the destruction of gut bacteria by antibiotic treatment could increase the honey bee's vulnerability to *N. ceranae* infection. *PLoS ONE* **2017**, *12*, e0187505. [CrossRef]

71. Baffoni, L.; Gaggìa, F.; Alberoni, D.; Cabbri, R.; Nanetti, A.; Biavati, B.; Di Gioia, D. Effect of Dietary Supplementation of *Bifidobacterium* and *Lactobacillus* Strains in *Apis mellifera* L. against *N. ceranae*. *Benef. Microbes* **2016**, *7*, 45–51. [CrossRef]

72. Corby-Harris, V.; Snyder, L.; Meador, C.A.D.; Naldo, R.; Mott, B.; Anderson, K.E. *Parasaccharibacter apium*, gen. nov., sp. nov., Improves Honey Bee (Hymenoptera: Apidae) Resistance to *N. ceranae*. *J. Econ. Entomol.* **2016**, *109*, 537–543. [CrossRef]

73. El Khoury, S.; Rousseau, A.; Lecoeur, A.; Cheaib, B.; Bouslama, S.; Mercier, P.L.; Demey, V.; Castex, M.; Giovenazzo, P.; Derome, N. Deleterious interaction between honeybees (*Apis mellifera*) and its microsporidian intracellular parasite *N. ceranae* was mitigated by administrating either endogenous or allochthonous gut microbiota strains. *Front. Ecol. Evol.* **2018**, *6*, 58. [CrossRef]

74. Arredondo, D.; Castelli, L.; Porrini, M.P.; Garrido, P.M.; Eguaras, M.J.; Zunino, P.; Antúnez, K. *Lactobacillus kunkeei* Strains Decreased the Infection by Honey Bee Pathogens *Paenibacillus larvae* and *N. ceranae*. *Benef. Microbes* **2018**, *9*, 279–290. [CrossRef] [PubMed]

75. Kazimierczak-Baryczko, M.; Szymaś, B. Improvement of the composition of pollen substitute for honey bee (*Apis mellifera* L.), through implementation of probiotic preparations. *J. Apic. Sci.* **2006**, *50*, 15–23.

76. Schmidt, K.; Engel, P. Probiotic Treatment with a Gut Symbiont Leads to Parasite Susceptibility in Honey Bees. *Trends Parasitol.* **2016**, *32*, 914–916. [CrossRef]

77. Rubanov, A.; Russell, K.A.; Rothman, J.A.; Nieh, J.C.; McFrederick, Q.S. Intensity of *N. ceranae* infection is associated with specific honey bee gut bacteria and weakly associated with gut microbiome structure. *Sci. Rep.* **2019**, *9*, 1–8. [CrossRef] [PubMed]

78. Vetvicka, V.; Fernandez-Botran, R. β-Glucan and Parasites. *Helminthologia* **2018**, *55*, 177–184. [CrossRef]

79. Saltykova, E.S.; Gaifullina, L.R.; Kaskinova, M.D.; Gataullin, A.R.; Matniyazov, R.T.; Poskryakov, A.V.; Nikolenko, A.G. Effect of Chitosan on Development of *N. apis* Microsporidia in Honey Bees. *Microbiology* **2018**, *87*, 738–743. [CrossRef]

80. Tlak Gajger, I.; Ribaric, J.; Matak, M.; Svecnjak, L.; Kozaric, Z.; Nejedli, S.; Smodis Skerl, I. Zeolite Clinoptilolite as a Dietary Supplement and Remedy for Honeybee (*Apis mellifera* L.) Colonies. *Vet. Med.* **2017**, *60*, 696–705. [CrossRef]

81. Higes, M.; Martín-Hernández, R.; Meana, A. *N. ceranae* in Europe: An Emergent Type C Nosemosis. *Apidologie* **2010**, *41*, 375–392. [CrossRef]

82. Fries, I.; Feng, F.; da Silva, A.; Slemenda, S.B.; Pieniazek, N.J. *N. ceranae* n. sp. (Microspora, Nosematidae), morphological and molecular characterization of a microsporidian parasite of the Asian honey bee *Apis cerana* (Hymenoptera, Apidae). *Eur. J. Protistol.* **1996**, *32*, 356–365. [CrossRef]

83. Medici, S.K.; Sarlo, E.G.; Porrini, M.P.; Braunstein, M.; Eguaras, M.J. Genetic Variation and Widespread Dispersal of *N. ceranae* in *Apis mellifera* Apiaries from Argentina. *Parasitol. Res.* **2012**, *110*, 859–864. [CrossRef] [PubMed]

84. Antúnez, K.; Invernizzi, C.; Mendoza, Y.; Zunino, P. Honeybee colony losses in Uruguay during 2013–2014. *Apidologie* **2017**, *48*, 364–370. [CrossRef]

85. Goblirsch, M. *N. ceranae* Disease of the Honey Bee (*Apis mellifera*). *Apidologie* **2018**, *49*, 131–150. [CrossRef]

86. Cristina, R.T.; Kovačević, Z.; Cincović, M.; Dumitrescu, E.; Muselin, F.; Imre, K.; Militaru, D.; Mederle, N.; Radulov, I.; Hădărugă, N.; et al. Composition and Efficacy of a Natural Phytotherapeutic Blend against Nosemosis in Honey Bees. *Sustainability* **2020**, *12*, 5868. [CrossRef]

87. Raymann, K.; Shaffer, Z.; Moran, N.A. Antibiotic Exposure Perturbs the Gut Microbiota and Elevates Mortality in Honeybees. *PLoS Biol.* **2017**, *15*, e2001861. [CrossRef]

88. Ptaszyńska, A.A.; Załuski, D. Extracts from *Eleutherococcus senticosus* (Rupr. et Maxim.) Maxim. roots: A new hope against honeybee death caused by nosemosis. *Molecules* **2020**, *25*, 4452. [CrossRef] [PubMed]

89. Schulz, M.; Łoś, A.; Grzybek, M.; Ścibior, R.; Strachecka, A. Piperine as a new natural supplement with beneficial effects on the life-span and defence system of honeybees. *J. Agric. Sci.* **2019**, *157*, 140–149. [CrossRef]

90. Strachecka, A.J.; Olszewski, K.; Paleolog, J. Curcumin stimulates biochemical mechanisms of *Apis mellifera* resistance and extends the apian life-span. *J. Apic. Sci.* **2015**, *59*, 129–141. [CrossRef]

91. Tasleem, F.; Azhar, I.; Ali, S.N.; Perveen, S.; Mahmood, Z.A. Analgesic and anti-inflammatory activities of *Piper nigrum* L. *Asian Pac. J. Trop. Med.* **2014**, *7*, S461–S468. [CrossRef]

92. Umar, S.; Sarwar, A.H.M.G.; Umar, K.; Ahmad, N.; Sajad, M.; Ahmad, S.; Katiyar, C.K.; Khan, H.A. Piperine ameliorates oxidative stress, inflammation and histological outcome in collagen induced arthritis. *Cell. Immunol.* **2013**, *284*, 51–59. [CrossRef] [PubMed]

93. Kurze, C.; Dosselli, R.; Grassl, J.; Le Conte, Y.; Kryger, P.; Baer, B.; Moritz, R.F.A. Differential proteomics reveals novel insights into *N. ceranae*–honey bee interactions. *Insect Biochem. Mol. Biol.* **2016**, *79*, 42–49. [CrossRef]

94. Cilia, G.; Cabbri, R.; Maiorana, G.; Cardaio, I.; Dall'Olio, R.; Nanetti, A. A novel TaqMan® assay for *Nosema ceranae* quantification in honey bee, based on the protein coding gene Hsp70. *Eur. J. Protistol.* **2018**, *63*, 44–50. [CrossRef] [PubMed]

95. Schwarz, R.S.; Moran, N.A.; Evans, J.D. Early Gut Colonizers Shape Parasite Susceptibility and Microbiota Composition in Honey Bee Workers. *Proc. Natl. Acad. Sci. USA* **2016**, *113*, 9345–9350. [CrossRef]

96. Daisley, B.A.; Chmiel, J.A.; Pitek, A.P.; Thompson, G.J.; Reid, G. Missing Microbes in Bees: How Systematic Depletion of Key Symbionts Erodes Immunity. *Trends Microbiol.* **2020**, *28*, 1010–1021. [CrossRef]

97. Paris, L.; Peghaire, E.; Moné, A.; Diogon, M.; Debroas, D.; Delbac, F.; El Alaoui, H. Honeybee Gut Microbiota Dysbiosis in Pesticide/Parasite Co-Exposures Is Mainly Induced by *N. ceranae*. *J. Invertebr. Pathol.* **2020**, *172*, 107348. [CrossRef] [PubMed]

98. Castelli, L.; Branchiccela, B.; Garrido, M.; Invernizzi, C.; Porrini, M.; Romero, H.; Santos, E.; Zunino, P.; Antúnez, K. Impact of Nutritional Stress on Honeybee Gut Microbiota, Immunity, and *N. ceranae* Infection. *Microb. Ecol.* **2020**, *80*, 908–919. [CrossRef]

99. Bravo, J.; Carbonell, V.; Sepúlveda, B.; Delporte, C.; Valdovinos, C.E.; Martín-Hernández, R.; Higes, M. Antifungal Activity of the Essential Oil Obtained from *Cryptocarya alba* against Infection in Honey Bees by *N. ceranae ceranae*. *J. Invertebr. Pathol.* **2017**, *149*, 141–147. [CrossRef]

100. Lee, J.K.; Kim, J.H.; Jo, M.; Rangachari, B.; Park, J.K. Anti-Nosemosis Activity of *Aster scaber* and *Artemisia dubia* Aqueous Extracts. *J. Apic. Sci.* **2018**, *62*, 27–38. [CrossRef]

101. Balamurugan, R.; Park, J.K.; Lee, J.K. Anti-Nosemosis Activity of Phenolic Compounds Derived from *Artemisia dubia* and *Aster scaber*. *J. Apic. Res.* **2020**, 1–11. [CrossRef]

102. Farjan, M.; Łopieńska-Biernat, E.; Lipiński, Z.; Dmitryjuk, M.; Żółtowska, K. Supplementing with vitamin C the diet of honeybees (*Apis mellifera carnica*) parasitized with *Varroa destructor*: Effects on antioxidative status. *Parasitology* **2014**, *141*, 770–776. [CrossRef]

103. Stanimirović, Z.; Glavinić, U.; Lakić, N.; Radović, D.; Ristanić, M.; Tarić, E.; Stevanović, J. Efficacy of Plant-Derived Formulation "Argus Ras" in *Varroa destructor* Control. *Acta Vet.* **2017**, *67*, 191–200. [CrossRef]

104. Aurori, A.C.; Bobiş, O.; Dezmirean, D.S.; Mărghitaş, L.A.; Erler, S. Bay Laurel (*Laurus nobilis*) as Potential Antiviral Treatment in Naturally BQCV Infected Honeybees. *Virus Res.* **2016**, *222*, 29–33. [CrossRef]

105. Fernández, N.J.; Damiani, N.; Podaza, E.A.; Martucci, J.F.; Fasce, D.; Quiroz, F.; Meretta, P.E.; Quintana, S.; Eguaras, M.J.; Gende, L.B. *Laurus nobilis* L. Extracts against *Paenibacillus larvae*: Antimicrobial Activity, Antioxidant Capacity, Hygienic Behavior and Colony Strength. *Saudi J. Biol. Sci.* **2019**, *26*, 906–912. [CrossRef]

106. Ferreira, T.P.; Oliveira, E.E.; Tschoeke, P.H.; Pinheiro, R.G.; Maia, A.M.S.; Aguiar, R.W.S. Potential Use of Negramina (*Siparuna guianensis* Aubl.) Essential Oil to Control Wax Moths and Its Selectivity in Relation to Honey Bees. *Ind. Crops Prod.* **2017**, *109*, 151–157. [CrossRef]

107. Stevanovic, J.; Stanimirovic, Z.; Simeunovic, P.; Lakic, N.; Radovic, I.; Sokovic, M.; Griensven, L.J.V. The Effect of *Agaricus brasiliensis* Extract Supplementation on Honey Bee Colonies. *An. Acad. Bras. Ciênc.* **2018**, *90*, 219–229. [CrossRef]

108. Lamontagne-Drolet, M.; Samson-Robert, O.; Giovenazzo, P.; Fournier, V. The Impacts of Two Protein Supplements on Commercial Honey Bee (*Apis mellifera* L.) Colonies. *J. Apic. Res.* **2019**, *58*, 800–813. [CrossRef]

109. Jehlík, T.; Kodrík, D.; Krištůfek, V.; Koubová, J.; Sábová, M.; Danihlík, J.; Tomčala, A.; Frydrychová, R.C. Effects of Chlorella sp. on Biological Characteristics of the Honey Bee *Apis mellifera*. *Apidologie* **2019**, *50*, 564–577. [CrossRef]

110. Higginson, A.D.; Gilbert, F.S.; Reader, T.; Barnard, C.J. Reduction of visitation rates by honeybees (*Apis mellifera*) to individual inflorescences of lavender (*Lavandula stoechas*) upon removal of coloured accessory bracts (Hymenoptera: Apidae). *Entomol. Gen.* **2007**, *29*, 165–178. [CrossRef]

111. Bekret, A.; Çankaya, S.; Silici, S. The Effects of Mixture of Plant Extracts and Oils are added to Syrup on Honey Bee Colony Development and Honey Yield. *TURJAF* **2015**, *3*, 365–370.

112. Utuk, A.E.; Piskin, F.C.; Girisgin, A.O.; Selcuk, O.; Aydin, L. Microscopic and molecular detection of *N. ceranae* spp. in honeybees of Turkey. *Apidologie* **2016**, *47*, 267–271. [CrossRef]

113. Martín-Hernández, R.; Meana, A.; Prieto, L.; Salvador, A.M.; Bailon, E.G.; Higes, M. Outcome of colonization of *Apis mellifera* by *Nosema ceranae*. *Appl. Environ. Microbiol.* **2007**, *73*, 6331–6338. [CrossRef] [PubMed]

114. Crozier, R.H.; Crozier, Y.C. The mitochondrial genome of the honeybee *Apis mellifera*: Complete sequence and genome organization. *Genetics* **1993**, *133*, 97–117. [CrossRef] [PubMed]

115. Delaplane, K.S.; Van Der Steen, J.; Guzman-Novoa, E. Standard methods for estimating strength parameters of *Apis mellifera* colonies. *J. Apic. Res.* **2013**, *52*, 1–12. [CrossRef]

Article

Amplicon Sequencing of Variable 16S rRNA from Bacteria and ITS2 Regions from Fungi and Plants, Reveals Honeybee Susceptibility to Diseases Results from Their Forage Availability under Anthropogenic Landscapes

Aneta A. Ptaszyńska [1,2,*,†], Przemyslaw Latoch [3,4], Paul J. Hurd [2], Andrew Polaszek [5], Joanna Michalska-Madej [6], Łukasz Grochowalski [6], Dominik Strapagiel [6], Sebastian Gnat [7], Daniel Załuski [8], Marek Gancarz [9,10], Robert Rusinek [9], Patcharin Krutmuang [11,12], Raquel Martín Hernández [13,14], Mariano Higes Pascual [13] and Agata L. Starosta [4,15]

Citation: Ptaszyńska, A.A.; Latoch, P.; Hurd, P.J.; Polaszek, A.; Michalska-Madej, J.; Grochowalski, Ł.; Strapagiel, D.; Gnat, S.; Załuski, D.; Gancarz, M.; et al. Amplicon Sequencing of Variable 16S rRNA from Bacteria and ITS2 Regions from Fungi and Plants, Reveals Honeybee Susceptibility to Diseases Results from Their Forage Availability under Anthropogenic Landscapes. *Pathogens* **2021**, *10*, 381. https://doi.org/10.3390/pathogens10030381

Academic Editor: Giovanni Cilia

Received: 22 January 2021
Accepted: 18 March 2021
Published: 22 March 2021

Publisher's Note: MDPI stays neutral with regard to jurisdictional claims in published maps and institutional affiliations.

[1] Department of Immunobiology, Faculty of Biology and Biotechnology, Institute of Biological Sciences, Maria Curie-Skłodowska University, Akademicka 19 Str., 20-033 Lublin, Poland

[2] School of Biological and Chemical Sciences, Queen Mary University of London, London E1 4NS, UK; p.j.hurd@qmul.ac.uk

[3] Polish-Japanese Academy of Information Technology, Koszykowa 86 Str., 02-008 Warsaw, Poland; przemyslawlatoch@gmail.com

[4] Laboratory of Gene Expression, ECOTECH-Complex, Maria Curie-Sklodowska University, ul. Gleboka 39, 20-612 Lublin, Poland; agata.starosta@umcs.pl

[5] Department of Life Sciences, Insects Division, Natural History Museum, London SW7 5BD, UK; a.polaszek@nhm.ac.uk

[6] Biobank Lab, Department of Molecular Biophysics, Faculty of Biology and Environmental Protection, University of Łódź, Pilarskiego 14/16, 90-231 Łódź, Poland; joanna.michalska@biol.uni.lodz.pl (J.M.-M.); lukasz.grochowalski@biol.uni.lodz.pl (Ł.G.); dominik.strapagiel@biol.uni.lodz.pl (D.S.)

[7] Department of Veterinary Microbiology, Faculty of Veterinary Medicine, Institute of Preclinical Veterinary Sciences, University of Life Sciences, Akademicka 12, 20-033 Lublin, Poland; sebastian.gnat@up.lublin.pl

[8] Department of Pharmaceutical Botany and Pharmacognosy, Ludwik Rydygier Collegium Medicum, Nicolaus Copernicus University, Marie Curie-Skłodowska 9, 85-094 Bydgoszcz, Poland; daniel_zaluski@onet.eu

[9] Institute of Agrophysics, Polish Academy of Sciences, Doświadczalna 4 Str., 20-290 Lublin, Poland; m.gancarz@ipan.lublin.pl (M.G.); r.rusinek@ipan.lublin.pl (R.R.)

[10] Faculty of Production and Power Engineering, University of Agriculture in Kraków, Balicka 116B, 30-149 Kraków, Poland

[11] Department of Entomology and Plant Pathology, Faculty of Agriculture, Chiang Mai University, Chiang Mai 50200, Thailand; patcharink26@gmail.com

[12] Research Center of Microbial Diversity and Sustainable Utilization, Faculty of Science, Chiang Mai University, Chiang Mai 50200, Thailand

[13] Centro de Investigación Apícola y Agroambiental (CIAPA), Laboratorio de Patología Apícola, IRIAF Instituto Regional de Investigación y Desarrollo Agroalimentario y Forestal, Consejería de Agricultura de la Junta de Comunidades de Castilla-La Mancha, Camino de San Martín s/n, 19180 Marchamalo, Spain; rmhernandez@jccm.es (R.M.H.); mhiges@jccm.es (M.H.P.)

[14] Instituto de Recursos Humanos para la Ciencia y la Tecnología (INCRECYT-FEDER), Fundación Parque Científico y Tecnológico de Castilla—La Mancha, 02006 Albacete, Spain

[15] Department of Molecular Biology, Institute of Biological Sciences, Maria Curie-Skłodowska University, Akademicka 19 Str., 20-033 Lublin, Poland

* Correspondence: aneta.ptaszynska@poczta.umcs.lublin.pl

† Senior author.

Abstract: European *Apis mellifera* and Asian *Apis cerana* honeybees are essential crop pollinators. Microbiome studies can provide complex information on health and fitness of these insects in relation to environmental changes, and plant availability. Amplicon sequencing of variable regions of the 16S rRNA from bacteria and the internally transcribed spacer (ITS) regions from fungi and plants allow identification of the metabiome. These methods provide a tool for monitoring otherwise uncultured microbes isolated from the gut of the honeybees. They also help monitor the composition of the gut fungi and, intriguingly, pollen collected by the insect. Here, we present data from amplicon sequencing of the 16S rRNA from bacteria and ITS2 regions from fungi and plants derived from honeybees collected at various time points from anthropogenic landscapes such as urban areas in Poland, UK, Spain, Greece, and Thailand. We have analysed microbial content of honeybee intestine as well as fungi and pollens. Furthermore, isolated DNA was used as the template for screening pathogens: *Nosema apis, N. ceranae, N. bombi*, tracheal mite (*Acarapis woodi*), any organism in the parasitic order Trypanosomatida, including Crithidia spp. (i.e., *Crithidia mellificae*), neogregarines including *Mattesia* and *Apicystis* spp. (i.e., *Apicistis bombi*). We conclude that differences between samples were mainly influenced by the bacteria, plant pollen and fungi, respectively. Moreover, honeybees feeding on a sugar based diet were more prone to fungal pathogens (*Nosema ceranae*) and neogregarines. In most samples *Nosema* sp. and neogregarines parasitized the host bee at the same time. A higher load of fungi, and bacteria groups such as Firmicutes (*Lactobacillus*); γ-proteobacteria, Neisseriaceae, and other unidentified bacteria was observed for *Nosema ceranae* and neogregarine infected honeybees. Healthy honeybees had a higher load of plant pollen, and bacteria groups such as: *Orbales, Gilliamella, Snodgrassella*, and Enterobacteriaceae. Finally, the period when honeybees switch to the winter generation (longer-lived forager honeybees) is the most sensitive to diet perturbations, and hence pathogen attack, for the whole beekeeping season. It is possible that evolutionary adaptation of bees fails to benefit them in the modern anthropomorphised environment.

Keywords: *Apis mellifera*; 16S rRNA; ITR2; NGS; *Nosema apis*; *Nosema ceranae*; *Nosema bombi*; *Acarapis woodi*; Trypanosomatida; *Crithidia* spp.; neogregarines; *Apicystis* spp.; antropocene; insectageddon; urban area; urban environment; bee biology

1. Introduction

Next-generation sequencing (NGS) is a culture-independent method often used for studying entire microbial communities, and helps to understand how microbes influence health and diseases of humans and animals including the honeybee [1–3]. Adult honeybees harbour a specialized gut microbiota of relatively low complexity with diet as a major factor influencing differences in bacterial loads [4]. Although honeybee microbiome core species composition is quite consistent regardless of environmental, geographical and genetic differences between specimens [2], some studies indicate that it can be sensitive to infection, changes in diet, malnutrition and many anthropogenic activities, such as extensive pesticide use and urban land-use changes [5–7].

Species within the *Apis* genus share fewer than 10 core species members, including *Lactobacillus, Bifidobacterium, Neisseria, Pasteurella, Gluconobacter, Snodgrassella* and *Gilliamella* [8–11]. Bacteria present in the honeybee gut provide numerous beneficial effects: they help digest and absorb necessary compounds and microelements, protect against mild poisonings with xenobiotics, acidify their environment which protects the gut from pathogenic microbes, e.g., *Paenibacillus larvae* that causes foulbrood, or *N. ceranae* that causes nosemosis [12,13]. They also have immunomodulation effects improving bees' immunity, strengthening the condition of the colony and prolonging bees' lives [11]. From the kingdom Fungi, yeasts are prevalent organisms in every environment in which bees conduct their life cycle, and can be isolated, for example, from honey and nectar. Honey microflora is composed of Gram-positive bacteria and yeasts, such as *Saccharomyces rouxi, S. mellis, S. bisporus, S. roesi, S. bailli, S. heterogenicus, Pichia (Hansenula) anomala*. The pollen reserves flora, which is dominated by bacteria from the genera *Pseudomonas* and *Lactobacillus*, and fungi from the genera Saccharomyces, Candida and Cryptococcus, far outnumbers the microflora of the honey [14–18]. Surprisingly, the gut flora of healthy, free-flying bees contains only a few yeasts if any, and diseases, malnutrition, antibiotics and insecticides cause an increase in the number of yeasts [16,19,20]. Therefore, an increased number of yeast colonies isolated from bees' guts may be considered as a stress indicator. However, recent work of Tauber et al. [5,21] suggested that, in general, yeasts are important during a younger bee's life, which includes in-house duties to feed the hive, and that the yeast community becomes less essential to the honeybee after foraging begins. Honeybee colonies are complex super-organisms where social immune defences, natural homeostatic mechanisms, microbiome diversity and function play a major role in disease resistance. However, there is still little known about connections between, and variation among bee pathogens, bee microbiota and anthropogenic changes of environment.

Currently, in developed countries, anthropogenic landscapes are the most impacting features and include those created either directly by human activity, or indirectly by natural processes triggered by human activity [22–24]. Not only has human activity influenced geological features, but it has also considerably affected flora and fauna [25]. The loss in biodiversity is often described as the sixth mass extinction, and a slump in insect mass and biodiversity is so spectacular that the term Insectaggedon has been used to describe the phenomenon [26–30].

Recent research shows insects to be dying out eight times faster than mammals, birds, or reptiles [26,31,32]. Most noteworthy factors behind the decline of insects are inappropriate application of pesticides, increased use of fertilizers and intense agronomic activities, highly intensive farming, insect malnutrition caused by farmland monocultures, parasites, long-term drought, long-term lack of sun, especially accompanied by low temperatures, as well as viral, bacterial, and fungal diseases [24,33]. Currently special concern is being paid to the decline of pollinators, largely because of their essential ecosystem services [20]. Therefore, both for educational purposes and as a way to preserve pollinator populations many urban pollinator initiatives have arisen recently (e.g.,"Life + Urbanbees" [34], "City Bees" [35], "Urban Beekeeping" [36]). Urban colonies were shown to be more productive than rural ones, as they had access to a greater number and variety of plant species, allowing honeybees to diversify nectar sources and produce honey at a higher rate [37,38]. On the other hand, wild pollinators such as *Bombus* and *Lasioglossum* spp. were negatively affected by urbanization, but increasing the abundance and richness of floral resources could partially compensate this effect [39]. Effects of urbanization on bees are complex, variable and not well-understood [40,41]. Therefore, the aim of this study was to use the amplicon sequencing of variable 16S rRNA and ITS2 to screen honeybee colonies originating from different urban areas, and to check if "the bees really love the city".

2. Results and Discussion

Adult honeybees harbour specialized gut microbiota of relatively low complexity with five core bacterial strains [3,4,9,42,43]: *Lactobacillus* Firm-4 and Firm-5 (Firmicutes), *Giliamella* (γ-proteobacteria, Orbales), *Snodgrassella* (γ-proteobacteria, Neisseriaceae), *Bifidobacterium* (Actinobacteria) and a number of elective bacterial strains, including *Frischella* (γ-proteobacteria, Orbales), *Bartonella* (α-proteobacteria, Rhizobiales), *Commensalibacter* (α-proteobacteria, Acetobacterales) and *Bombella* (α-proteobacteria, Acetobacterales), which was also confirmed in this study (Figures 1 and 2, Supplementary Materials Tables S1 and S2).

NGS was used successfully for taxonomic assessment of pollen and plants from many ecological and palynological studies, and to determine plant–pollinator interactions, or to confirm the floral composition of honey [3,44–47].

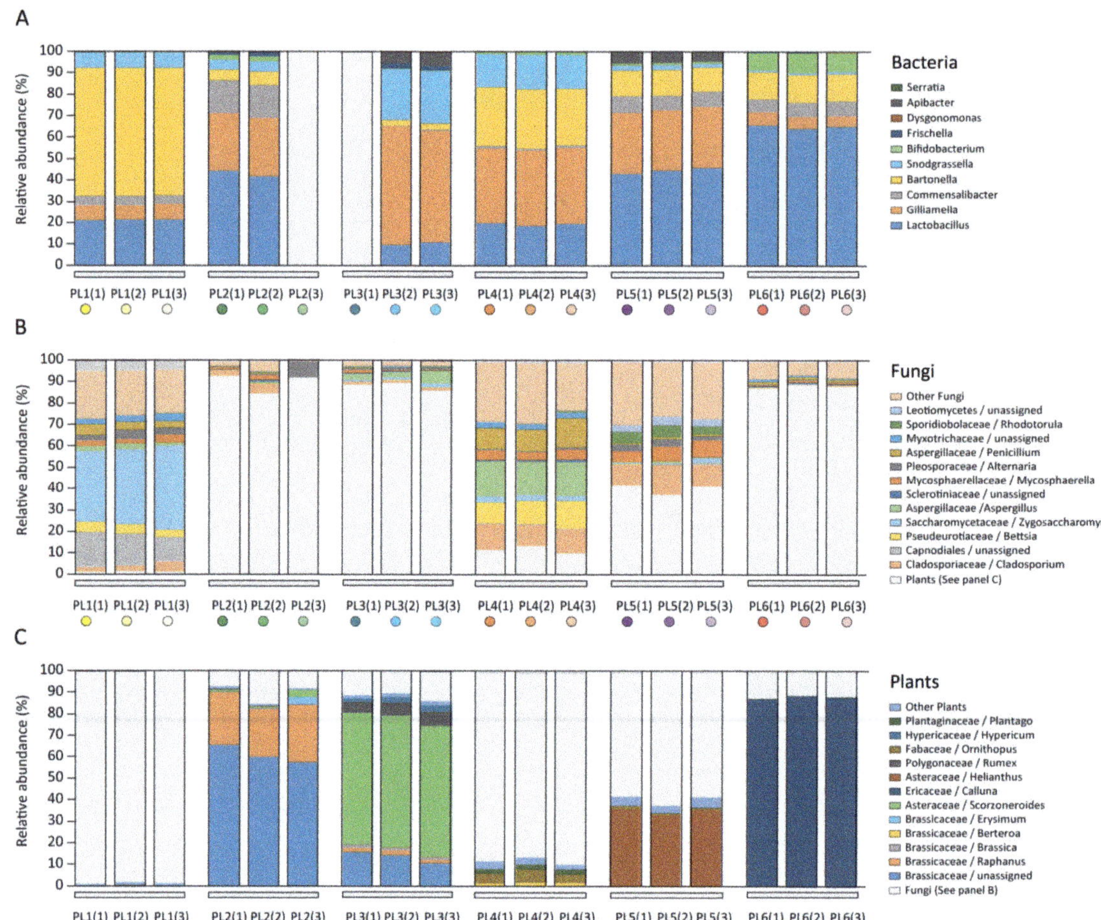

Figure 1. Composition of bacteria (**A**), fungi (**B**) and pollen (**C**) from Polish honeybee samples.

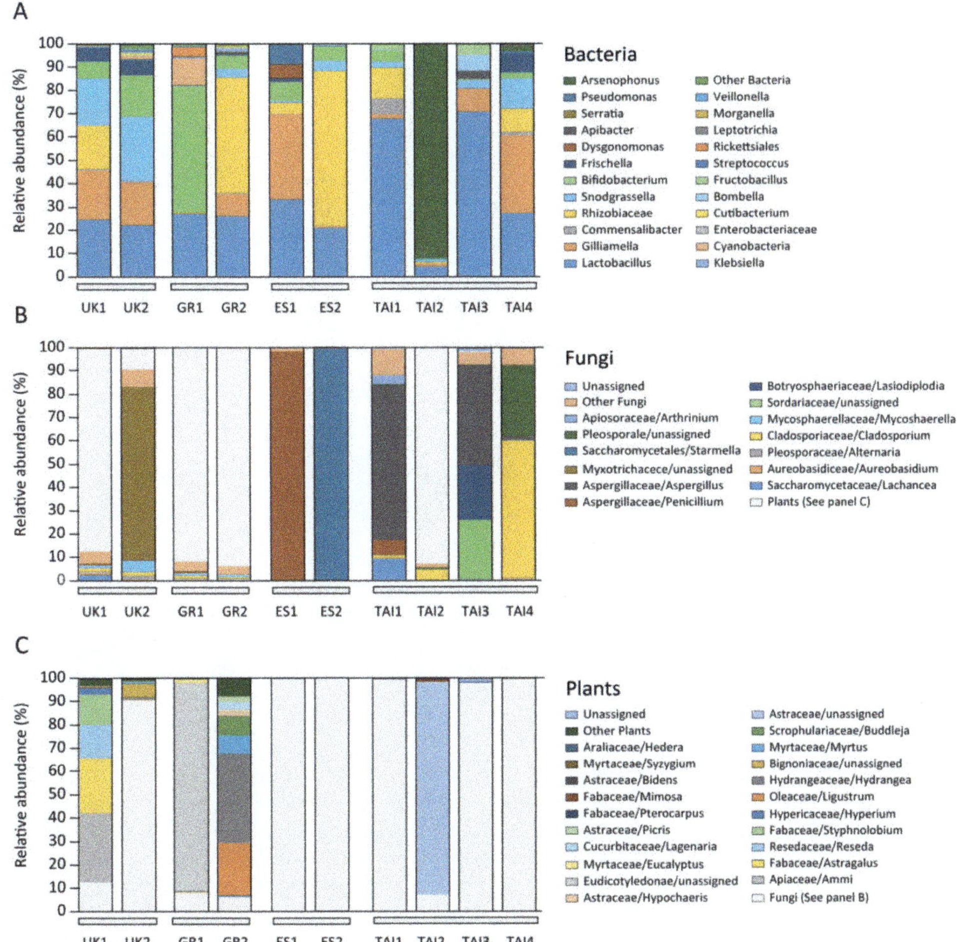

Figure 2. Composition of bacteria (**A**), fungi (**B**) and pollen (**C**) from UK, Greek, Spanish, and Thai bee samples.

2.1. Microbiome and Pollen Composition of Honeybees from Poland (Differences over the Vegetation Season)

Following NGS analysis, it was possible to assign samples to the time that they were collected by comparing them with vegetation periods of nectar- and pollen-rich plants. From one location, 3 specimens (forager honeybees) were taken, as the representative and consistent number for each group (data adequacy confirmed by the principal component analysis (PCA) analysis of amplicons of the 16S rRNA from bacteria and the ITSs region from fungi and plants, Supplementary Materials Figure S1). Microbiome down to genera analysis enabled division of Polish honeybee samples into 6 sub-groups: PL1, PL2, PL3, PL4, PL5, and PL6 (Supplementary Materials Table S1).

PL1 forager honeybees were collected in April and 16 taxa were identified in total from 16S amplicon analysed and 164 taxa in ITS2 analyses. In PL1 a low number of pollen types was observed (1.81%, SD = 0.512 belonging primarily to the *Betula* sp. and *Urtica* sp. genera) and a high number of fungi (93.83%, SD = 0.554). The low number of pollen types may relate to the low availability of flowering resources at the start of the flowering season, when the bees have to consume the stored feeding solution in the colony, or after feeding by the beekeeper later in the season. A sugar-based diet can lead to higher yeast numbers as observed in PL1 (35.92%, SD = 3.208) (Supplementary Materials Tables S1 and S2). Furthermore, the pathogenic fungus *Nosema ceranae* was detected in the PL1 (Supplementary Materials Table S3).

PL2 forager honeybees were collected in May and 16 taxa were identified in total from 16S amplicons analysed and 147 taxa in ITS2. *Lactobacillus* spp. were the dominant bacteria in PL2 (43.02%, SD = 1.704) (Supplementary Materials Table S1). At the same time, the percentage of fungi was moderate (9.59%; SD = 4.476) with the prevalence of *Cladosporium* and a small content of fungi from other genera (Supplementary Materials Table S2). The dominant pollen content in the PL2 with 87.54% (SD = 3.886) came from plants from the *Brassicaceae*, mainly unassigned, and speciesof *Raphanus* (Supplementary Materials Table S1B) known for its high protein pollen content [31]. No pathogens were detected in PL2 (Supplementary Materials Table S3).

PL3 forager honeybees were collected in June and 18 taxa were identified in total from 16S amplicons analysed and 177 taxa in ITS2. This group shows a higher number of γ-proteobacteria from *Gilliamella* and *Snodgrassella* (54.03%, SD = 2.333) (Supplementary Materials Table S1). PL3 honeybees collected pollen from Polygonaceae, which contain moderate amounts of amino acids [48]. In this group, fungal load was moderate (11.25%, SD = 1.847) with *Aspergillus*, *Cladosporium* and *Mycosphaerella* (mainly transferred as bioaerosols by wind in the air) at the level of 3.76% (SD = 1.9486), 1.21% (SD = 0.0961), and 0.81% (SD = 0.624), respectively (Supplementary Materials Table S1B). No pathogens were detected in PL3 (Supplementary Materials Table S3).

PL4 forager honeybees were collected in July and 16 taxa were identified in total from 16S amplicons analysed and 322 taxa in ITS2. This group was differentiated on the basis of the highest fungal DNA loads (87.31%, SD = 1.680) and trace amounts of plant DNA (Supplementary Materials Table S1B). The fungi were mainly *Aspergillus* (15.98%, SD = 0.503), *Cladosporium* (10.91%, SD = 1.048), *Penicillum* (11.07%, SD = 2.190), and *Betsia* (11.20%, SD = 1.572) spp. The spores of the three first genera commonly appeared in the air (transferred as bioaerosols), but the presence of *Betsia* species is suggestive of poor health of the honeybee colony. The pollen mould (*B. alvei*) is a saprophyte that lives on the pollen stored combs, especially in temperate regions [49]. Furthermore, pathogens belonging to *N. ceranae* and neogregarines were detected in the PL4 (Supplementary Materials Table S3). Moreover, a high fungal load and a small amount of plant pollen indicates that honeybees most likely gained sugar based diet. At the same time, PL4 honeybees had high amounts of γ-proteobacteria, Orbales, *Gilliamella* (35.32%, SD = 0.370) α-*proteobacteria Rhizobiales*, *Bartonella* (27.06%, SD = 0.560), known honeybee gut symbionts (Supplementary Materials Table S1A). *Gilliamella apicola* was found to be a dominant gut bacterium in honeybees and bumble bees, and this bacterium simultaneously utilizes glucose, fructose and mannose, and has the ability to break down other potentially toxic carbohydrates [50].

PL5 forager honeybees were collected in August and 27 taxa were identified in total from 16S amplicons analysed and 284 taxa in ITS2. PL5 diet was mainly based on *Helianthus* sp. (34.98%, SD = 1.616). PL5 had a higher number of *Lactobacillus* (44.31%, SD = 1.456) and *Gilliamella* (28.46%, SD = 0.129) species (Supplementary Materials Table S1A). The fungal load was medium (59.16%, SD = 2.225) with bioaerosol *Cladosporium* (11.23%, SD = 2.225) and *Mycosphaerella* (6.69%, SD = 1.358) as the main representatives (Supplementary Materials Table S1B). No pathogens were detected in PL5 (Supplementary Materials Table S3).

PL6 forager honeybees were collected in September, and 28 taxa were identified in total from 16S amplicons analysed and 191 taxa in ITS2. PL6 had the highest level of *Firmicutes* and *Lactobacillus* with 65.37%, SD = 0.731, and a slightly higher number of bacteria from the Comensalibacter group (Supplementary Materials Table S1A). PL6 honeybee diet was rich in *Calluna* pollen (88.26%, SD = 0.585), which displays low protein content and is considered to be a poor source of food for bees, and thus destructive for colony development [51,52]. The fungal load was moderate (11.26%, SD = 0.718) with *Penicillium* as a dominant genus with 0.96%, SD = 0.104 (Supplementary Materials Table S2). Pathogens belonging to *N. ceranae* and neogregarines were detected in PL6 (Supplementary Materials Table S3).

Generally, PCA analysis of stores from spring honeybees (PL1 and PL2), summer honeybees (PL3 and PL4) and autumn honeybees (PL5 and PL6) involved splitting data into five major components which accounted for 100% of the variation. It can be concluded from the PCA analysis that PL3 clearly differed from the others, and was mainly influenced by bacteria (b) and plants (p). PL2 and PL6 were similar to each other, bacteria (b) and plants (p) having the greatest impact as well. PL1 and PL5 were similar to each other. The greatest, albeit low, influence was caused by bacteria (b) and fungi (f). PL4, similarly to PL3, clearly stood out from the others, and was mainly influenced by plants (p) and fungi (f). It could also be generally seen that bacteria had the greatest influence on the variability of the plant pollen-bacteria-fungi system, followed by the plants, and then the fungi. The first two components accounted for over 53% of the variability of the entire system. Positive PC2 values may describe the summer months, and negative PC2 values the months closer to spring and autumn (Supplementary Materials Figure S1).

PC1 and PC3, the two main components, account for approximately 50% of the system variability (Supplementary Materials Figure S1a,b). It can be concluded that positive PC3 also described spring and autumn values, and negative PC3 values described the summer, which clearly differs from the others. Therefore, the third component described the season.

To summarise, we observed a prevalence of *Lactobacillus* and *Bartonella* species in honeybees collected during spring (April PL1, May PL2) and autumn months (September PL6), while in summer months, (June PL3, July PL4, August PL5) microbiome analyses showed the prevalence of *Gilliamella*, which is in agreement with previous findings [4,9,43]. In late July (PL4) the physiology of honeybees changes due to their adaptations to overwintering, and a role from two types of bacterial groups, i.e., *Lactobacillus* and *Giliamella* is suggested to be played in this process. However, increased *Lactobacillus* and *Giliamella* occurrence may simply be a consequence of a protein-rich diet. We observed fluctuations in the microbiome composition correlating with changes in the protein-richness of the pollen available in the environment. During spring and autumn, it is common for honeybees' diet to be based on sugars ingested from honeydew. These high sugar diets can lead to fungal infections observed in April (PL1), May (PL2) and June (PL3) in honeybee samples. Probably, the detected pathogens were present in the honeybee colonies throughout the season but owing to the colony biology and well-balanced diet observed during May (PL4), June (PL3) and August (PL5), infections were less frequent among foragers and may have gone unnoticed in the whole colony screening tests.

2.2. Microbiome and Pollen Composition of Honeybees from UK, Greece, Spain, and Thailand

Forager honeybees from the UK were collected on the roof of the Fogg Building at Queen Mary University, London (UK1) and in the garden of the Natural History Museum, London (UK2), in July 2019. The mean day temperature in the first half of July was 21.93 °C (71.47 °F) with average humidity equal to 65% [53]. Their bacterial microbiota was mainly composed of *Firmicutes* (*Lactobacillus*), γ-proteobacteria (*Orbales, Gilliamella*), α-proteobacteria (*Rhizobiales, Bartonella*), γ-proteobacteria (*Neisseriaceae, Snodgrassella*), and *Actinobacteria* (*Bifidobacterium*) (Supplementary Materials Table S2A). In total, 42 and 93 taxa were identified (species level) from the 16S amplicon analysis, and 70 and 94 taxa (species level) in the ITS2 in the UK1 and UK2 groups, respectively. The UK1 sample contained modest amounts of fungi (12.59%) with the prevalence of plant pollen (87.41%) derived from Apiaceae (*Ammi*), Fabaceae (*Astragalus*), Resedaceae (*Reseda*), and Fabaceae (*Styphnolobium*). UK2 honeybees foraged on Hydrangeaceae (*Hydrangea*), and Bignoniaceae plants (Supplementary Materials Table S2B). The UK2 sample was dominated by fungi (90.80%), mainly from the Myxotrichaceae, which is reported to be a common hive fungus in Europe [54,55]. Moreover, the fungal pathogen *N. ceranae* was detected in the UK2 (Supplementary Materials Table S3).

Forager honeybees from Greece (GR1, GR2) were collected in November from two colonies inhabiting the garden of the Agricultural University of Athens. The mean day temperature in the first half of November was 18.53 °C (65.35 °F) with average humidity of 25% [56]. In total, 80 and 31 taxa were identified from the 16S amplicon analysis, and 42 and 96 taxa in the ITS2 in the GR1 and GR2 group respectively. Although the colonies were close, their microbiota and food preferences differed. GR1 microbiota were mainly Actinobacteria *(Bifidobacterium)*, Firmicutes *(Lactobacillus)*, and Cyanobacteria which reached 54.41%, 26.84%, and 11.66%, respectively. Cyanobacteria indicated some colony health problems most probably connected with the contamination of water used by the honeybee colony (Supplementary Materials Table S2A). Moreover, pathogens belonging to *N. ceranae* and neogregarines were detected in the GR1 (Supplementary Materials Table S3). The fungal load was miniscule (7.94%) and contained mainly fungi present in the air transferred as bioaerosols by wind, such as Mycosphaerella and *Cladosporium*. GR1 honeybees foraged mainly on Eudicotyledonae plants. GR2 microbiome contained α-proteobacteria (Rhizobiales, *Bartonella*), Firmicutes *(Lactobacillus)*, and γ-proteobacteria (Orbales, *Gilliamella*) with 49.29%, 25.84%, and 9.78%, respectively. The amount of fungi was miniscule (6.03%) with the dominant taxon of Mycosphaerella reaching 1.09%. GR2 honeybees foraged mainly on Oleaceae *(Ligustrum)*, Hydrangeaceae *(Hydrangea)*, Myrtaceae *(Myrtus)*, and Scrophulariaceae *(Buddleja)* (Supplementary Materials Table S2B).

Forager honeybees from Spain (ES1, ES2) were collected in November from experimental colonies located near Marchamalo. The mean day temperature in the first half of November was 18 °C (64.40 °F) with average humidity equal to 51% [57]. In total, 34 and 25 taxa were identified from the 16S amplicon analysis, and 31 and 13 taxa in the ITS2 in the ES1 and ES2 groups, respectively. ES1 contained γ-proteobacteria (Orbales, *Gilliamella*), and Firmicutes *(Lactobacillus)*, at the level of 36.63%, and 33.11%, respectively (Supplementary Materials Table S2A). These honeybees most probably gained sugar based diet, since pollen DNA was hardly detected (Araliaceae *(Hedera)* 0.12% for ES1 and 0.04% for ES2). ES1 and ES2 had dominant fungal fraction containing spores transferred as bioaerosols by wind in the air, such as *Penicillium, Cladosporium* and *Mycosphaerella* (Supplementary Materials Table S2B). Additionally, pathogens belonging to *N. ceranae* and neogregarines were detected in ES2 (Supplementary Materials Table S3).

Forager honeybees from Thailand consisted of both: western honeybee *(Apis mellifera* TAI1 and TAI2) and Asian honeybee *(Apis cerana* TAI3 and TAI4). In Eastern Asia, these two bee genera inhabit the same locations, resulting in the transfer of pathogens from *Apis cerana* to *Apis mellifera*, as described for *Varroa destructor* and *Nosema ceranae*. Thai samples were collected in February, the best month in the year for honeybee colonies in Thailand. The mean day temperature in the first half of Februry was 33.33 °C (91.99 °F) with average humidity equal to 59% [58].

In western honeybee *(Apis mellifera* TAI1 and TAI2), in total 29 and 22 taxa were identified from the 16S amplicon analysis, and 50 and 94 taxa in the ITS2 in the TAI1, TAI2 groups, respectively. TAI1 contained high loads of Firmicutes *(Lactobacillus)* and *Bartonella* (Supplementary Materials Table S2A), and a high quantity of fungi such as *Aspergillus* (66.42%), Saccharomycetaceae (9.09%), and *Peniclilium* (6.53%) (Supplementary Materials Table S2B). A trace amount of plant pollen was detected in TAI1 *(Pterocarpus* with 0.16%, *Mimosa* 0.08%) indicating sugar based diet to have been the main source of forage (Supplementary Materials Table S2B). Bacterial microbiota of TAI2 was mainly composed of *Arsenophonus* bacteria (γ-proteobacteria, Enterobacteriales) 92.11% of which were insects' intracellular symbionts. *Arsenophonus* species showed a broad spectrum of symbiotic relationships varying from parasitic son-killers to coevolving mutualists [59]. Moreover, pathogens belonging to *N. ceranae* and neogregarines were detected in TAI1 (Supplementary Materials Table S3). TAI2 honeybees foraged mainly on Asteraceae pollen

(91.33%) and contained only a miniscule amount of fungi transferred as bioaerosols by wind in the air, as *Cladosporium* 4.32%.

From Asian honeybee (*Apis cerana*) in total 19 and 42 taxa were identified from the 16S amplicon analysis, and 59 and 85 taxa in the ITS2 in the TAI3 and TAI4 groups, respectively. *Apis cerana* samples contained 70.74% of *Lactobacillus* genus for TAI3, 27.17% for TAI4, *Gilliamella*, 33.45% for TAI4, and *Snodgrassella* with 3.60% and 13.13% for TAI3 and TAI4, respectively. Fungi present in the samples were related with the air bioaerosols, including *Aspergillus* 42.77% for TAI3, *Cladosporium* 59.03% and *Pleosporales* 30.38% for TAI4. The amount of plant pollen was miniscule (*Pterocarpus* with 0.42% for TAI3, *Mimosa* 0.15% for TAI4) indicating sugar based diet as the main forage for Thai *A. cerana* bees (Supplementary Materials Table S2). Moreover, pathogens belonging to *N. ceranae* and neogregarines were detected in TAI4 (Supplementary Materials Table S3).

PCA analysis of the honeybees' stores split data into five major components which accounted for 100% of the variation. PC1 and PC2 components accounted for nearly 57% of the variation, respectively 32.83% and 23.65 (Figure 3, figures a and b should be considered simultaneously).

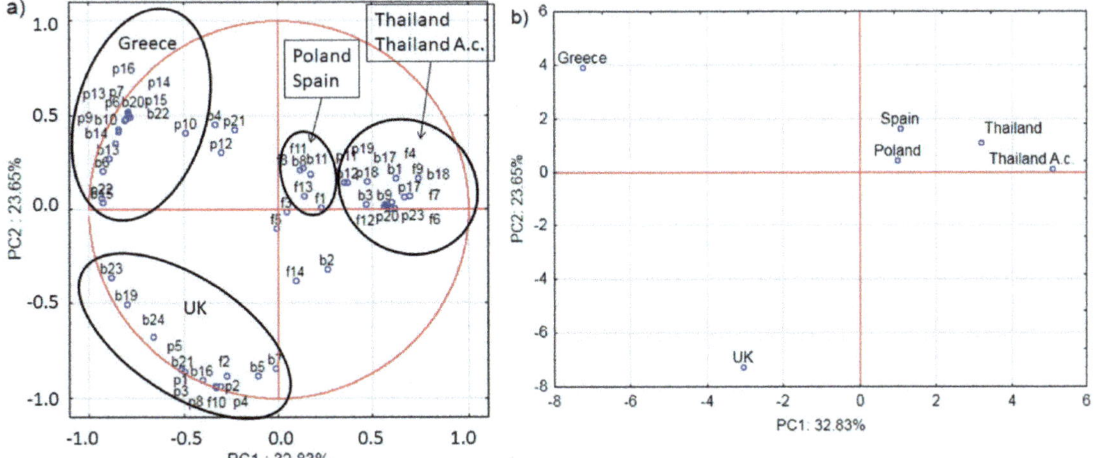

Figure 3. Loading plot (**a**) and score plot (**b**) of the principal components' analysis (PC1 and PC2) carried out on the analytical data of the taxonomy detected in the world bees (Thailand A.c.—*A. cerana*). Small letters on loading plot (a): b—data obtained from bacteria NGS analysis, f—data obtained from fungal NGS analysis, p—data obtained from plant NGS analysis.

The PCA analysis allowed us to determine the differences between bees from different countries. Four different areas were distinguished. Clear differences can be seen between bee samples from Greece, UK, Spain, Poland, and the two samples of bees from Thailand (Figure 3b).

PCA analysis shows Greek samples to markedly differ from the others, which is mainly influenced by bacteria (b) and plants (p). During the sample gathering, the weather conditions in Greece were most similar to those of Spain. Therefore, there must have been reasons for the Greek samples differences other than the weather. Most probably, nutrition was of the greatest importance in this case. The PCA analysis findings for UK bee samples also clearly differ from the others, mainly due to bacteria (b), then plants (p) and finally fungi (f). One can also distinguish samples from Spain and Poland from the other samples, which is mostly influenced by bacteria (b) and fungi (f), however, these are similar to the results for the samples from Thailand which are mostly influenced by bacteria (b), plants (p) and fungi (f). Bacteria, and, to a lesser extent, plants and fungi can be said to have the greatest influence on the variability of the system (Figure 3a,b).

Furthermore, ANOVA analysis confirms the correlation between the health status of honeybees and some of their bacterial microbiota (Supplementary Materials Table S4 and [60]). Bacterial groups such as Firmicutes (*Lactobacillus*); γ-proteobacteria, *Orbales*, *Gilliamella*, γ-proteobacteria, Neisseriaceae, *Snodgrassella*, Enterobacteriaceae, and other unidentified bacteria had significantly different loads in healthy and in *Nosema ceranae* and neogreagarine infected honeybees. The load of fungi was always higher in infected honeybees (p = 0.002429) whereas the load of plant pollens was always higher in healthy honeybees (p = 0.030446).

Foragers are worker honeybees of a similar age, when they collect water, nectar, and pollen as well as supplements necessary for the colony to survive. All forager bees have similar function and physiological processes [61] and similar microbiota. Therefore, they should share similar microbiota. However, studies indicated that forager honeybees have a "contingent microbiome" dependent mainly on the food they forage [2,10,11]. This carries a danger, because with poor food resources, the microbiota will be inappropriate and non-functioning [62]. Currently, many factors influence bee microbiota e.g.,: monocultures, nutritional stress, pesticide exposure and agrochemicals, many of which exhibit antimicrobial properties, and thus contribute greatly to reductions in honeybee stress tolerance and disease resistance, leading to higher honeybee mortality, and a high rate of colony loss [5,63], pathogens which trigger bee malnutrition [64], changes in the composition of their microelements [65] and yeast content [19].

Intestinal pathogens such as microsporidia and neogragarines can strongly interfere with bee microbiota (Supplementary Materials Table S4). During *Nosema*-infection the honeybee intestine is covered by a layer of mature spores which is the cause of deprivation of the physiological function of the bee alimentary tract for food absorption [64]. Recent studies have revealed the *N. ceranae* infection course, showing a spring peak, and a subsequent decline in summer and autumn [19,66,67]. These studies also confirmed the seasonal pattern of *Nosema* infection, as in samples taken during April (PL1) and July (PL4, UK2) *N. ceranae* was detected. However, the presence of the parasite in Autumn samples (GR1, ES2) may indicate the colony health problem, and may pose a threat to overwintering. It is worth emphasizing that *N. ceranae* was found both in *Apis mellifera* (TAI1) and *A. ceranae* (TAI4) Thai samples. Besides *Nosema*-infection in samples PL4, PL6, GE1, ES2, TAI1, TAI4 also harboured other intestinal parasite such as neogregarines. Neogregarines, since their first detection in *A. mellifera* and *Bombus* sp. described in 1992 were linked to declines in bee populations [68–70]. Neogregarines, inhabit the intestines of many invertebrates and lead to the impoverishment of the host's organism. Neogregarines' interaction with their bee hosts had not been deciphered in full detail and still more studies need to be undertaken in regard to nutrient uptake and malnutrition, facilitating susceptibility to other diseases, etc.

Pollination is a crucial process for the maintenance of plant-based food supplies [71]. To maintain the health of honeybees, it is favourable to prevent the spread of disease, prevent exposure to insecticides and pesticides and provide a variety of plants to maintain optimal nutrition and microbiome throughout the season [72]. The use of NGS techniques in the identification of the pollen pool preferentially chosen by honeybees can provide strategies to maintain healthy colonies of bees. This technique can greatly expand and supplement knowledge based on long term observation for the most efficient "pollinator-friendly" plants [73–75]. This should help to create more effective pollinator beneficial plant species composition for "pollinator-friendly" gardens or to enrich the plant species on flower strips.

3. Conclusions

Honeybee dietary preferences, developed during the course of evolution, may not be currently favourable for honeybees. It is an urgent issue that should be carefully studied to aid bees to survive in the anthropogenic biosphere.

In our research, the composition of honeybee microbiomes was mainly determined by dietary preferences and forage availability. Even when colonies originated from one apiary, for instance from Spain or Greece, honeybees chose different plants to forage on. UK bees from a single highly urbanised area (London) exhibited particularly high diversity, chose different food sources and were otherwise prone to diseases. Furthermore, honeybees choosing a sugar based diet were more susceptible to pathogens (*Nosema ceranae* and neogregarines). The period when honeybees switch to the winter generation (longer-lived forager honeybees) was the time when the whole colony proved most sensitive to dietary perturbations.

Our findings are in line with limited other reports, that suggested honeybees from varying apiaries make independent decisions on the choice of pollen and nectar they forage upon [76]. As observed, colonies of bees located in close proximity may have different pollen composition in their gut. It remains unclear how bees decide which pollen to forage. However, flower structure, nectar volume, sugar content and composition were indicated to play a role in attracting bees [51,52,73]. Some studies indicated that honeybees primarily chose pollen rich in essential amino acids [51,52,73]. On the other hand, honeybees, because of their considerable energy need and having being weakened by disease, were unable to undertake forage flights to collect good quality pollen, and collected pollen of poorer nutritional quality. It is possible that evolutionary adaptation of bees has failed to benefit them in the modern anthropomorphised environment.

4. Highlights

1. The composition of honeybee microbiomes is mainly determined by their forage availability.
2. Honeybees were more susceptible to pathogens if they did not receive a well-balanced diet, and especially honeybees on sugar based diet were more prone to fungal pathogens (*Nosema ceranae*) and neogregarines. In most samples *Nosema* sp. and neogregarines parasitized the host bee at the same time.
3. The period when honeybees switch to the winter generation (longer-lived forager honeybees) is the most sensitive to diet perturbations, and hence pathogen attack, for the whole beekeeping season.

5. Materials and Methods

5.1. Honeybee Collection and DNA Isolation

Forager honeybees were recognized as bees returning to the hive and captured at the hive entrance about the midday. Forager honeybees from Poland were collected from one location in Lublin [51°15′ N 22°34′ E] each month from April to September 2018 (PL1-PL6). Forager honeybees from UK were collected from the roof of the Fogg Building of Queen Mary University, London (UK2) [51°52′ N 0°03′ W] and in the garden of the Natural History Museum, London (UK1) [51°29′ N 0°10′ W] in July 2019. Greek (GR1, GR2) samples of forager honeybees were collected in November 2017 from two colonies inhabiting the garden of The Agricultural University of Athens [37°59′ N 23°42′ E]. Forager honeybees from Spain (ES1, ES2) were collected in November 2017 from experimental colonies located at Marchamalo (Centro Apícola y Agroambiental de Marchamalo (CIAPA-IRIAF), Marchamalo, Spain [40°68′ N 3°21′ W]. Thai samples of forager bees consisting of both western honeybee (*Apis mellifera*) and Asian honeybee (*A. cerana*) were collected in February 2018 in the proximity of Chiang Mai University [18°50′ 98°58′ E], *A. mellifera* samples were marked TAI1 and TAI2, and *A. cerana* as TAI3 and TAI4. Genomic DNA was extracted from whole honeybees using QIAamp DNA Kit according to manufacturer's instructions. Isolates were sent to the Biobank, Poland for NGS analysis.

5.2. NGS

NGS sequencing and the analysis of the 16S rRNA bacterial gene amplicon was based on the V3-V4 region and the ITS2 eukaryotic region for bee DNA samples. Amplicon libraries, were prepared using the *16S Metagenomic Sequencing Library Preparation, Preparing 16S Ribosomal RNA Gene Amplicons for the Illumina MiSeq System* (Illumina® San Diego,

CA, USA) protocol. Information about primers sequences, PCR conditions is shown in Supplementary Materials Table S5.

All data are available at https://www.ncbi.nlm.nih.gov/bioproject/PRJNA686953 (Submission Registration date: 21 December 2020).

5.3. Positive, Negative Control

The positive quality control for the V3-V4 region of the 16s rRNA gene was the DNA isolate derived from an ear swab. For the ITS2 region, it was DNA isolated from the *Saccharomyces cerevisiae* strain. PCR grade water was the negative quality control for both kinds of amplicons.

5.4. Purification, Clean-Up

The amplicons obtained were purified using magnetic beads (AMPure XP beads; Beckman Coulter Brea, CA, USA) according to Illumina® protocol.

5.5. Library Pooling—Concentration, Normalization

Before pooling samples for libraries, the concentration was measured. The concentration [ng/uL] was measured using the NanoDrop™ 2000/c Spectrophotometer (Thermo Fisher Scientific Vienna; Austria) for each amplicon. Samples were diluted (PCR grade water) to the same concentration and pooled. To determine the final library concentration in [nM], the *NEBNext Ultra DNA Library Prep Kit for Illumina* (New England Biolabs® Inc. Ipswich, MA; USA) protocol was followed. The final concentration of pooled libraries for sequencing was 8 pM.

5.6. Sequencing

Prepared libraries were sequenced on an Illumina MiSeq platform, 2×300 sequence reading in paired ends mode. The run contained PhiX libraries (PhiX Control Kit v3, Illumina® San Diego, CA, USA), to serve as an internal positive quality control.

5.7. 16S rRNA Bacterial Gene Analyses

Reads from the sequencing run were imported into the QIIME 2 version 2019.10 artifact [77]. Then sequences were trimmed at first 21 bp for forward and reverse reads and truncated to 250 for forward reads and 240 for reverse reads. Reads were then denoised with DADA2 (Divisive Amplicon Denoising Algorithm v. 2) [78] and merged together. Sequences were aligned with MAFFT (Multiple Alignment using Fast Fourier Transform) [79] and used to construct a phylogeny with fasttree [80]. Rarefaction was performed with at least 14,096 sequences per sample for subsequent stages of the analysis. Taxonomic assignments of representative sequences were conducted using q2-feature-classifier with the sklearn classifier [81] trained on SILVA 132 database at 99% similarity level [82].

5.8. ITS2 Region Analyses

Analogous steps as for "*16S rRNA bacterial gene analyses*" were performed for the ITS2 analysis. Reads were trimmed and denoised separately, they were then merged for further analysis. Reads were then trimmed at first 21 bp for forward and reverse reads and truncated to 300 for forward reads and 215 for reverse reads. Taxonomic assignments were conducted analogous to 16S analysis with classifier trained on ITS gene clustered at 99% similarities within UNITE database released 04.02.2020 containing all eukaryotes [83]. Sampling depth was set to 34,100 sequences for the diversity analyses.

5.9. Analyses of Amplicon from Honeybee Samples

Amplicons for the 16S region and ITS2 were sequenced using the Illumina MiSeq platform. Data were trimmed and merged. For 16S analyses only full-length reads over 229 bp with medium length of all sequences at 414 bp were used. Sequences were assigned to taxonomy using classifier trained on SILVA 132 database with minimum similarity 90%

of read matching to the reference. For ITS2 analyses only full-length reads over 269 bp with medium length of all sequences at 337 bp were used. Sequences were assigned to taxonomy using classifier trained on all eukaryotes UNITE database v8.2 with the minimum similarity of 90% of the read matching to the reference [84,85].

5.10. Screening for Pathogen Infected Honeybee Samples

Isolated DNA was used as the template for screening pathogens: *Nosema apis*, *Nosema ceranae*, *Nosema bombi*, tracheal mite (*Acarapis woodi*), any organism in the parasitic order Trypanosomatida, including Crithidia spp. (i.e., *Crithidia mellificae*), neogregarines including *Mattesia* and *Apicystis* spp. (i.e., *Apicistis bombi*), using PCR techniques described earlier [67,86–88]. Primers used for pathogen detection are listed in the Supplementary Materials Table S5. Detection of the pathogens in honeybee samples.

5.11. Statistical Analysis

Analyses of correlations and Principal Component Analysis (PCA) were performed using software Statistica (version 12.0, StatSoft Inc., Oklahoma; USA) at the significance level of $\alpha = 0.05$. The analysis was used to determine the relationships between the bee sample and the bacterial group, plant group, and fungi group. The optimum number of principal components obtained in the PCA analysis was established based on Cattell's criterion. The data matrix for the PCA of the Polish bees had 37 columns and 6 rows and of the world samples of bees had 61 columns and 6 rows (UK, Spain, Greek, Thailand and Poland) had 61 columns and 6 rows. The input matrix was auto-scaled. One-way ANOVA was performed to establish the correlation between honeybees' health status and the bacteria, fungi and plant pollen detected. For the ANOVA test, the level of statistical significance was assumed to be $\alpha = 0.05$, and the same level of statistical significance was used in all comparisons. The results for which p values are equal to, or less than, 0.05 were obtained differ significantly from each other.

Supplementary Materials: The following are available online at https://www.mdpi.com/2076-0817/10/3/381/s1, **Table S1.** Taxonomy analysis of 16S and ITS amplicon sequencing. **Table S1. A.** Taxonomy of 16S amplicon sequencing. Six sub-groups of three bees each were analysed. Table presents relative abundance of each of the listed bacterium (numbers as percentages) in relation to the entire amplicon. **Table S1. B.** Taxonomy of ITS2 amplicon sequencing. Six subgroups of three bees each were analysed. Hits for plants and fungi were grouped into separate fractions. % for either Plant or Fungi is a sum of all counts for each fraction (highlighted in grey). **Table S2. A.** Taxonomy of 16S amplicon sequencing. Four subgroups of two bees (UK, GR, ES) or four bees (TAI) each were analysed. Table presents relative abundance of each of the listed bacterium (numbers as percentages) in relation to the entire amplicon. **Table S2. B.** Taxonomy of ITS2 amplicon sequencing. Four subgroups of two bees (UK, GR, ES) or four bees (TAI) each were analysed. Hits for plants and fungi were grouped into separate fractions. % for either Plant or Fungi is a sum of all counts for each fraction (highlighted in grey). **Table S3.** Detection of the pathogens in honeybee samples. **Table S4.** The correlation between honeybees' health status and the detected bacteria. Data with significant differences are written in bold. The ANOVA test, $\alpha = 0.05$; $p \leq 0.05$. **Figure S3.** Loading plot (a) and score plot (b) of the principal components analysis (PC1 and PC2) carried out on the analytical data of the taxonomy detected in all Polish bees (PL1 to PL6). **Table S5.** The list of primers and PCR conditions.

Author Contributions: A.A.P. (senior author), designed the experiments, analysed data, and wrote the paper. P.L., analysed data, prepared figures, supplemental information, and methods. P.J.H., analysed data, especially of metabiome and parasites, co-wrote the paper. A.P. co-wrote the paper. J.M.-M., conducted laboratory work for sequencing library preparation, sequencing, and detection of pathogens. Ł.G., analysed raw data from metabiom sequencing, prepared tables, co-wrote the paper. D.S., analysed data from metabiom sequencing, co-wrote the paper. S.G., performed genetic analyses. D.Z., drafted and made a correction of the manuscript. M.G., and R.R., analysed data using PCA, interpreted the results, co-wrote the paper. P.K., analysed Thai data. R.M.H., and M.H.P., analysed UK data, co-wrote the paper. A.L.S., analysed data, prepared figures and tables and co-wrote

Pathogens **2021**, *10*, 381

the paper. All authors read and approved the final manuscript. All authors have read and agreed to the published version of the manuscript.

Funding: The publication of the article was financed by the Polish National Agency for Academic Exchange under the Foreign Promotion Programme (NAWA), for AAP (bee-research.umcs.pl; Api Lab UMCS PPI/PZA/2019/1/00039). Honeybees were collected during the Miniatura 2 project ID 418332 founded by the NCN for AAP and the first research plans were possible thanks to EU: GB-TAF-7137 SYNTHESYS project for AAP. Work in the Starosta lab was funded by EMBO Installation Grants 2017, and POIR. 04.04.00-00-3E9C/17-00 for ALS. Work in the Hurd lab was funded by the BBSRC (BB/L023164/1) and granted to PJH.

Institutional Review Board Statement: In Europe, the EU Directive 2010/63/EU on the protection of animals used for scientific purposes laid the down the ethical framework for the use of animals in scientific experiments. The scope of this directive also includes specific invertebrate species, i.e. cephalopods, but no insects. Thus, according to European legislation no specific permits were required for the described studies.

Informed Consent Statement: Not applicable.

Data Availability Statement: The data presented in this study are openly available in https://www.ncbi.nlm.nih.gov/bioproject/PRJNA686953 (accessed on 21 March 2021).

Data Citation: Gancarz: M., Hurd, P.J., Latoch, P., Polaszek, A., Michalska-Madej, J., Grochowalski, Ł., Strapagiel, D., Gnat, S., Załuski, D., Rusinek, R., Starosta, A.L.., Krutmuang, P., Martín Hernández, R., Higes Pascual, M., Ptaszyńska, A.A. (2021). Dataset of the next-generation sequencing of variable 16S rRNA from bacteria and ITS2 regions from fungi and plants derived from honeybees kept under anthropogenic landscapes. Data in Brief.

Lead Contact: Further information and requests for resources and reagents should be directed to and will be fulfilled by the Lead Contact, Aneta A. Ptaszyńska (aneta.ptaszynska@poczta.umcs.lublin.pl).

Conflicts of Interest: The authors declare no competing interest.

References

1. Yao, R.; Xu, L.; Lu, G.; Zhu, L. Evaluation of the Function of Wild Animal Gut Microbiomes Using Next-Generation Sequencing and Bioinformatics and its Relevance to Animal Conservation. *Evol. Bioinform.* **2019**, *15*, 1176934319848438. [CrossRef]
2. Romero, S.; Nastasa, A.; Chapman, A.; Kwong, W.K.; Foster, L.J. The honey bee gut microbiota: Strategies for study and characterization. *Insect Mol. Biol.* **2019**, *28*, 455–472. [CrossRef]
3. Hawkins, J.; De Vere, N.; Griffith, A.; Ford, C.R.; Allainguillaume, J.; Hegarty, M.J.; Baillie, L.; Adams-Groom, B. Using DNA Metabarcoding to Identify the Floral Composition of Honey: A New Tool for Investigating Honey Bee Foraging Preferences. *PLoS ONE* **2015**, *10*, e0134735. [CrossRef]
4. Kešnerová, L.; Emery, O.; Troilo, M.; Liberti, J.; Erkosar, B.; Engel, P. Gut microbiota structure differs between honeybees in winter and summer. *ISME J.* **2020**, *14*, 801–814. [CrossRef]
5. Tauber, J.P.; Nguyen, V.; Lopez, D.; Evans, J.D. Effects of a Resident Yeast from the Honeybee Gut on Immunity, Microbiota, and Nosema Disease. *Insects* **2019**, *10*, 296. [CrossRef] [PubMed]
6. McNeil, D.J.; McCormick, E.; Heimann, A.C.; Kammerer, M.; Douglas, M.R.; Goslee, S.C.; Grozinger, C.M.; Hines, H.M. Bumble bees in landscapes with abundant floral resources have lower pathogen loads. *Sci. Rep.* **2020**, *10*, 1–12. [CrossRef] [PubMed]
7. Patel, V.; Pauli, N.; Biggs, E.; Barbour, L.; Boruff, B. Why bees are critical for achieving sustainable development. *Ambio* **2021**, *50*, 49–59. [CrossRef]
8. Kwong, W.K.; A Moran, N. Evolution of host specialization in gut microbes: The bee gut as a model. *Gut Microb.* **2015**, *6*, 214–220. [CrossRef]
9. Ahn, J.H.; Hong, I.P.; Bok, J.I.; Kim, B.Y.; Song, J.; Weon, H.Y. Pyrosequencing analysis of the bacterial communi-ties in the guts of honey bees Apis cerana and Apis mellifera in Korea. *J. Microbiol.* **2012**, *50*, 735–745. [CrossRef] [PubMed]
10. Khan, K.A.; Al-Ghamdi, A.A.; Ghramh, H.A.; Ansari, M.J.; Ali, H.; Alamri, S.A.; Kahtani, S.N.A.-; Adgaba, N.; Qasim, M.; Hafeez, M. Structural diversity and functional variability of gut microbial communities associated with honey bees. *Microb. Pathog.* **2020**, *138*, 103793. [CrossRef] [PubMed]
11. Kwong, W.K.; Moran, N.A. Gut microbial communities of social bees. *Nat. Rev. Genet.* **2016**, *14*, 374–384. [CrossRef]
12. Raymann, K.; Bobay, L.M.; Moran, N.A. Antibiotics reduce genetic diversity of core species in the honeybee gut micro-biome. *Mol. Ecol.* **2018**, *27*, 2057–2066. [CrossRef] [PubMed]
13. Pachla, A.; Ptaszyńska, A.A.; Wicha, M.; Kunat, M.; Wydrych, J.; Oleńska, E.; Małek, W. Insight into probiotic properties of lactic acid bacterial endosymbionts of Apis mellifera L. derived from the Polish apiary. *Saudi J. Biol. Sci.* **2021**, *28*, 1890–1899. [CrossRef] [PubMed]
14. Gilliam, M.; Prest, D.B. Microbiology of feces of the larval honey bee, Apis mellifera. *J. Invertebr. Pathol.* **1987**, *49*, 70–75. [CrossRef]

15. Gilliam, M.; Valentine, D.K. Bacteria isolated from the intestinal contents of foraging worker honey bees, Apis mellifera: The genus Bacillus. *J. Invertebr. Pathol.* **1976**, *28*, 275–276. [CrossRef]
16. Gilliam, M. Are yeasts present in adult worker honey bees as a consequence of stress? *Ann. Enlomal. Soc. Am.* **1973**, *66*, 1176. [CrossRef]
17. Gilliam, M. Identification and roles of non-pathogenic microflora associated with honey bees. *FEMS Microbiol. Lett.* **1997**, *155*, 1–10. [CrossRef]
18. Gilliam, M.; Taber, S. Diseases, pests, and normal microflora of honeybees Api smellifera, from feral colonies. *J. Invertebr. Pathol.* **1991**, *58*, 286–289. [CrossRef]
19. Ptaszyńska, A.A.; Paleolog, J.; Borsuk, G. Nosema ceranae Infection Promotes Proliferation of Yeasts in Honey Bee Intestines. *PLoS ONE* **2016**, *11*, e0164477. [CrossRef]
20. Gilliam, M.; Prest, D.B. Fungi isolated from the intestinal contents of foraging worker honey bees, Apis mellifera. *J. Invertebr. Pathol.* **1972**, *20*, 101–103. [CrossRef]
21. Tauber, J.P.; McMahon, D.; Ryabov, E.; Kunat, M.; Ptaszynska, A.A.; Evans, J.D. Honeybee intestines retain low yeast titers, but no bacterial mutualists, at emergence. *Yeast* **2021**, in press.
22. Zalasiewicz, J.; Waters, C.; Williams, M. Chapter 32: The Anthropocene. In *A Geologic Time Scale*; Gradstein, F., Ogg, J., Schmitz, M., Ogg, G., Eds.; Cambridge University Press: Cambridge, UK, 2020. [CrossRef]
23. Howard, J. Anthropogenic Landforms and Soil Parent Materials. In *Progress in Soil Science*; Springer: Cham, Switzerland, 2017; Volume 23, pp. 25–51.
24. Le Conte, Y.; Navajas, M. Climate change: Impact on honey bee populations and diseases. *Rev. Sci. Tech.* **2008**, *27*, 485–497, 499–510. [CrossRef] [PubMed]
25. IUCN Red List, IUCN Species Survival Commission. Guidelines for Application of IUCN Red List criteria at regional and national levels. Table 3; Last Updated: Dec. 2020 List version 2020-3. Available online: https://www.iucnredlist.org/statistics (accessed on 15 December 2020).
26. Fox, R. The decline of moths in Great Britain: A review of possible causes. *Insect. Conserv. Divers.* **2012**, *6*, 5–19. [CrossRef]
27. Benton, T.G.; Bryant, D.M.; Cole, L.; Crick, H.Q. Linking agricultural practice to insect and bird populations: A his-torical study over three decades. *J. Appl. Ecol.* **2002**, *39*, 673–687. [CrossRef]
28. Thomas, J.A.; Telfer, M.G.; Roy, D.B.; Preston, C.D.; Greenwood, J.J.D.; Asher, J.; Fox, R.; Clarke, R.T.; Lawton, J.H. Comparative Losses of British Butterflies, Birds, and Plants and the Global Extinction Crisis. *Science* **2004**, *303*, 1879–1881. [CrossRef] [PubMed]
29. Hallmann, C.A.; Foppen, R.P.B.; Van Turnhout, C.A.M.; De Kroon, H.; Jongejans, E. Declines in insectivorous birds are associated with high neonicotinoid concentrations. *Nat. Cell Biol.* **2014**, *511*, 341–343. [CrossRef]
30. Pimm, S.L.; Jenkins, C.N.; Abell, R.; Brooks, T.M.; Gittleman, J.L.; Joppa, L.N.; Raven, P.H.; Roberts, C.M.; Sexton, J.O. The biodi-versity of species and their rates of extinction, distribution, and protection. *Science* **2014**, *344*, 1246752. [CrossRef]
31. Sánchez-Bayo, F.; Wyckhuys, K.A. Worldwide decline of the entomofauna: A review of its drivers. *Biol. Conserv.* **2019**, *232*, 8–27. [CrossRef]
32. Marshman, J.; Blay-Palmer, A.; Landman, K. Anthropocene Crisis: Climate Change, Pollinators, and Food Security. *Environments* **2019**, *6*, 22. [CrossRef]
33. EU Pollinators Initiative. Brussels, 1.6.2018. COM(2018) 395 final. Available online: https://eur-lex.europa.eu/legal-content/EN/TXT/PDF/?uri=CELEX:52018DC0395&from=PL (accessed on 15 December 2020).
34. LIFE+ URBANBE. Available online: https://ec.europa.eu/environment/life/project/Projects/index.cfm?fuseaction=home.showFile&rep=file&fil=URBANBEES_Management_Plan.pdf (accessed on 15 December 2020).
35. City Bees. Available online: https://www.eea.europa.eu/atlas/eea/city-bees/story/article (accessed on 15 December 2020).
36. Urban Beekeeping. Available online: https://beeproject.ca/urban-beekeeping (accessed on 15 December 2020).
37. Garnett, T. Urban agriculture in London: Rethinking our food economy. In *Growing Cities, Growing Food*; German Foundation for International Development: Feldafing, Germany, 2000; pp. 477–500.
38. Wilson, C.J.; Jamieson, M.A. The effects of urbanization on bee communities depends on floral resource availability and bee functional traits. *PLoS ONE* **2019**, *14*, e0225852. [CrossRef]
39. Cariveau, D.P.; Winfree, R. Causes of variation in wild bee responses to anthropogenic drivers. *Curr. Opin. Insect Sci.* **2015**, *10*, 104–109. [CrossRef]
40. Hernandez, J.L.; Frankie, G.W.; Thorp, R.W. Ecology of Urban Bees: A Review of Current Knowledge and Directions for Future Study. *Cities Environ.* **2009**, *2*, 1–15. [CrossRef]
41. Ellegaard, K.M.; Engel, P. Genomic diversity landscape of the honey bee gut microbiota. *Nat. Commun.* **2019**, *10*, 1–13. [CrossRef] [PubMed]
42. Cilia, G.; Fratini, F.; Tafi, E.; Mancini, S.; Turchi, B.; Sagona, S.; Cerri, D.; Felicioli, A.; Nanetti, A. Changes of Western honey bee Apis mellifera ligustica (Spinola, 1806) ventriculus microbial profile related to their in-hive tasks. *J. Apic. Res.* **2021**, *60*, 198–202. [CrossRef]
43. Laha, R.C.; De Mandal, S.; Ralte, L.; Ralte, L.; Kumar, N.S.; Gurusubramanian, G.; Satishkumar, R.; Mugasimangalam, R.; Kuravadi, N.A. Meta-barcoding in combination with palynological inference is a potent diagnostic marker for hon-ey floral composition. *AMB Express* **2017**, *7*, 132. [CrossRef]
44. Gous, A.; Swanevelder, D.Z.H.; Eardley, C.D.; Willows-Munro, S. Plant-pollinator interactions over time: Pollen metabarcoding from bees in a historic collection. *Evol. Appl.* **2019**, *12*, 187–197. [CrossRef]
45. Keller, A.; Danner, N.; Grimmer, G.; Ankenbrand, M.; Von Der Ohe, K.; Rost, S.; Härtel, S.; Steffan-Dewenter, I. Evaluating multiplexed next-generation sequencing as a method in palynology for mixed pollen samples. *Plant Biol.* **2014**, *17*, 558–566. [CrossRef] [PubMed]

46. Baksay, S.; Pornon, A.; Burrus, M.; Mariette, J.; Andalo, C.; Escaravage, N. Experimental quantification of pollen with DNA metabarcoding using ITS1 and trnL. *Sci. Rep.* **2020**, *10*, 1–9. [CrossRef]
47. Szczęsna, T. Protein content and amino acid composition of bee-collected pollen from selected botanical origins. *J. Agric. Sci.* **2006**, *50*, 81–90.
48. Wynns, A.A. Convergent evolution of highly reduced fruiting bodies in Pezizomycotina suggests key adaptations to the bee habitat. *BMC Evol. Biol.* **2015**, *15*, 1–11. [CrossRef]
49. Zheng, H.; Nishida, A.; Kwong, W.K.; Koch, H.; Engel, P.; Steele, M.I.; Moran, N.A. Metabolism of Toxic Sugars by Strains of the Bee Gut Symbiont Gilliamella apicola. *mBio* **2016**, *7*, e01326-16. [CrossRef] [PubMed]
50. Vanderplanck, M.; Vereecken, N.J.; Grumiau, L.; Esposito, F.; Lognay, G.; Wattiez, R.; Michez, D. The importance of pollen chemistry in evolutionary host shifts of bees. *Sci. Rep.* **2017**, *7*, 43058. [CrossRef] [PubMed]
51. Vanderplanck, M.; Moerman, R.; Rasmont, P.P.; Lognay, G.C.; Wathelet, B.; Wattiez, R.; Michez, D. How Does Pollen Chemistry Impact Development and Feeding Behaviour of Polylectic Bees? *PLoS ONE* **2014**, *9*, e86209. [CrossRef] [PubMed]
52. Past Weather in London, England, United Kingdom—July 2019. Available online: https://www.timeanddate.com/weather/uk/london/historic?month=07&year=2019 (accessed on 15 December 2020).
53. Rodríguez-Andrade, E.; Stchigel, A.M.; Terrab, A.; Guarro, J.; Cano-Lira, J.F. Diversity of xerotolerant and xerophilic fungi in honey. *IMA Fungus* **2019**, *10*, 1–30. [CrossRef]
54. Burnside, C.E. Saprophytic Fungi Associated with the Honey Bee. *Bee World* **1929**, *10*, 42. [CrossRef]
55. Batra, L.R.; Batra, S.W.T.; Bohart, G.E. The mycoflora of domesticated and wild bees (Apoidea). *Mycopathologia et Mycologia Applicata* **1973**, *49*, 13–44. [CrossRef]
56. Past Weather in Athens, Greece—November 2017. Available online: https://www.timeanddate.com/weather/greece/athens/historic?month=11&year=2017 (accessed on 15 December 2020).
57. Past Weather in Madrid, Spain—November 2017. Available online: https://www.timeanddate.com/weather/spain/madrid/historic?month=11&year=2017 (accessed on 15 December 2020).
58. Past Weather in Chiang Mai, Thailand—February 2018. Available online: https://www.timeanddate.com/weather/thailand/chiang-mai/historic?month=02&year=2018) (accessed on 15 December 2020).
59. Nováková, E.; Hypša, V.; A Moran, N. Arsenophonus, an emerging clade of intracellular symbionts with a broad host distribution. *BMC Microbiol.* **2009**, *9*, 143. [CrossRef]
60. Gancarz, M.; Hurd, P.J.; Latoch, P.; Polaszek, A.; Michalska-Madej, J.; Grochowalski, Ł.; Strapagiel, D.; Gnat, S.; Załuski, D.; Rusinek, R.; et al. Dataset of the next-generation sequencing of variable 16S rRNA from bacteria and ITS2 regions from fungi and plants derived from honeybees kept under anthropogenic landscapes. Data in Brief. [CrossRef]
61. Abou-Shaara, H. The foraging behaviour of honey bees, Apis mellifera: A review. *Veterinární Medicína* **2014**, *59*, 1–10. [CrossRef]
62. Daisley, B.A.; Chmiel, J.A.; Pitek, A.P.; Thompson, G.J.; Reid, G. Missing Microbes in Bees: How Systematic Depletion of Key Symbionts Erodes Immunity. *Trends Microbiol.* **2020**, *28*, 1010–1021. [CrossRef]
63. Kakumanu, M.L.; Reeves, A.M.; Anderson, T.D.; Rodrigues, R.R.; Williams, M.A. Honey Bee Gut Microbiome Is Altered by In-Hive Pesticide Exposures. *Front. Microbiol.* **2016**, *7*, 1255. [CrossRef]
64. A Ptaszyńska, A.; Borsuk, G.; Mułenko, W.; Demetraki-Paleolog, J. Differentiation of Nosema apis and Nosema ceranae spores under Scanning Electron Microscopy (SEM). *J. Apic. Res.* **2014**, *53*, 537–544. [CrossRef]
65. Ptaszyńska, A.A.; Gancarz, M.; Hurd, P.J.; Borsuk, G.; Wiacek, D.; Nawrocka, A.; Strachecka, A.; Załuski, D.; Paleolog, J. Changes in the bioelement content of summer and winter western honeybees (*Apis mellifera*) induced by Nosema ceranae infection. *PLoS ONE* **2018**, *13*, e0200410. [CrossRef]
66. Emsen, B.; De la Mora, A.; Lacey, B.; Eccles, L.; Kelly, P.G.; Medina-Flores, C.A.; Guzman-Novoa, E. Seasonality of Nosema ceranae Infections and Their Rela-tionship with Honey Bee Populations, Food Stores, and Survivorship in a North Ameri-can Region. *Veter. Sci.* **2020**, *7*, 131. [CrossRef]
67. Martín-Hernández, R.; Meana, A.; Prieto, L.; Salvador, A.M.; Garrido-Bailón, E.; Higes, M. Outcome of Colonization of Apis mellifera by Nosema ceranae. *Appl. Environ. Microbiol.* **2007**, *73*, 6331–6338. [CrossRef]
68. Lipa, J.; Triggiani, O. A newly recorded neogregarine (Protozoa, Apicomplexa), parasite in honey bees (Apis mellifera) and bumble bees (Bombus spp). *Apidologie* **1992**, *23*, 533–536. [CrossRef]
69. Cepero, A.; Ravoet, J.; Gómez-Moracho, T.; Bernal, J.L.; Del Nozal, M.J.; Bartolomé, C.; Maside, X.; Meana, A.; González-Porto, A.V.; De Graaf, D.C.; et al. Holistic screening of collapsing honey bee colonies in Spain: A case study. *BMC Res. Notes* **2014**, *7*, 1–10. [CrossRef]
70. Graystock, P.; Ng, W.H.; Parks, K.; Tripodi, A.D.; Muñiz, P.A.; Fersch, A.A.; Myers, C.R.; McFREDERICK, Q.S.; McArt, S.H. Dominant bee species and floral abundance drive parasite temporal dynamics in plant-pollinator communities. *Nat. Ecol. Evol.* **2020**, *4*, 1–10. [CrossRef] [PubMed]
71. Costanza, R.; d'Arge, R.; de Groot, R.; Farber, S.; Grasso, M.; Hannon, B.; Limburg, K.; Naeem, S.; O'Neill, R.V.; Paruelo, J.; et al. The value of the world's ecosystem services and natural capital. *Nature* **1997**, *387*, 253–260. [CrossRef]
72. Goulson, D.; Nicholls, E.; Botías, C.; Rotheray, E.L. Bee declines driven by combined stress from parasites, pesticides, and lack of flowers. *Science* **2015**, *347*, 1255957. [CrossRef]
73. Mayer, C.; Adler, L.; Armbruster, S.; Dafni, A.; Eardley, C.; Huang, S.-Q.; Kevan, P.G.; Ollerton, J.; Packer, L.; Ssymank, A. Pollination ecology in the 21st Century: Key questions for future research. *J. Pollinat. Ecol.* **2011**, *3*, 8–23. [CrossRef]

74. De Vere, N.; Jones, L.E.; Gilmore, T.; Moscrop, J.; Lowe, A.; Smith, D.; Hegarty, M.J.; Creer, S.; Ford, C.R. Using DNA metabarcoding to investigate honey bee foraging reveals limited flower use despite high floral availability. *Sci. Rep.* **2017**, *7*, 42838. [CrossRef] [PubMed]
75. Ptaszyńska, A.A.; Załuski, D. Extracts from *Eleutherococcus senticosus* (Rupr. et Maxim.) Maxim. Roots: A New Hope Against Honeybee Death Caused by Nosemosis. *Molcules* **2020**, *25*, 4452. [CrossRef] [PubMed]
76. Vaudo, A.D.; Tooker, J.F.; Grozinger, C.M.; Patch, H.M. Bee nutrition and floral resource restoration. *Curr. Opin. Insect Sci.* **2015**, *10*, 133–141. [CrossRef] [PubMed]
77. Bolyen, E.; Rideout, J.; Dillon, M.; Bokulich, N.; Abnet, C.; Al-Ghalith, G.; Alexander, H.; Alm, E.; Arumugam, M.; Asnicar, F.; et al. Reproducible, interactive, scalable and extensible microbiome data science using QIIME 2. *Nat. Biotechnol.* **2019**, *37*, 852–857. [CrossRef] [PubMed]
78. Callahan, B.J.; Mcmurdie, P.J.; Rosen, M.J.; Han, A.W.; Johnson, A.J.A.; Holmes, S.P. DADA2: High-resolution sample inference from Illumina amplicon data. *Nat. Methods* **2016**, *13*, 581–583. [CrossRef]
79. Katoh, K.; Standley, D. MAFFT multiple sequence alignment software version 7: Improvements in performance and usability. *Mol. Biol. Evol.* **2013**, *30*, 772–780. [CrossRef]
80. Price, M.N.; Dehal, P.S.; Arkin, A.P. FastTree 2 – Approximately Maximum-Likelihood Trees for Large Alignments. *PLoS ONE* **2010**, *5*, e9490. [CrossRef]
81. Fabian, P.; Gaël, V.; Alexandre, G.; Vincent, M.; Bertrand, T.; Olivier, G.; Mathieu, B.; Peter, P.; Ron, W.; Vincent, D.; et al. Scikit-learn: Machine learning in python. *J. Machine Learn. Res.* **2011**, *12*, 2825–2830.
82. Quast, C.; Pruesse, E.; Yilmaz, P.; Gerken, J.; Schweer, T.; Yarza, P.; Peplies, J.; Glöckner, F.O. The SILVA ribosomal RNA gene database project: Improved data processing and web-based tools. *Open. External Link New Window Nucl. Acids Res.* **2013**, *41*, D590–D596. [CrossRef]
83. Abarenkov, K.; Nilsson, R.H.; Larsson, K.-H.; Alexander, I.J.; Eberhardt, U.; Erland, S.; Høiland, K.; Kjøller, R.; Larsson, E.; Pennanen, T.; et al. The UNITE database for molecular identification of fungi - recent updates and future perspectives. *New Phytol.* **2010**, *186*, 281–285. [CrossRef]
84. Klindworth, A.; Pruesse, E.; Schweer, T.; Peplies, J.; Quast, C.; Horn, M.; Glöckner, F.O. Evaluation of general 16S ribosomal RNA gene PCR primers for classical and next-generation sequencing-based diversity studies. *Nucleic Acids Res.* **2012**, *41*, e1. [CrossRef] [PubMed]
85. White, T.J.; Bruns, T.D.; Lee, S.B.; Taylor, J.W.; Innis, M.A.; Gelfand, D.H.; Sninsky, J. Amplification and Direct Sequencing of Fungal Ribosomal RNA Genes for Phylogenetics. In *PCR Protocols: A Guide to Methods and Applications*; Innis, M.A., Gelfand, D.H., Sninsky, J.J., White, T.J., Eds.; Academic Press: San Diego, CA, USA, 1990; pp. 315–322.
86. Klee, J.; Tay, W.T.; Paxton, R.J. Specific and sensitive detection of Nosema bombi (Microsporidia: Nosematidae) in bumble bees (Bombus spp.; Hymenoptera: Apidae) by PCR of partial rRNA gene sequences. *J. Invertebr. Pathol.* **2006**, *91*, 98–104. [CrossRef] [PubMed]
87. Yang, B.; Peng, G.; Li, T.; Kadowaki, T. Molecular and phylogenetic characterization of honey bee viruses, Nosemamicrosporidia, protozoan parasites, and parasitic mites in China. *Ecol. Evol.* **2013**, *3*, 298–311. [CrossRef] [PubMed]
88. Meeus, I.; De Graaf, D.; Jans, K.; Smagghe, G. Multiplex PCR detection of slowly-evolving trypanosomatids and neogregarines in bumblebees using broad-range primers. *J. Appl. Microbiol.* **2009**, *109*, 107–115. [CrossRef] [PubMed]

Article

To Treat or Not to Treat Bees? Handy VarLoad: A Predictive Model for *Varroa destructor* Load

Hélène Dechatre [1,2], Lucie Michel [3], Samuel Soubeyrand [3], Alban Maisonnasse [2,4], Pierre Moreau [5], Yannick Poquet [1,2], Maryline Pioz [1,2], Cyril Vidau [2,6], Benjamin Basso [1,2,6], Fanny Mondet [1,2,*,†] and André Kretzschmar [3,*,†]

[1] INRAE (Institut National de la Rechernche Agronomique et de l'Environnement), Abeilles et Environnement, 84914 Avignon, France; helene.dechatre@gmail.com (H.D.); yannick.poquet@hotmail.com (Y.P.); maryline.pioz@inrae.fr (M.P.); benjamin.basso@inrae.fr (B.B.)

[2] UMT PrADE (Unité Mixte Technologique: Protection de l'Abeille dans l'Environnement), 84914 Avignon, France; a.maisonnasse.adapi@free.fr (A.M.); cyril.vidau@itsap.asso.fr (C.V.)

[3] INRAE (Institut National de la Rechernche Agronomique et de l'Environnement), BioSP, 84914 Avignon, France; lucie.michel@inrae.fr (L.M.); samuel.soubeyrand@inrae.fr (S.S.)

[4] ADAPI (Association Pour le Développement de l'Apiculture en Provence), 13626 Aix en Provence, France

[5] Beekeeper, 07270 Empurany, France; pierre-pm.moreau@laposte.net

[6] ITSAP (Institut Technique et Scientifique de l'Abeille et de la Pollinisation), Institut de l'Abeille, 84914 Avignon, France

* Correspondence: fanny.mondet@inrae.fr (F.M.); andre.kretzschmar@inrae.fr (A.K.); Tel.: +33-432-722-175 (A.K.)

† Equal contribution.

Citation: Dechatre, H.; Michel, L.; Soubeyrand, S.; Maisonnasse, A.; Moreau, P.; Poquet, Y.; Pioz, M.; Vidau, C.; Basso, B.; Mondet, F.; et al. To Treat or Not to Treat Bees? Handy VarLoad: A Predictive Model for *Varroa destructor* Load. *Pathogens* **2021**, *10*, 678. https://doi.org/10.3390/pathogens10060678

Academic Editor: Giovanni Cilia

Received: 21 April 2021
Accepted: 26 May 2021
Published: 30 May 2021

Abstract: The parasitic *Varroa destructor* is considered a major pathogenic threat to honey bees and to beekeeping. Without regular treatment against this mite, honey bee colonies can collapse within a 2–3-year period in temperate climates. Beyond this dramatic scenario, Varroa induces reductions in colony performance, which can have significant economic impacts for beekeepers. Unfortunately, until now, it has not been possible to predict the summer Varroa population size from its initial load in early spring. Here, we present models that use the Varroa load observed in the spring to predict the Varroa load one or three months later by using easily and quickly measurable data: phoretic Varroa load and capped brood cell numbers. Built on 1030 commercial colonies located in three regions in the south of France and sampled over a three-year period, these predictive models are tools designed to help professional beekeepers' decision making regarding treatments against Varroa. Using these models, beekeepers will either be able to evaluate the risks and benefits of treating against Varroa or to anticipate the reduction in colony performance due to the mite during the beekeeping season.

Keywords: *Apis mellifera*; *Varroa destructor*; treatment; predictive model; beekeeping; decision-making tool

1. Introduction

The parasite *Varroa destructor* is considered a major pathogenic threat to honey bees [1] and to beekeeping. This mite is an ectoparasite affecting both adult bees and broods. Female mites have two distinct stages: a phoretic stage on adult bees and a reproductive stage, which takes place inside a capped brood during bee metamorphosis. The Varroa threat is not new for the beekeeping community, but with colony importations and the commerce of bees, this threat continues to increase. Indeed, these circumstances favor the Varroa spread throughout territories and the world's apiaries. This threat is all the more important given that the parasite spread is rapid [2]. Thus, bees and beekeepers cannot adapt and respond efficiently; on the contrary, Varroa, with continuous exposure to miticide treatments, responds with mechanisms of resistance [3–5]. Consequently, the current challenge is to develop new methods to limit Varroa numbers inside colonies.

Without regular efficient treatment against this mite, honey bee colonies can collapse within a 2–3-year period in temperate climates. Varroa feeding on pupal hemolymph can induce a decrease in adult bee body weight and malformations as well as reducing their life spans, thus weakening their immune systems [6–8]. Thus, it seems logical that infested colonies are less productive and efficient than healthy colonies, which can have significant economic impacts for beekeepers [9,10]. Beyond a threshold of 3 phoretic Varroa mites per 100 bees, the decrease in performance is correlated with the Varroa load [10]. According to this study, a colony with more than 3 phoretic Varroa mites per 100 bees produces, on average, 2.65 kg less honey than a colony below this threshold. Unfortunately, until now, it has not been possible to predict, from the mite population size in the spring, the population load in the summer, despite studies by Arechavaleta-Velasco and Guzman-Novoa (2001), Harris et al. (2003), and Lodesani et al. (2002), confirmed significant correlations between the amount of brood and/or the fertility of the mites [11–13] and population growth [1].

Models of Varroa dynamics have been previously established but mainly carried theoretical descriptions and only allowed for the evaluation of the instantaneous Varroa load. Wilkinson and Smith's model [14] was built from virtual colonies, and DeGrandi-Hoffman and Curry's model [15] was based on the BEEPOP honey bee colony population dynamics model [16]; a BEEHAVE Varroa unit was developed by Becher et al. [17]. These models were based on parametric values available from previous studies [16,18–26]. As these models primarily work with mathematical extrapolation, instead of being data-derived, we assumed that the resulting parametric values could be revised. Additionally, in the twenty years since these models were published, Varroa biology may have coevolved with its host. The coevolution between Varroa and honey bees has been reported by Kurze et al. [27] and includes host resistance behaviors, which involve a decrease in the Varroa reproduction rate as well as perturbations in the biological cycle of the mite. Moreover, previous studies serving as the basis for model construction were based on honey bees with different European origins and on Africanized honey bees [19]. Honey bee origins affect Varroa reproduction [28] and, consequently, Varroa population sizes. To increase its predictability, here, we used a model based on empirical data.

The most important information for a beekeeper is not the Varroa load at the time of honey flow because most treatment compounds, even some labeled "natural" (e.g., formic acid or thymol), are banned or not recommended during honey flow [29]. The aim of this study was therefore to predict the Varroa load one or three months later, from its baseline level in early spring, to anticipate colony performance for honey flow, knowing that the reduced performance threshold is 3 phoretic Varroa mites per 100 bees. Aimed as a useful tool for beekeepers, the model Handy Varload is based on inexpensive, accessible, and quickly measurable data in the field.

2. Results

2.1. Variable Selection (for Variable Definitions, See Materials and Methods, Statistical Analysis)

The variable "phoretic Varroa" measured at $t = 0$ was continuous with 25% of zeros, 27% of the data in [0, 1], and 48% of the data in [1,30]. The zero-inflated beta distribution is similar to the beta distribution but allows zeros as response values in which the ν parameter models the probability of obtaining zero. The distribution features of the variable "phoretic Varroa" (Vp_t) require dividing by 100 in order to fit the data to the interval [0, 1]. We then modeled this new response variable by a zero-inflated beta distribution, with parameter variation depending on covariates. A first model selection was performed to choose the best variables to model μ (see AICc comparisons in Table 1; more details are provided in Supplementary Materials Table S1). At the end of this preliminary selection, two models including the apiary random factor were retained, one for the 1-month adjustment and the other for the 3-month adjustment, noted (*) and (**) in Table 1, respectively.

Table 1. Comparisons of the tested models investigating the influence of phoretic Varroa numbers (per 100 bees) at $t = 0$, capped brood cell numbers, varbrood, and date of predicted phoretic Varroa numbers as a function of the estimation length, using the AICc criterion. N = 867 for data adjustment at one month (x = 1) and N = 93 for data adjustment at three months (x = 3).

Model	Adjustment for x = 1 AICc	Adjustment for x = 3 AICc
phoretic Varroa	−2477.5	−268.3
capped brood cells	−2320.6	−261.8
varbrood (**)	**−2540.1**	**−296.9**
date	−2413.0	−260.9
phoretic Varroa + capped brood cells	−2488.0	−271.1
phoretic Varroa + date	−2561.7	−266.3
phoretic Varroa + varbrood	−2538.2	−295.7
capped brood cells + date	−2412.1	−261.4
capped brood cells + varbrood	−2580.1	−297.7
date + varbrood	−2618.2	−294.7
phoretic Varroa + capped brood cells + date	−2564.9	−269.1
phoretic Varroa + capped brood cells + varbrood	−2582.4	−296.0
phoretic Varroa + date + varbrood	−2616.2	−293.8
capped brood cells + date + varbrood (*)	**−2645.5**	**−295.5**
phoretic Varroa + capped brood cells + varbrood + date	−2647.8	−293.9
(*) and (**) + apiary random effect	**−2651.1**	**−316.9**

The final models (A and B, see below) were obtained after a second variable selection based on AICc comparisons, using the modeled σ and ν added to the (*) and (**) preliminary models. The number of phoretic Varroa present at t was modeled by the following zero-inflated beta models (BEZI in "*gamlss*"):

$$Vp_t \sim \text{BEZI}\ (\mu, \sigma) \qquad \text{with } (1 - \nu) \text{ probability}$$
$$Vp_t = 0 \qquad \text{with } \nu \text{ probability}$$

For data adjustment at one month: (A)

$$\mu = \text{logit}^{-1}\ (\alpha_0 + \alpha_1 Vb_{t-x} + \alpha_2 Cp_{t-x} + \alpha_3 D_t + Ap)$$
$$\sigma = \exp\ (\beta_0 + \beta_1 Vb_{t-x} + \beta_2 D_t + Ap)$$
$$\nu = \text{logit}^{-1}\ (\gamma_0 + \gamma_1 Vb_{t-x} + \gamma_2 Cp_{t-x} + \gamma_3 D_t + Ap)$$

For data adjustment at three months: (B)

$$\mu = \text{logit}^{-1}\ (\alpha_0 + \alpha_1 Vb_{t-x} + Ap)$$
$$\sigma = \exp\ (\beta_0 + Ap)$$
$$\nu = \text{logit}^{-1}\ (\gamma_0 + \gamma_1 Vb_{t-x} + \gamma_2 Vp_{t-x})$$

where the α, β, and γ parameters are coefficients used to model μ, σ, and ν, respectively. As a consequence of this second variable selection, the final AICc was −3179.2 (A) and −343.5 (B) (see details in Tables S2 and S3). Varbrood, which was retained by model selection in all cases except for σ of model B, appeared as the most important explanatory variable.

2.2. Goodness of Fit and Prediction Evaluation

2.2.1. Parameter Uncertainty

For models A and B, the α, β, and γ parameters associated with μ, σ, and ν were estimated and their $CI_{95\%}$s were computed (see Table 2). Based on the intercept, we can note that varbrood had the largest influence on the data adjustment for each parameter of model A. Moreover, the order of influence of model covariates was the same regardless of the parameter: varbrood > date > capped brood cells. For ν of model B, phoretic Varroa had a larger influence than varbrood. The CI$_{95}$ range as positively correlated with covariate weights, i.e., the greater the weight, the larger the uncertainty.

Table 2. Estimated coefficient and 95% confidence interval (CI$_{95\%}$) of models A and B investigating the influence of varbrood, capped brood cells, phoretic Varroa, and date on the number of phoretic Varroa mites for mu, sigma, and nu parameters.

Model	Parameter	Covariate	Estimated Coefficient	Lower 95% CI	Upper 95% CI
A	Mu	Intercept	−5.830	−6.021	−5.640
		varbrood	0.025	0.021	0.028
		capped brood cells	0.002	0.001	0.003
		date	0.014	0.012	0.015
	Sigma	Intercept	6.579	6.233	6.925
		varbrood	−0.023	−0.030	−0.016
		date	−0.018	−0.021	−0.015
	Nu	Intercept	2.073	1.475	2.672
		varbrood	−0.063	−0.087	−0.039
		capped brood cells	−0.003	−0.006	−0.001
		date	−0.032	−0.039	−0.025
B	Mu	Intercept	−3.982	−4.175	−3.790
		varbrood	0.023	0.019	0.027
	Sigma	Intercept	4.460	4.140	4.779
	Nu	Intercept	−0.701	−1.468	0.065
		varbrood	−0.077	−0.167	0.012
		phoretic Varroa	−3.786	−10.747	3.176

Moreover, the apiary effect depended on the horizon of prediction. Thus, the mean apiary effect was zero with varying estimated standard deviations depending on the data adjustment; at one month, the estimated standard deviation was 0.285, with a standard deviation of this estimate of 0.798, and at three months, the estimated standard deviation was 0.681, with a standard deviation of this estimate of 0.914 (see Supplementary Materials Tables S2 and S3).

2.2.2. Prediction Quality

The prediction quality can be evaluated using confidence intervals and error rates of models. Table 3 shows that for cross-validation, 97.6% (N = 4999) of sampled phoretic Varroa mites were in their CI$_{95\%}$ with model A and 97.3% (N = 2328) with model B. These coverage rates are heterogeneous with respect to Vp_t: they overestimate the targeted values (95%, 70%, or 50%) when $Vp_t \leq 3$, they are consistent when $3 < Vp_t \leq 10$, and they are significantly lower than the targeted values when $Vp_t > 10$, which roughly corresponds to only 5–10% of the hives. These results hold approximately for all tackled cases (cross-validation and training validation; models A and B).

Table 3. Coverage rates of confidence intervals ($CI_{95\%}$, $CI_{70\%}$, and $CI_{50\%}$) of Vp_t for both approaches, cross-validation and training validation, for models A and B. The coverage rate provides the proportion of times that the CI contains the true value of Vp_t. For each method and each model, numbers of observed hives are reported for each class of Vp_t.

Cross-Validation								
Model A	Model B	Observed Vp_t	Model A			Model B		
Observed Colony Numbers			$CI_{95\%}$	$CI_{70\%}$	$CI_{50\%}$	$CI_{95\%}$	$CI_{70\%}$	$CI_{50\%}$
4999	2328	all	97.6	83.6	67.7	97.3	83.1	67.8
4027	1700	≤3	99.6	91.5	76.3	99.7	97.9	87.9
724	526	>3 and ≤10	92.7	53.7	34.5	99.8	51.1	16
248	102	>10	80.6	42.3	24.2	45.1	2	0

Training Validation								
Model A	Model B	Observed Vp_t	Model A			Model B		
Observed Colony Numbers			$CI_{95\%}$	$CI_{70\%}$	$CI_{50\%}$	$CI_{95\%}$	$CI_{70\%}$	$CI_{50\%}$
1438	749	all	92.6	75.3	61.8	57.8	39	26
1140	546	≤3	95.9	82.6	69	61.2	44.1	29.9
229	137	>3 and ≤10	82.1	49.8	37.6	60.6	29.9	17.5
69	66	>10	72.5	39.1	21.7	24.2	15.2	12.1

Predicted quantiles were used as an indicator of the accuracy of the prediction aimed by the model, i.e., the proportion of hives to be treated against Varroa. Predicting values by simulation may be seen as minimizing the risk of an incorrect prediction (the risk of unnecessarily increasing the number of hives to be treated) or may be necessary to more accurately target the correctly predicted value (the risk of ignoring a proportion of hives which should be treated and which will not be). For model B, outputs are based on the average Varroa load in April of 0.7 phoretic Varroa mites per 100 bees [30] (quoted Vp_{t-x}) and the threshold of 3 phoretic Varroa mites per 100 bees at the beginning of summer [10] (quoted Vp_t). The model indicates for each colony whether or not to treat (prediction that the threshold will exceed three Varroa mites). Figure 1 describes two extreme situations that correspond to two treatment strategies. The first two strategies, represented by Q97.5 and Q85, are no-risk situations because the model indicated that all colonies are to be treated, and thus no risks are taken of having a colony that exceeds the threshold of three. In these cases, the input costs are great, and 73% of colonies are unnecessarily treated. The second strategy (Q50) is an attempt to justify no treatment, and it estimates the respective risk; it provides reasons not to treat 71% of colonies at the risk of not treating the 24% of colonies that need treatment. This could be seen as the price to pay for engaging in a process of decreasing inputs. Intermediate quantiles allow beekeepers to find correct indicators based on calculated trade-offs. For example, considering indicators for Q72 (or Q71.5), 27% of colonies observed exceed the threshold of three; the model predicted to treat 11% when necessary (10% for Q71.5) and 17% when not necessary (16% for Q71.5). In these cases, there were as many colonies that were treated when not necessary (17%—in orange) as colonies untreated when necessary (16%—in red) for Q72, and the inverse occurred for Q71.5 (Figure 1).

The first and third cases are the hives that are necessary to treat. The percentages of these four categories are provided for each level of risk.

This figure is based partly on Table S4 of Supplementary Materials; Tables S4 and S5 show all results for models A and B of the two model evaluations (cross-validation and training validation). For both models, the smaller the quantile, the lower the global error rate. For larger quantiles (Q97.5 and Q85), models predicted better Vp_t when the phoretic Varroa number exceeded the threshold of three Varroa mites at t. Model predictions of Vp_t were relatively good when the earlier phoretic Varroa number was at three, the maximum. However, models failed to produce correct predictions when the mite number at t-x was higher than three for model A and higher than 0.7 for model B.

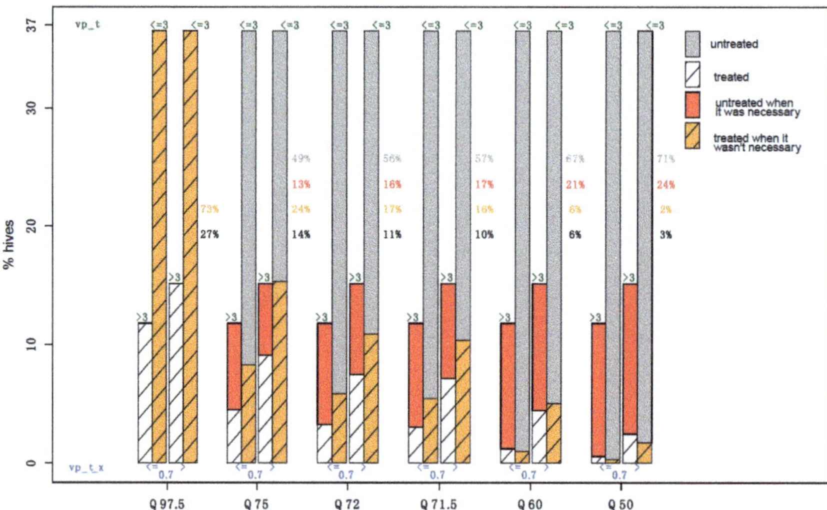

Figure 1. In this figure, 5 scenarios are presented with increasing risk (from left to right) taken by the beekeeper to not treat when the model predicts it was necessary or to treat when it was unnecessary. The risk is inversely proportional to the measure of quantile Q. For each level of risk, four cases are represented: (1) Hives with vp_t_x (i.e., Vp at $t = 0$) < = 0.7 and vp_t (i.e., Vp three months later) > 3; (2) Hives with vp_t_x (i.e., Vp at $t = 0$) < = 0.7 and vp_t (i.e., Vp three months later) < = 3; (3) Hives with vp_t_x (i.e., Vp at $t = 0$) > 0.7 and vp_t (i.e., Vp three months later) < = 3; (4) Hives with vp_t_x (i.e., Vp at $t = 0$) > 0.7 and vp_t (i.e., Vp three months later) < = 3.

3. Discussion

3.1. Selected Variables

The Handy VarLoad (HVL) model allowed for the prediction of the Varroa load at a given moment t, as a function of the previously observed Varroa load and of the available area for their reproduction, i.e., the number of honey bee brood cells.

Seasons influence the Varroa load, but only in the short term. This could be explained by the fact that, in one month, a beekeeper management intervention or a particular climatic event can have an effect on one or two Varroa generations, as the generation interval of capped brood is 12 days. The mite population growth rate is exponential during short periods (three months) and when mite populations are low; in contrast, Varroa population growth follows a logistic dynamic over longer periods (covering the entire production period) when density-dependent factors influence population growth [12]. Consequently, an event which increases or decreases Varroa reproduction may change the short-term Varroa load but have an insignificant influence on the long-term Varroa load. For example, disruption of honey bee colony broods could be offset by the Varroa population growth itself. Conversely, if colony brood disruption speeds up Varroa reproduction, the mite population size eventually stabilizes due to density dependence [25,31]. Moreover, during a three-month period, colonies undergo a series of favorable and unfavorable disruptions for Varroa development, particularly climatic, which will balance each other out. Finally, the apiary effect acts regardless of the delay between two phoretic Varroa measurements. Thus, the biological variability between colonies, the differences in management strategy between beekeepers, year, and region (climate) influence the Varroa load of the colony [32–35].

Contrary to previous mathematical models on Varroa load, the HVL model allows one to obtain a prediction with a measure of uncertainty, as well as the associated uncertainty for each parameter. The model uncertainty includes variability at the inter-apiary scale, in beekeeping management strategies, and in year and region effects. The apiary effect included in the model induces a large amount of prediction uncertainty, but, at the same time, it assimilates the sampling diversity related to apiary characteristics (management, year, and region).

3.2. Beekeepers' Interest

The model presented here allows one to have a representation of the risk beekeepers take by not treating the apiary, according to the percentage of colonies that exceed the threshold of three Varroa mites. Different quantiles propose different decision-making indicators for beekeepers taking into account trade-offs between cost, time, and environmental effects of treatments, on the one hand, and the risk of losing infested colonies, on the other.

Moreover, beside economic trade-offs, Varroa treatments are not without consequences and, indeed, may induce acaricide resistance in Varroa [3–5], which is why beekeepers should treat only when economic risks are real. It is worth noting that treatments during the beekeeping season are not efficient over the long term [36]. These types of treatment must be used only when the aim is to temporarily decrease the Varroa load to optimize honey flow performance. Thus, this model takes into account integrated pest management.

The model can also be used to determine which apiaries should be given priority on lavender and sunflower fields if the spot number is limited. However, despite the fact that managing colonies at the apiary scale is more efficient, as honey bee colony performances are highly dependent on the characteristics of any apiary (Kretzschmar et al., unpublished data), beekeepers may want to manage Varroa at the colony scale and thus strictly follow the model prediction.

3.3. Limits and Prospects of the Model

The choice to use only easy-to-measure variables in the field impairs the model's goodness of fit and, consequently, the estimation/prediction accuracy. Taking into account other variables (Varroa foundress density, Varroa infestation rate in the capped brood, natural death of Varroa mites measured on sticky boards, etc.) would have allowed better predicting the Varroa load. Including these additional variables in the present model could have easily improved its prediction power. Nevertheless, it would be far too long and complex to collect that type of data in the actual schedule of a beekeeper. If the sampling plan is unrealistic and impracticable at a large scale, the HVL model will be worthless. However, such improved models could be developed for researchers or technicians who work on a smaller scale and need to have better precision in their experimental frameworks. Another limit of this study is the sampled colony number: the more hives sampled, the better the estimation. In the present study, as the number of repetitions for each factor (management, year, and region) is limited, our sampling variation increased model uncertainty. Nevertheless, the Handy VarLoad model will be improved by the accumulation of data issued from the numerous experiments in which the two handy variables it uses (phoretic Varroa load and capped brood area) are commonly collected. As the database on which the model is based increases, the effect of covariates (apiary, region, season, beekeeping practices, etc.) can be better integrated.

4. Materials and Methods

4.1. Data Sampling

Data were collected from 310 colonies from 2014 to 2016 in three regions of France (PACA, AURA, and Occitanie; "dataset1") and from 720 colonies in 2018 in three regions of France (PACA, Nouvelle Aquitaine, and Centre; "dataset2"). Most of the colonies were kept on 10-frame Dadant hives and contained hybrid *Apis mellifera* L. queens. Colonies belonged to commercial beekeepers and thus displayed different sizes, dynamics, and management styles, which allowed us to take into account the variability which exists between beekeepers and apiaries. No treatment against Varroa was applied during the sampling periods.

At each sampling point, the amount of capped brood (noted *Cb*) was determined according to the ColEval method [37], and the phoretic mite load was estimated by sampling around three hundred bees (or 45 g) from a frame containing an uncapped brood. Sampled bees were washed with a detergent solution and the number of Varroa mites retrieved (noted Vp) was counted [38]. Finally, to take into account seasonality, a "date" variable (noted D) was also created in which days were reported on a perpetual calendar with day 1 starting on 15 March of each year. This variable described the number of days ran from an initial time, which corresponds to the beginning of the measurable increase in the Varroa population after wintering. In our case, it corresponded approximately to the middle of March.

Sampling points were repeated at 30-day intervals, except for apiaries R16 to R18 ("dataset1"), in which measurements were sometimes performed every 12 days to mimic the generation interval of capped broods.

4.2. Statistical Methods

4.2.1. Distribution Adjustment on "dataset1"

All statistics were performed using the statistical software R version 3.3.0 [39]. Estimation of model parameters was carried out using the "gamlss" function of the eponymous package (Rigby and Stasinopoulos, 2005). The response variable (number of observed Varroa mites per 100 bees) was modeled with a generalized additive model for location, scale, and shape (GAMLSS). GAMLSS is an extension of the generalized linear model and the generalized additive model. It is a distribution-based approach to semiparametric regression models, in which all the parameters of the assumed distribution for the response can be modeled as additive functions of the explanatory variables, such as the *location* (e.g., *mean* μ), the *scale* (e.g., variance σ^2), the *shape* (skewness and kurtosis), and some *inflation* (e.g., at zero, ν). Moreover, we chose to use GAMLSS because it offers numerous choices for the distribution of the response variable and is suitable for time series data (Rigby and Stasinopoulos, 2001). GAMLSS was fitted to data using maximum (penalized) likelihood estimation implemented with the RS algorithm, which does not require accurate starting values for μ, σ, and ν to ensure convergence in comparison with the CG algorithm [40,41]. The most parsimonious model with the lowest corrected Akaike's information criterion (AICc) [42], was selected; models with differences in AICc values lower than or equal to two were considered to be equivalent. We chose this selection criterion because, it is the most suitable criterion to model selection in predictive models for ecology and time series applications including forecasting [43]. Thus, it allows for the selection of the model that will best predict the response variable, i.e., the model with the best predictive accuracy.

Variables, which were described above, were transformed as follows to comply with the scaling conditions during model fitting:

$$Cb = \frac{Cb0}{100} \tag{1}$$

$$Vp = \frac{Vp0 * 100 * 0.14}{sw} \tag{2}$$

$$Vb = \log\left(\frac{Vp}{Cb + 130} * 100 + 1\right) * 50 \tag{3}$$

where Cb (Equation (1)) is a scaled value of the number of capped brood cells $Cb0$; Vp (Equation (2)) is the normalized rational number of Varroa mites for 100 honey bees (called "phoretic Varroa" in the present study), knowing that the weight per bee is 0.14 g, and sw in Equation (2) is the sampling weight of bees; Vb is a variable called "varbrood", built to take into account the role of the amount of brood in the regulation of Varroa reproduction, and, more specifically, to integrate the fact that the more spread out the capped brood, the harder it is to capture phoretic Varroa mites hidden in the capped brood. The varbrood variable was thus obtained by taking the Neperian logarithm of the number of phoretic Varroa and dividing it by the number of capped brood cells. In Equation (3), 130 corresponds to the Cb median, 100 and 50 multipliers are necessary for the scale, and +1 is used to avoid obtaining log(0). These three quantitative variables were mathematically reduced to the same scale, in order to be able to compare their respective weights during model adjustment. The date (measured as a number of days after the first measurement) was used without transformation.

The rational number of phoretic Varroa mites present at t (Vp_t) was modeled in the GAMLSS framework by a zero-inflated beta distribution with mean μ, standard deviation σ, and inflation at zero ν. Different specifications for μ, σ, and ν were used (see Results section). Our models were designed to predict Vp_t from explanatory variables typically collected at time $t-x$. Two horizons of prediction x were considered: a short-term horizon ($x = 1$ month, noted model A hereafter) and a long-term horizon ($x = 3$ months, noted model B hereafter). For $x = 1$ (model A), all data were used to fit the models (867 observations), whereas for $x = 3$ (model B), all the data providing this interval were used to avoid the use of time-overlapping pairs of observations (93 observations). Phoretic Varroa numbers, capped brood cell numbers, and varbrood present at $t-x$, as well as the date at t, were exploited as fixed factors; they are denoted by Vp_{t-x}, Cb_{t-x}, Vb_{t-x}, and D_t, respectively. Moreover, an «apiary» factor (noted Ap) was used as a random factor and includes the variability of the apiary, beekeeping management strategy, and year and region effects.

4.2.2. Goodness of Fit and Prediction Including "dataset2"

To assess the goodness of fit of the selected models, we explored the uncertainty of parameters and the prediction quality by comparing the predicted and observed values of Vp_t. We evaluated the prediction quality using two methods: cross-validation and training validation. In both methods, model performance was evaluated on data not included in the sample used to estimate model parameters.

For increasing the domain where the uncertainty of the parameters and the prediction quality of the models could be explored, a larger dataset ("dataset2") was added to the first dataset ("dataset1") with which the model parameters were estimated.

In the cross-validation method, observations of the hives of a given apiary were removed from the database, the model was fitted to the remaining data, and estimated parameters were plugged in to predict Vp_t for the hives of the apiary whose observations were removed (this case corresponds to predicting Vp_t for a new apiary based on observations collected from other apiaries). This procedure was repeated for each apiary of the dataset (i.e., 54 times for model A and 40 times for model B) and allowed us to provide averaged cross-validation assessments of the prediction performance.

The prediction performance was assessed with respect to the two following criteria:

- The actual coverage of 95%, 70%, and 50% confidence intervals of Vp_t (denoted by $CI_{95\%}$, $CI_{70\%}$, and $CI_{50\%}$), providing the proportion of times that the true value of Vp_t is contained within the CI;
- The use of different predicted quantiles of Vp_t (namely, $Q_{97,5\%}$, $Q_{85\%}$, $Q_{75\%}$, and $Q_{50\%}$) to evaluate the risk that the actual Vp_t exceeds the problematic threshold of 3 Varroa mites for 100 bees.

These criteria (CI and quantiles) were empirically calculated from 1000 simulations of the zero-inflated beta distribution in which the estimated values of μ, σ, and ν were inserted (random factors incorporated into μ, σ, and ν were randomly drawn at each simulation from centered normal distributions with standard deviations equal to their estimated values). Note that estimation uncertainty was neglected in this simulation procedure; this choice may lead to un-calibrated confidence intervals and quantiles. The comparisons of quantiles $Q_{97.5\%}$, $Q_{85\%}$, $Q_{75\%}$, and $Q_{50\%}$ with the threshold of 3 Varroa mites for 100 bees can be used as indicators to assess whether Vp_t will exceed this problematic threshold. The efficiency of these indicators was assessed with the error rate $\tau^{error}(\alpha)$ calculated as the ratio between (i) the number of hives for which the predicted quantile $Q_\alpha(Vp_t)$ is less than or equal to 3 at time t, whereas the actual observation Vp_t is greater than 3, and (ii) the number of observations Vp_t greater than 3:

$$\tau^{error}(\alpha) = \frac{\sum_{R=1}^{K} \sum_{r=1}^{N_R} 1\left(Vp_{t,R,r} > 3, \ Q_{\alpha}^{-R}(Vp_{t,R,r}) \leq 3 \right)}{\sum_{R=1}^{K} \sum_{r=1}^{N_R} 1\left(Vp_{t,R,r} > 3 \right)},$$

(4)

where K is the number of apiaries (54 for model A and 40 for model B), N_R is the number of hives in the apiary R (which ranges between 7 and 51), $Vp_{t,R,r}$ is the observed value of phoretic Varroa at time t for the hive r of the apiary R, and $Q_{\alpha}^{-R}(Vp_{t,R,r})$ is the predicted quantile at $\alpha\%$ of $Vp_{t,R,r}$ for the hives of the apiary R whose observations were removed $(-R)$. The indicator function $E \rightarrow 1I$ takes the value of 1 if event E is true, or otherwise 0.

Equation (4) presents the case in which observations are greater than 3 and predictions are less than or equal to 3, and $\tau^{error}(\alpha)$ was also calculated when observations are less than or equal to 3 and predictions are greater than 3. The error rate $\tau^{error}(\alpha)$ can be computed in other specific conditions, for example, conditions related to the number of phoretic Varroa at $t-x$ $(Vp_{t-x,R,r})$.

In the training validation method, observations of the hives of all apiaries after a specific date, t_A = 31 (15 April) for model A, and t_B = 118 (11 July) for model B, were removed from the database, the model was fitted to remaining data, and estimated parameters were plugged in to predict Vp_t for the hives of all apiaries after t_A or t_B (this case corresponds to predicting Vp_t for an apiary already installed, based on observations collected beforehand from this apiary). This procedure was repeated for each year of the dataset (i.e., from 2014 to 2016 and 2018 for both models A and B) and allowed us to provide averaged training validation assessments of the prediction performance already introduced in the cross-validation approach.

Supplementary Materials: The following are available online at https://www.mdpi.com/article/10.3390/pathogens10060678/s1, Table S1: Comparisons of the tested models investigating the influence of phoretic Varroa mite (per 100 bees) numbers, capped brood cell numbers, varbrood, and date on next phoretic Varroa numbers as a function of the estimation length, using the AICc criterion. N = 867 for data adjustment at one month (x = 1) and N = 93 for data adjustment at three months (x = 3), Table S2: R graphic output of model A summary, Table S3: R graphic output of model B summary, Table S4: Performance comparisons between different quantiles (Q97.5, Q85, Q75, Q50) for models A and B, depending on the observed phoretic Varroa numbers at t−x and the observed phoretic Varroa numbers at t. Error rates represent the colony percentage for which the Varroa load at the horizon of prediction was badly predicted with the cross-validation method. For each quantile, the number of hives to treat depends on the error rate for which their percentages were reported, Table S5: Performance comparisons between different quantiles (Q97.5, Q85, Q75, Q50) for models A and B, depending on the observed phoretic Varroa numbers at t-x and the observed phoretic Varroa numbers at t. Error rates represent the colony percentage for which the Varroa load at the horizon of prediction was badly predicted with the training validation method. For each quantile, the number of hives to treat depends on the error rate for which their percentages were reported.

Author Contributions: Conceptualization, F.M. and A.K.; methodology, F.M., A.K., L.M., H.D. and S.S.; investigation, P.M., A.M., C.V., B.B., F.M. and A.K.; data curation, H.D., L.M., S.S., Y.P., M.P. and A.K.; writing—original draft preparation, H.D., F.M. and A.K.; supervision, F.M. and A.K. All authors have read and agreed to the published version of the manuscript.

Funding: This research received no external funding.

Institutional Review Board Statement: Not applicable.

Informed Consent Statement: Not applicable.

Data Availability Statement: All data are available upon request to the corresponding author.

Acknowledgments: We wish to thank all beekeepers that gave us access to their apiaries to record data.

Conflicts of Interest: The authors declare no conflict of interest.

References

1. Rosenkranz, P.; Aumeier, P.; Ziegelmann, B. Biology and control of Varroa destructor. *J. Invertebr. Pathol.* **2010**, *103*, 96–119. [CrossRef]
2. Denmark, H.A.; Cromroy, H.L.; Cutts, L. *Varroa Mite. Varroa Jacobsoni Oudemans: (Acari: Varroidae)*; Fla. Department Agric. & Consumer Serv., Division of Plant Industry: Gainesville, FL, USA, 1991.
3. Baxter, J.; Eischen, F.; Pettis, J.; Wilson, W.T.; Shimanuki, H. Detection of fluvalinate-resistant Varroa mites in US honey bees. *Am. Bee J.* **1998**, *138*, 291.
4. Elzen, P.J.; Westervelt, D. Detection of coumaphos resistance in Varroa destructor in Florida. *Am. Bee J.* **2002**, *142*, 291–292.
5. Pettis, J.S. A scientific note on Varroa destructor resistance to coumaphos in the United States. *Apidologie* **2004**, *35*, 91–92. [CrossRef]
6. De Jong, D.; De Jong, P.H.; Goncalves, L.S. Weight loss and other damage to developing worker honeybees from infestation with *V. jacobsoni. J. Apic. Res.* **1982**, *21*, 165–216. [CrossRef]
7. Schneider, P.; Drescher, W. Einfluss der Parasitierung durch die Milbe *Varroa Jacobsoni* aus das Schlupfgewicht, die Gewichtsentwicklung, die Entwicklung der Hypopharynxdrusen und die Lebensdauer von *Apis mellifera. Apidologie* **1987**, *18*, 101–106. [CrossRef]
8. Yang, X.; Cox-Foster, D.L. Impact of an ectoparasite on the immunity and pathology of an invertebrate: Evidence for host immunosuppression and viral amplification. *Proc. Natl. Acad. Sci. USA* **2005**, *102*, 7470–7475. [CrossRef]
9. Emsen, B.; Guzman-Novoa, E.; Kelly, P.G. Honey production of honey bee (Hymenoptera: Apidae) colonies with hight and low Varroa destructor (Acari: Varroidae) infestation rates in eastern Canada. *Can. Entomol.* **2014**, *146*, 236–240. [CrossRef]
10. Kretzschmar, A.; Maisonnasse, A.; Dussaubat, C.; Cousin, M.; Vidau, C. Performances des colonies vues par les observatoires de ruchers. *Innov. Agron.* **2016**, *53*, 81–93.
11. Arechavaleta-Velasco, M.E.; Guzman-Novoa, E. Relative effect of four characteristics that restrain the population growth of the mite *Varroa destructor* in honey bee (*Apis mellifera* L.) colonies. *Apidologie* **2001**, *32*, 157–174. [CrossRef]
12. Harris, J.W.; Harbo, J.R.; Villa, J.D.; Danka, R.G. Variable population growth of *Varroa destructor* (Mesostigmata: Varroidae) in colonies of honey bees hymenoptera: Apidae) during a 10-year period. *Environ. Entomol.* **2003**, *32*, 1305–1312. [CrossRef]
13. Lodesani, M.; Crailsheim, C.; Moritz, R.F.A. Effect of some characters on the population growth of mite *Varroa jacobsoni* in *Apis mellifera* L. colonies and results of a bi-directional selection. *J. Appl. Entomol.* **2002**, *126*, 130–137. [CrossRef]
14. Wilkinson, D.; Smith, G.C. A model of the mite parasite, *Varroa destructor*, on honeybees (*Apis mellifera*) to investigate parameters important to mite population growth. *Ecol. Model.* **2002**, *148*, 263–275. [CrossRef]
15. DeGrandi-Hoffman, G.; Curry, R. A mathematical model of Varroa mite (*Varroa destructor* Anderson and Trueman) and honeybee (*Apis mellifera* L.) population dynamics. *Int. J. Acarol.* **2004**, *30*, 259–274. [CrossRef]
16. DeGrandi-Hoffman, G.; Roth, S.A.; Loper, G.L.; Erickson, E.H., Jr. BEEPOP: A honeybee population dynamics simulation model. *Ecol. Model.* **1989**, *45*, 133–150. [CrossRef]
17. Becher, M.A.; Grimm, V.; Thorbek, P.; Horn, J.; Kennedy, P.J.; Osborne, J.L. BEEHAVE: A systems model of honeybee colony dynamics and foraging to explore multifactorial causes of colony failure. *J. Appl. Ecol.* **2014**, *51*, 470–482. [CrossRef] [PubMed]
18. Jay, S.C. The development of honeybees in their cells. *J. Apic. Res.* **1963**, *2*, 117–134. [CrossRef]
19. De Jong, D.; De Jong, P.H. Longevity of Africanized honey bees (Hymenoptera: Apidae) infested by *Varroa jacobsoni* (Parasitiformes: Varroidae). *J. Econ. Entomol.* **1983**, *76*, 766–768. [CrossRef]
20. Kovac, H.; Crailsheim, K. Lifespan of *Apis mellifera carnica* Pollm. infested by *Varroa jacobsoni* Oud. In relation to season and extent of infestation. *J. Apic. Res.* **1988**, *27*, 230–238. [CrossRef]
21. Fries, I.; Camazine, S.; Sneyd, J. Population dynamics of *Varroa jacobsoni:* A model and a review. *Bee World* **1994**, *75*, 4–28. [CrossRef]
22. Donzé, G.; Herrmann, M.; Bachofen, B.; Guerin, P.M. Effect of mating frequency and brood cell infestation rate on the reproductive success of the honeybee parasite Varroa jacobsoni. *Ecol. Entomol.* **1996**, *21*, 17–26. [CrossRef]
23. Martin, S.J.; Kemp, D. Average number of reproductive cycles performed by *Varroa jacobsoni* in honey bee (*Apis mellifera*) colonies. *J. Apic. Res.* **1997**, *36*, 113–123. [CrossRef]
24. Calis, J.N.M.; Fries, I.; Ryrie, S.C. Population modeling of *Varroa jacobsoni* Oud. *Apidologie* **1999**, *30*, 111–124. [CrossRef]
25. Martin, S.J. A population model for the ectoparasitic mite *Varroa jacobsoni* in honey bee (*Apis mellifera*) colonies. *Ecol. Model.* **1998**, *109*, 267–281. [CrossRef]
26. Martin, S.J. The role of Varroa and viral pathogens in the collapse of honey bee colonies: A modeling approach. *J. Appl. Entomol.* **2001**, *38*, 1082–1093.
27. Kurze, C.; Routtu, J.; Moritz, R.F. Parasite resistance and tolerance in honeybees at the individual and social level. *Zoology* **2016**, *119*, 290–297. [CrossRef]
28. Camazine, S. Differential reproduction of the mite *Varroa jacobsoni* (Mesostigmata: Varroidae), on Africanized and European honey bees (Hymenoptera: Apidae). *Ann. Entomol. Soc. Am.* **1986**, *79*, 801–803. [CrossRef]
29. Bogdavov, S.; Charrière, J.-D.; Imdorf, A.; Kilchenmann, V.; Fluri, P. Determination of residues in honey after treatments with formic and oxalic acid under field conditions. *Apidologie* **2002**, *33*, 399–409. [CrossRef]

30. Maisonnasse, A.; Frontero, L.; Kretzschmar, A. Mesurer le taux de VP/100ab (Varroa phorétique pour 100 abeilles) dans les ruchers pour optimiser la gestion et la production. In Proceedings of the 5ème Journée de la Recherche Apicole, Paris, France, 8–9 February 2017.

31. Sumpter, D.J.T.; Broomhead, D.S. Relating individual behaviour to population dynamics. *Proc. R. Soc. Lond. B Biol. Sci.* **2001**, *268*, 925–932. [CrossRef]

32. Eguaras, M.; Marcangeli, J.; Oppedisano, M.; Fernández, N. Seasonal changes in *Varroa jacobsoni* reproduction in temperate climates of Argentina. *Bee Sci.* **1994**, *3*, 120–123.

33. Garcia-Fernandez, P.; Rodriguez, R.B.; Orantesbermejo, F.J. Influence of climate on the evolution of the population-dynamics of the *Varroa* mite on honeybees in the south of Spain. *Apidologie* **1995**, *26*, 371–380.

34. Kraus, B.; Velthuis, H.H.W. High humidity in the honey bee (*Apis mellifera* L.) brood nest limits reproduction of the parasitic mite *Varroa jacobsoni* Oud. *Naturwissenschaften* **1997**, *84*, 217–218. [CrossRef]

35. Currie, R.W.; Tahmasbi, G.H. The ability of high-and low-grooming lines of honey bees to remove the parasitic mite Varroa destructor is affected by environmental conditions. *Can. J. Zool.* **2008**, *86*, 1059–1067. [CrossRef]

36. Mondet, F.; Maisonnasse, A.; Kretzschmar, A.; Alaux, C.; Vallon, J.; Basso, B.; Dangleant, A.; Le Conte, Y. Varroa: Son impact, les méthodes d'évaluation de l'infestation et les moyens de lutte. *Innov. Agron.* **2016**, *53*, 63–80.

37. Hernandez, J.; Maisonnasse, A.; Cousin, M.; Beri, C.; Le Quintrec, C.; Bouetard, A.; Castex, D.; Decante, D.; Servel, E.; Buchwalder, G.; et al. ColEval: Honeybee COLony Structure EVALuation for Field Surveys. *Insects* **2020**, *11*, 41. [CrossRef] [PubMed]

38. Dietemann, V.; Nazzi, F.; Martin, S.J.; Anderson, D.L.; Locke, B.; Delaplane, K.S.; Wauquiez, Q.; Tannahill, C.; Frey, E.; Ziegelmann, B.; et al. *BEEBOOK Volume II. Standard Methods for Varroa Research 4.2.3.1.2.3*; University of Bern: Bern, Switzerland, 2012.

39. R Core Team. *R: A Language and Environment for Statistical Computing*; R Foundation for Statistical Computing: Vienna, Austria, 2021; Available online: https://www.R-project.org/ (accessed on 25 May 2021).

40. Rigby, R.A.; Stasinopoulos, D.M. Generalized additive models for location, scale and shape, (with discussion). *Appl. Stat.* **2005**, *54*, 507–554. [CrossRef]

41. Rigby, R.A.; Stasinopoulos, D.M. The GAMLSS project: A flexible approach to statistical modelling. In *New Trends in Statistical Modelling, Proceedings of the 16th International Workshop on Statistical Modelling, Odense, Denmark, 2–6 July 2001*; Klein, B., Korsholm, L., Eds.; Odense University Press: Odense, Denmark, 2001; pp. 249–256.

42. Hurvich, C.M.; Tsai, C.-L. Regression and Time Series Model Selection in Small Samples. *Biometrika* **1989**, *76*, 297–307. [CrossRef]

43. Aho, K.; Derryberry, D.; Peterson, T. Model selection for ecologists: The worldviews of AIC and BIC. *Ecology* **2014**, *95*, 631–636. [CrossRef]

MDPI

Article

Genomic Sequencing and Comparison of Sacbrood Viruses from *Apis cerana* and *Apis mellifera* in Taiwan

Ju-Chun Chang [1,2], Zih-Ting Chang [1], Chong-Yu Ko [1], Yue-Wen Chen [1,*] and Yu-Shin Nai [2,*]

[1] Department of Biotechnology and Animal Science, National Ilan University, Yilan 260, Taiwan; c1887g@gmail.com (J.-C.C.); a0923653853@gmail.com (Z.-T.C.); ricardo7677@gmail.com (C.-Y.K.)
[2] Department of Entomology, National Chung-Hsing University, Taichung 402, Taiwan
* Correspondence: chenyw@niu.edu.tw (Y.-W.C.); ysnai@nchu.edu.tw (Y.-S.N.)

Abstract: Sacbrood virus (SBV) was the first identified bee virus and shown to cause serious epizootic infections in the population of *Apis cerana* in Taiwan in 2015. Herein, the whole genome sequences of SBVs in *A. cerana* and *A. mellifera* were decoded and designated AcSBV-TW and AmSBV-TW, respectively. The whole genomes of AcSBV-TW and AmSBV-TW were 8776 and 8885 bp, respectively, and shared 90% identity. Each viral genome encoded a polyprotein, which consisted of 2841 aa in AcSBV-TW and 2859 aa in AmSBV-TW, and these sequences shared 95% identity. Compared to 54 other SBVs, the structural protein and protease regions showed high variation, while the helicase was the most highly conserved region among SBVs. Moreover, a 17-amino-acid deletion was found in viral protein 1 (VP1) region of AcSBV-TW compared to AmSBV-TW. The phylogenetic analysis based on the polyprotein sequences and partial VP1 region indicated that AcSBV-TW was grouped into the SBV clade with the AC-genotype (17-aa deletion) and was closely related to AmSBV-SDLY and CSBV-FZ, while AmSBV-TW was grouped into the AM-genotype clade but branched independently from other AmSBVs, indicating that the divergent genomic characteristics of AmSBV-TW might be a consequence of geographic distance driving evolution, and AcSBV-TW was closely related to CSBV-FZ, which originated from China. This 17-amino-acid deletion could be found in either AcSBV or AmSBV in Taiwan, indicating cross-infection between the two viruses. Our data revealed geographic and host specificities between SBVs. The amino acid difference in the VP1 region might serve as a molecular marker for describing SBV cross-infection.

Keywords: sacbrood virus; sacbrood disease; *Apis cerana*; *Apis mellifera*

Citation: Chang, J.-C.; Chang, Z.-T.; Ko, C.-Y.; Chen, Y.-W.; Nai, Y.-S. Genomic Sequencing and Comparison of Sacbrood Viruses from *Apis cerana* and *Apis mellifera* in Taiwan. *Pathogens* **2021**, *10*, 14. https://dx.doi.org/10.3390/pathogens10010014

Received: 1 December 2020
Accepted: 22 December 2020
Published: 28 December 2020

Publisher's Note: MDPI stays neutral with regard to jurisdictional claims in published maps and institutional affiliations.

1. Introduction

Sacbrood virus (SBV) is a single-stranded, positive-sense RNA virus that belongs to the *Iflaviridae* family [1–3]. The particles of SBV are 28 nm in diameter, nonenveloped, icosahedral [4,5]. SBV is a common honeybee virus that exhibits a high prevalence of infection mainly in early larval stage of honeybees. This condition affects the broods of honeybees, and the specific symptoms can be easily identified in dead deformed larvae in hives with fluid-filled sacs [6–9].

Sacbrood disease, which is caused by SBV infection, was first reported and verified in *Apis mellifera* in 1964 [7,10]. The ectoparasitic mite *V. destructor* plays a role in SBV transmission [11]. Infection with SBV (AmSBV) is now commonly found in *A. mellifera* worldwide and does not usually result in *A. mellifera* colony loss [12–16]. However, according to a previous report from 1976, the SBV found in *A. cerana* (AcSBV) has large impacts on *A. cerana* in several Asian countries, including China, Korea, India, Vietnam, and Thailand [17–22]. During 1991–1992, an outbreak of sacbrood disease caused up to 90% colony losses in Thailand [23–30]. As mentioned above, AcSBV infection usually causes a high rate of *A. cerana* larvae death and may even lead to whole-colony collapse [31,32].

The *A. cerana* is an indigenous honeybee species in Taiwan. The natural fitness of *A. cerana* is better than *A. mellifera*. The *A. cerana* has a higher tolerance for low temperature and better performance on the pollination in the mountain regions than *A. mellifera*. Therefore, *A. cerana* contributes to the pollination of mountainous orchards, including plums and peaches, etc., which are counted for 1.3% of Taiwan agricultural production [33]. Since AcSBV was first detected in Taiwan in 2015, many beekeepers in Taiwan have reported significant *A. cerana* larval death with symptoms caused by AcSBV in *A. cerana* colonies [34,35]. The virus was found to have spread from southern Taiwan to northern and then eastern Taiwan in 2016. Based on long-term surveillance data, the prevalence rate of AcSBV in *A. cerana* colonies had dramatically increased from 47% to ~70% at the end of 2016 and continued to increase to 72% in 2017 [35]. More than 90% of *A. cerana* colonies were influenced by AcSBV infection from 2016–2019, and the colony collapse does have impacts on the sale market of *A. cerana* colonies, price of honey, and the productions of mountainous orchards. The prevalence of SBV in the population of *A. cerana* in Taiwan is now irreversible.

To better understand the relationship between each SBV strain among Asian countries, the analysis of genome sequences from different geographic areas could provide an accurate and reliable method of detecting variations within the same type of genome based on molecular comparisons. Several studies have examined the whole genome sequences of either AmSBV or AcSBV worldwide. In Korea, six AmSBVs (AmSBV-Kor1 [KP296800.1], AmSBV-Kor2 [KP296801.1], AcSBV-Kor3 [KP296802.1], AcSBV-Kor4 [KP296803.1], AmSBV-Kor19 [JQ390592.1], and AmSBV-Kor21 [JQ390591.1]) from *A. mellifera* were sequenced and further compared (Table 1) [24,36]. In 2017, a comparative genomic analysis among nine SBVs of *A. cerana* and *A. mellifera* was performed in Vietnam [30]. These reports identified different genomic features and revealed the genetic diversity among these SBVs, suggesting that viral cross-infections might occur between AcSBV and AmSBV.

Table 1. Information on the sacbrood virus (SBV) strains used in this study.

No.	Name	Host	Total Size (bp)	Location	Accession No.	Reference
1	AmSBV-TW	*Apis mellifera*	8885	Taiwan	MN082651	This study
2	AcSBV-TW	*Apis cerana*	8776	Taiwan	MN082652	This study
3	AcSBV-IndTN-1	*Apis cerana*	8740	India	KX663835.1	[37]
4	AmSBV-Kor21	*Apis mellifera*	8855	Korea	JQ390591.1	[24]
5	AmSBV-Kor19	*Apis mellifera*	8784	Korea	JQ390592.1	[24]
6	South Australia_1	*Apis mellifera*	8821	Australia	KY887697.1	[38]
7	South Australia_2	*Apis mellifera*	8831	Australia	KY887698.1	[38]
8	South Australia_3	*Apis mellifera*	8848	Australia	KY887699.1	[38]
9	SBV-UK	*Apis mellifera*	8832	UK	NC_002066.1	[6]
10	AcSBV-Viet-LDst	*Apis cerana*	8832	Viet Nam	KJ959613.1	-
11	CSBV-SXYL-2015	*Apis cerana* (bee larvae)	8776	China	KU574662.1	[39]
12	Korean	*Apis cerana*	8792	Korea	HQ322114.1	-
13	CSBV-BJ	*Apis cerana* (bee larvae)	8857	China	KF960044.1	-
14	MD1	*Apis mellifera*	8861	USA	MG545286.1	[40]
15	MD2	*Apis mellifera*	8861	USA	MG545287.1	[40]
16	AmCSBV-SDLY	*Apis mellifera*	8794	China	MG733283.1	[40]
17	CSBV-LNQY-2009	*Apis cerana* (bee larvae)	8863	China	HM237361.1	[39]
18	CSBV-JLCBS-2014	*Apis cerana* (bee larvae)	8794	China	KU574661.1	[39]
19	AcSBV-Viet-SBM2	*Apis cerana*	8854	Viet Nam	KC007374.1	[41]
20	AcSBV-India-II10	*Apis cerana*	8550	India	JX194121.1	-
21	AcSBV-India-II2	*Apis cerana*	8680	India	JX270795.1	-
22	AcSBV-India-K1A	*Apis cerana*	8743	India	JX270796.1	-
23	AcSBV-India-K5B	*Apis cerana*	8700	India	JX270797.1	-
24	AcSBV-India-K3A	*Apis cerana*	8756	India	JX270798.1	-
25	AcSBV-India-S2	*Apis cerana*	8741	India	JX270799.1	-
26	AcSBV-India-II9	*Apis cerana*	8740	India	JX270800.1	-
27	AmSBV-Kor1	*Apis mellifera*	8837	Korea	KP296800.1	[36]
28	AmSBV-Kor2	*Apis mellifera*	8834	Korea	KP296801.1	[36]
29	AcSBV-Kor3	*Apis cerana*	8787	Korea	KP296802.1	[36]
30	AcSBV-Kor4	*Apis cerana*	8786	Korea	KP296803.1	[36]

Table 1. *Cont.*

No.	Name	Host	Total Size (bp)	Location	Accession No.	Reference
31	AcSBV-Viet1	*Apis cerana*	8787	Viet Nam	KM884990.1	[30]
32	AcSBV-Viet2	*Apis cerana*	8786	Viet Nam	KM884991.1	[30]
33	AcSBV-Viet3	*Apis cerana*	8787	Viet Nam	KM884992.1	[30]
34	AmSBV-Viet4	*Apis mellifera*	8787	Viet Nam	KM884993.1	[30]
35	AcSBV-Viet5	*Apis cerana*	8784	Viet Nam	KM884994.1	[30]
36	AmSBV-Viet6	*Apis mellifera*	8836	Viet Nam	KM884995.1	[30]
37	VN3	*Apis mellifera*	8820	Australia	KY465673.1	[42]
38	VN2	*Apis mellifera*	8832	Australia	KY465674.1	[42]
39	VN1	*Apis mellifera*	8835	Australia	KY465675.1	[42]
40	QLD	*Apis mellifera*	8835	Australia	KY465678.1	[42]
41	SA	*Apis mellifera*	8823	Australia	KY465677.1	[42]
42	WA2	*Apis mellifera*	8832	Australia	KY465671.1	[42]
43	WA1	*Apis mellifera*	8832	Australia	KY465672.1	[42]
44	NT	*Apis mellifera*	8830	Australia	KY465679.1	[42]
45	TAS	*Apis mellifera*	8835	Australia	KY465676.1	[42]
46	AcSBV-Viet-BP	*Apis cerana*	8831	Viet Nam	KX668139.1	-
47	AcSBV-Viet-NA	*Apis cerana*	8791	Viet Nam	KX668140.1	-
48	AcSBV-Viet-BG	*Apis cerana*	8784	Viet Nam	KX668141.1	-
49	CSBV-SXnor1	*Apis cerana*	8705	China	KJ000692.1	-
50	CSBV-FZ	*Apis cerana*	8800	China	KM495267.1	[43]
51	CSBV-GZ	*Apis cerana* (bee larvae)	8740	China	AF469603.1	[5]
52	SBV-Brno	*Apis mellifera*	8832	Czech Republic	KY273489.1	-
53	SBV-Hynor	*Apis cerana* (bee larvae)	8779	Viet Nam	KJ959614.1	-
54	SBV-Sydney	*Apis mellifera*	8833	Australia	MF623170.1	-
55	SBV_MS	*Apis mellifera*	8828	Sweden	MH267698.1	-
56	SBV_MR	*Apis mellifera*	8830	Sweden	MH267697.1	-

-: unpublished.

According to our previous data, cross-infection might occur between AcSBV and AmSBV in Taiwan [33,35]. However, the available information on the whole genome sequences of AcSBV and AmSBV in Taiwan is insufficient. Therefore, this study attempted to determine and analyze two complete genome sequences of AcSBV and AmSBV in Taiwan. This is the first complete genome sequences of AcSBV and AmSBV from Taiwan. Phylogenetic analysis based on conserved viral proteins and similarity comparisons of the genomic sequences with those of 54 other SBV strains worldwide were also performed, these results may contribute to better understanding the variation of other SBV strains.

2. Results

2.1. Genomic Sequences and Analysis of SBV Strains in Taiwan

The whole genomes of AmSBV and AcSBV from *A. mellifera* and *A. cerana* in Taiwan were sequenced. The complete genome sequences of the two SBV strains were deposited in GenBank under the accession numbers MN082651 for AmSBV-TW and MN082652 for AcSBV-TW. The genomes of AmSBV-TW and AcSBV-TW were annotated by using NCBI ORF finder, and the numbers of RNAs encoded by AmSBV-TW and AcSBV-TW were 8885 and 8776, respectively. The 5′ and 3′ untranslated regions (UTRs) of AmSBV-TW were 212 and 115 nt, respectively; for AcSBV-TW, the 5′ and 3′ UTRs were 174 and 272 nt, respectively. Only one open reading frame (ORF) was predicted in the genomic RNA sequence of AmSBV-TW, which extended from nt 213 to 8792, encoding a putative polyprotein of 2859 amino acids. The genomic RNA of AcSBV-TW also encoded one ORF, from nt 175 to nt 8700, encoding a putative polyprotein of 2841 amino acids (Figure 1). Two structural domains were identified as rhv-like domains in the 5′ region of AmSBV-TW and AcSBV-TW, and three nonstructural domains, including *helicase*, *protease*, and *RNA-dependent RNA polymerase* (*RdRp*), were located at the 3′ regions of both AmSBV-TW and

AcSBV-TW. The analysis of the protein domain arrangement and genomic structures of AmSBV-TW and AcSBV-TW revealed characteristics of family *Iflaviridae* (Figure 1).

(A)

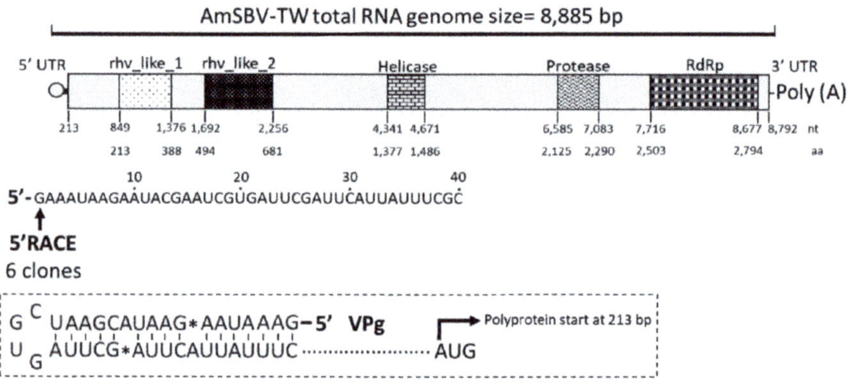

(B)

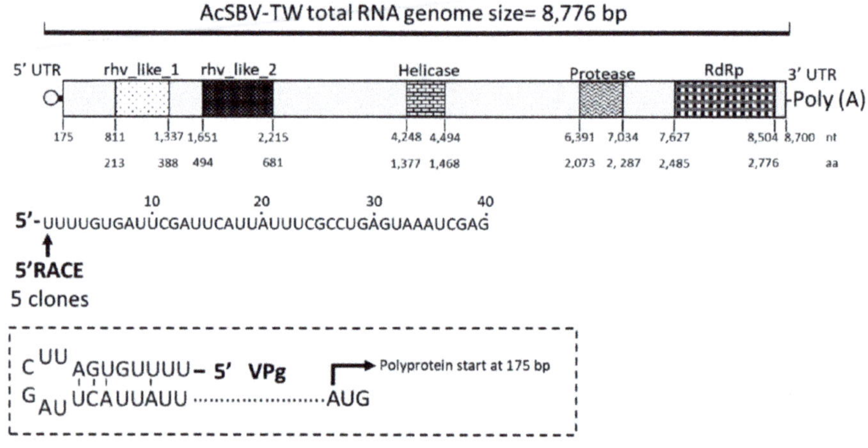

Figure 1. The genomic maps of (**A**) AmSBV-TW (accession number: MN082651) and (**B**) AcSBV-TW (accession number: MN082652). The full-length sequences were obtained using a combination of RT-PCR amplification and rapid amplification of 5′ and 3′ cDNA ends (5′ RACE and 3′ RACE). The nucleotide (nt) and amino acid (aa) positions of each domain was indicated below the schematic of AmSBV-TW and AcSBV-TW, respectively. The 5′ terminal sequences of AmSBV-TW and AcSBV-TW were determined by 5′ RACE, and the prediction of the 5′ secondary structure of AmSBV-TW and AcSBV-TW was performed on the RNAfold WebServer (http://rna.tbi.univie.ac.at/cgi-bin/RNAWebSuite/RNAfold.cgi) and presented in the dotted box. VPg = viral protein genomic linked. * Mismatch nucleotide base.

2.2. Comparisons of SBV Strains

The sequences of AmSBV-TW and AcSBV-TW were first compared to each other (Table 2). The results of nt sequence comparisons showed that the full-length genomic RNA and ORF regions were highly conserved between AmSBV-TW and AcSBV-TW, sharing 90% identity, while the 5′ and 3′ UTRs showed high variation between AmSBV-TW and AcSBV-TW, presenting 68% and 73% identity, respectively (Table 2). In the amino acid sequence comparisons, the helicase protein domain exhibited the highest identity (99%) and was the most conserved protein domain between AmSBV-TW and AcSBV-TW, followed by rhv_like_2 (97%), RdRp (96%), polyprotein, and rhv_like_1 (95%), while the nonstructural protein protease showed low identity (75%) between AmSBV-TW and AcSBV-TW (Table 2).

Table 2. Comparison of genomic sequences and protein regions of AcSBV and AmSBV in Taiwan.

Genomic Region/Protein Region	Region Name	AmSBV-TW	AcSBV-TW	Identity (%)
Genomic region (nt)	Full length	8885	8776	90%
	5′ UTR	211	174	68%
	ORF region	8580	8526	90%
	3′ UTR	96	75	73%
Protein region (aa)	Polyprotein	2859	2841	95%
	rhv_like_1	176	176	95%
	rhv_like_2	188	188	97%
	Helicase	110	110	99%
	Protease	166	215	75%
	RdRp	292	292	96%

The genomic regions of AmSBV-TW and AcSBV-TW were further compared to those of SBV strains from other countries (Table 3). The nucleotide sequences of the whole AmSBV-TW and AcSBV-TW genomes shared identities of 87% (AcSBV-India-II10) to 92% (AcSBV-Viet-SBM2) and 88% (South Australia_1, 2, 3, SBV_MR, MD1, 2 and AcSBV-India-II10) to 96% (AcSBV-Viet1, 2, AmCSBV-SDLY and CSBV-FZ), respectively, with other SBVs (Table 1; Table 3). The identities of the 5′ and 3′ UTRs showed high variation among SBVs; for AmSBV-TW, the 5′ and 3′ UTRs shared 30% (CSBV-SXnor1) to 78% (Korean strain) and 10% (AcSBV-India-K5B) to 94% (MD1 strain) identities, respectively, with those of other SBVs, while AcSBV-TW showed 42% (CSBV-SXnor1) to 93% (AcSBV-Viet3) identity for the 5′ UTR and 11% (AcSBV-India-K5B) to 85% (NT strain) identity for the 3′UTR (Table 1; Table 3).

Table 3. Comparison of the nucleotide sequence homology (%) of different genomic regions of AcSBV-TW, AmSBV-TW, and 54 other SBV strains.

SBV Strains	Full Length (nt)		5′ UTR		ORF Region		3′ UTR	
	AcSBV-TW	AmSBV-TW	AcSBV-TW	AmSBV-TW	AcSBV-TW	AmSBV-TW	AcSBV-TW	AmSBV-TW
AcSBV-TW	-	90	-	68	-	90	-	70
AmSBV-TW	90	-	68	-	90	-	70	-
VN3	89	90	85	72	89	90	88	77
SBV-Sydney	89	90	84	71	88	90	84	71
VN1	89	90	85	72	89	90	88	77
AmSBV-Kor21	89	90	86	73	89	90	71	90
VN2	89	90	86	73	89	91	89	78
South Australia_1	88	90	87	71	88	90	76	63
South Australia_3	88	90	79	76	88	90	85	71
South Australia_2	88	89	87	70	89	90	85	71
QLD	89	90	86	72	89	90	89	78
AmSBV-Kor1	89	90	85	72	89	90	87	79
SA	89	90	85	72	89	90	89	78
WA2	89	90	87	73	89	90	89	78
WA1	89	90	86	72	89	91	89	78
NT	89	91	85	72	89	91	85	72
TAS	89	90	87	73	89	90	88	77
SBV_MS	89	90	83	75	89	90	77	64
SBV_MR	88	90	84	75	89	90	80	66
MD1	88	90	82	77	89	90	75	94
MD2	88	90	82	78	89	90	75	93
SBV-Brno	89	90	86	72	89	91	87	77

Table 3. *Cont.*

SBV Strains	Full Length (nt)		5′ UTR		ORF Region		3′ UTR	
	AcSBV-TW	AmSBV-TW	AcSBV-TW	AmSBV-TW	AcSBV-TW	AmSBV-TW	AcSBV-TW	AmSBV-TW
SBV-UK	89	90	84	70	89	90	87	77
AcSBV-Viet1	96	90	92	71	96	90	93	73
AcSBV-Viet2	96	90	89	69	96	90	93	73
AcSBV-Viet3	95	90	93	72	96	90	92	73
AmSBV-Viet4	95	90	89	71	95	90	92	73
SBV-Hynor	95	90	89	71	95	90	85	66
AcSBV-Viet5	95	90	88	69	96	90	85	70
AmCSBV-SDLY	96	90	88	77	96	90	94	70
CSBV-JLCBS-2014	95	90	84	76	95	91	88	76
AmSBV-Kor2	93	91	81	72	93	91	91	75
Korean	93	90	90	78	94	91	88	76
AcSBV-Kor4	94	90	84	70	94	91	90	75
AcSBV-Kor3	93	90	84	71	93	91	90	75
AmSBV-Kor19	93	90	86	73	93	91	88	76
CSBV-LNQY-2009	91	90	61	55	92	91	50	58
CSBV-FZ	96	91	91	72	97	91	92	73
CSBV-GZ	93	90	65	50	94	92	23	20
AcSBV-India-II10	88	87	-	-	91	91	-	-
AcSBV-India-II2	90	89	47	36	91	91	27	22
AcSBV-India-K1A	90	90	66	52	91	91	39	32
AcSBV-India-K5B	90	90	65	51	91	91	11	10
AcSBV-IndTN-1	90	90	68	53	91	91	47	37
AcSBV-India-K3A	90	90	61	47	91	91	50	41
AcSBV-India-S2	89	89	65	51	90	91	23	20
AcSBV-India-II9	90	90	67	52	91	91	50	41
AcSBV-Viet-BP	90	90	81	70	91	91	78	65
AmSBV-Viet6	91	91	85	71	91	91	84	76
AcSBV-Viet-LDst	91	91	83	72	91	91	82	68
AcSBV-Viet-NA	93	90	89	71	94	91	77	70
AcSBV-Viet-BG	92	90	83	70	93	90	66	54
AcSBV-Viet-SBM2	92	92	84	72	92	92	70	88
CSBV-BJ	91	91	82	69	91	91	68	87
CSBV-SXnor1	90	89	42	30	92	91	27	23
CSBV-SXYL-2015	91	90	81	68	91	91	23	20

-: Non comparable.

The amino acid identities among the SBV strains were similar to the variations in the nucleotide identities. In the comparison of polyprotein amino acid sequences, AmSBV-TW was most similar to SBV-UK, with 98% aa identity, and AcSBV-TW was most similar to AcSBV-Viet1 and 2, sharing 98% aa identity. (Table 4). Among the structural proteins (rhv_like_1 and rhv_like_2), AmSBV-TW shared 81% (AmSBV-Viet6) to 98% (NT strain) and 94% (AcSBV-Viet-NA) to 99% (AmSBV-Kor1) identities with those of other SBVs, and AcSBV-TW shared 81% (AmSBV-Viet6) to 98% (CSBV-FZ) and 94% (AcSBV-Viet-BP) to 100% (AmCSBV-SDLY, CSBV-JLCBS-2014 and AcSBV-Viet-BG) identities with those of other SBVs (Table 4). The identities of the nonstructural proteins, including helicase, protease and RdRp, between AmSBV-TW and other SBVs showed greater variation than those of the structural proteins, ranging from 60–100%, 70–98%, and 88–98%, respectively, while the corresponding values were 60–100%, 67–99%, and 89–98% for AcSBV-TW (Table 4).

Table 4. Amino acid sequence homology (%) of AcSBV-TW, AmSBV-TW, and 54 other SBV strains.

SBVs	Polyprotein		rhv_like_1		rhv_like_2		Helicase		Protease		RdRp	
	AcSBV-TW	AmSBV-TW	AcSBV-TW	AmSBV-TW	AcSBV-TW	AmSBV-TW	AcSBV-TW	AmSBV-TW	AcSBV-TW	AmSBV-TW	AcSBV-TW	AmSBV-TW
AcSBV-TW	-	95	-	95	-	97	-	99	-	75	-	96
AmSBV-TW	95	-	95	-	97	-	99	-	75	-	96	-
VN3	94	97	96	97	96	98	99	100	88	85	96	96
SBV-Sydney	94	97	95	96	96	98	99	100	88	85	97	97
VN1	94	97	96	97	96	98	60	60	88	85	96	96
AmSBV-Kor21	94	97	90	91	96	98	98	99	88	85	97	97
VN2	94	97	96	97	96	98	99	100	88	85	97	97
South Australia_1	94	97	95	96	96	98	99	100	88	85	96	96
South Australia_3	94	97	95	96	96	98	99	100	88	85	97	97
South Australia_2	94	97	95	96	96	98	99	100	88	85	97	97
QLD	94	97	96	97	96	98	99	100	87	83	97	97
AmSBV-Kor1	94	97	96	97	96	99	99	100	88	84	96	96
SA	94	97	96	97	96	98	99	100	89	83	97	97
WA2	94	97	96	97	96	98	99	100	88	85	96	96
WA1	94	97	96	97	96	98	99	100	88	84	97	96
NT	95	97	96	98	96	98	99	100	98	77	97	97
TAS	94	97	95	96	96	98	99	100	88	85	97	96
SBV_MS	94	97	96	97	96	98	99	99	88	85	96	96
SBV_MR	94	97	95	96	96	98	99	99	88	85	96	96
MD1	95	97	96	97	96	98	99	99	84	88	97	97
MD2	95	97	96	97	96	98	99	99	88	85	97	97
SBV-Brno	95	97	96	97	96	98	98	99	98	77	97	97
SBV-UK	95	98	96	97	96	98	99	100	98	77	97	97
AcSBV-Viet1	98	96	92	91	98	96	100	99	76	98	97	96
AcSBV-Viet2	98	96	92	91	98	96	100	99	76	98	97	96
AcSBV-Viet3	97	95	92	91	98	96	99	98	98	75	97	96
AmSBV-Viet4	97	96	94	95	98	96	100	99	99	75	97	96
SBV-Hynor	97	95	92	91	99	96	100	99	96	75	93	92
AcSBV-Viet5	97	95	92	90	99	96	100	99	76	98	93	92
AmCSBV-SDLY	97	95	92	91	100	97	100	99	76	98	97	96
CSBV-JLCBS-2014	97	95	94	95	100	97	100	99	91	83	97	96
AmSBV-Kor2	96	96	95	97	97	96	100	99	88	85	93	93
Korean	96	96	95	97	97	96	100	99	88	85	97	97
AcSBV-Kor4	97	96	94	96	97	96	99	98	88	84	94	92
AcSBV-Kor3	96	96	94	96	97	96	99	98	88	84	94	94
AmSBV-Kor19	96	96	94	95	97	96	100	99	88	85	96	96
CSBV-LNQY-2009	94	95	94	96	96	97	98	99	79	73	97	96
CSBV-FZ	97	95	98	97	99	96	100	99	67	86	98	97
CSBV-GZ	95	96	94	97	97	97	99	98	79	75	98	98
AcSBV-India-II10	93	94	84	84	96	98	99	100	96	75	96	97
AcSBV-India-II2	94	95	93	93	95	96	99	100	96	75	96	96

Table 4. *Cont.*

SBVs	Polyprotein		rhv_like_1		rhv_like_2		Helicase		Protease		RdRp	
	AcSBV-TW	AmSBV-TW	AcSBV-TW	AmSBV-TW	AcSBV-TW	AmSBV-TW	AcSBV-TW	AmSBV-TW	AcSBV-TW	AmSBV-TW	AcSBV-TW	AmSBV-TW
AcSBV-India-K1A	93	95	93	93	96	98	99	100	97	76	96	96
AcSBV-India-K5B	94	95	93	91	97	97	99	100	97	75	96	96
AcSBV-IndTN-1	94	95	92	93	96	97	99	100	97	76	95	95
AcSBV-India-K3A	94	96	90	90	96	96	99	100	98	76	93	94
AcSBV-India-S2	93	95	90	90	97	97	97	98	91	70	89	88
AcSBV-India-II9	94	95	92	93	96	97	99	100	97	76	95	96
AcSBV-Viet-BP	93	95	82	82	94	95	96	97	95	74	94	95
AmSBV-Viet6	94	96	81	81	95	95	99	100	97	75	96	97
AcSBV-Viet-LDst	94	96	84	84	95	95	99	100	97	75	96	97
AcSBV-Viet-NA	96	96	92	91	95	94	99	100	96	75	93	93
AcSBV-Viet-BG	95	95	85	85	100	97	99	100	96	75	94	95
AcSBV-Viet-SBM2	95	97	95	96	97	97	99	100	73	97	96	97
CSBV-BJ	94	95	81	82	96	96	98	99	78	75	96	96
CSBV-SXnor1	94	95	95	96	96	97	98	99	78	75	97	96
CSBV-SXYL-2015	94	95	94	95	97	97	96	97	87	85	96	96

-: Non comparable.

Comparisons of nucleotide and amino acid sequences revealed the deletion of 51 base pairs (17 amino acids) in AcSBV-TW (from amino acid positions 712–730 (VP1 region) in the ORF region) compared to AmSBV-TW, and the same deletion (AC-genotype) was found in most of the other SBVs from *A. cerana*, including AcSBV-Viet1, 2, 3, 5, AcSBV-Hynor, AmCSBV-SDLY, CSBV-JLCBS-2014, AcSBV-Korean, AcSBV-Kor3, 4, AcSBV-Viet-NA, AcSBV-Viet-BG, except AmSBV-Kor19 and AmSBV-Viet4 (Figure 2). Another 10–13-amino-acid deletion was found in six SBVs from *A. cerana* in India and two SBVs from *A. cerana* in China, including AcSBV-India-II2, -II9, -II10, -K1A, -K5B, -TN-1, CSBV-LNQY-2009, and CSBV-FZ (Figure 2). However, a less than 10-amino-acid deletion was found in the SBVs from *A. mellifera* in Australia, including AmSBV-VN3 and SA (Figure 2). Similar to other SBVs from *A. mellifera*, AmSBV-TW showed no deletion in the 712–730 amino acid region of the ORF (AM-genotype), and same to the AcSBVs from China (CSBV-GZ, -BJ, -SXnor1, and SXYL-2015), India (AcSBV-India-K3A and -S2), and Vietnam (AcSBV-Viet-BP, -LDst, and -SBM2) also lacked the 17-amino-acid deletion (Figure 2).

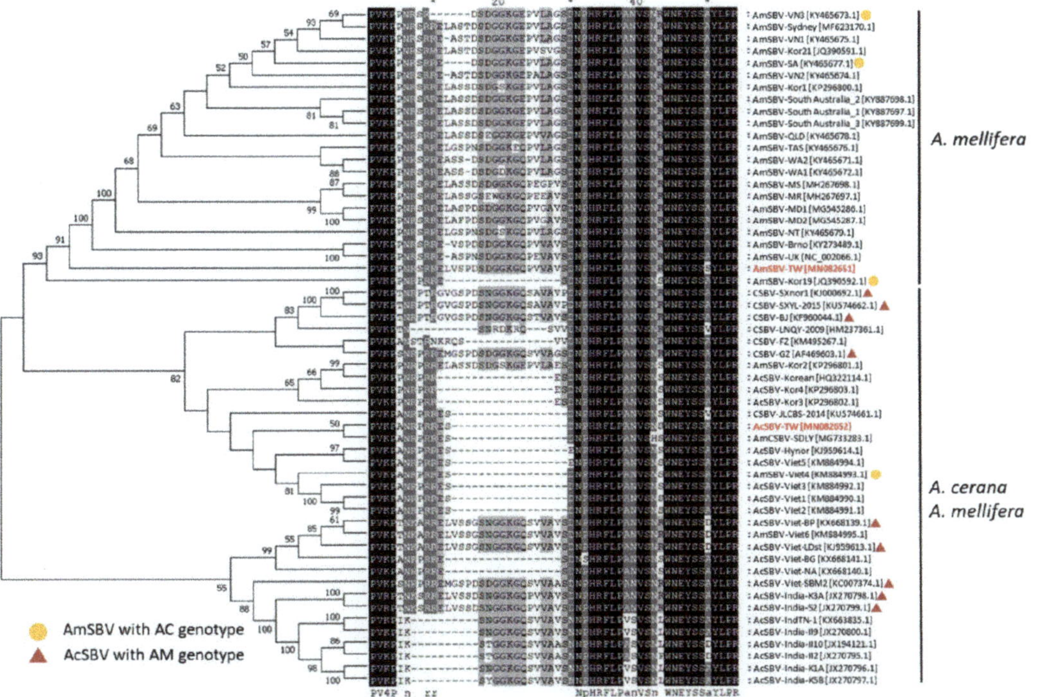

Figure 2. Phylogenetic tree constructed based on the polyprotein amino acid sequences of AmSBV-TW and AcSBV-TW and 54 other SBV strains from the NCBI database. The phylogenetic tree was constructed using the neighbor-joining (NJ) method and 1000 bootstrap replications. The pairwise alignment indicated the deletion patterns in the VP1 region of SBV strains. The round shape symbols indicated AmSBVs with AC genotype (deletion in VP1 region), and the red triangle shape symbols indicated AcSBVs with AM genotype (non-deletion in the VP1 region). red font: The Taiwan strains from this study. *: A note for every 10 bases.

2.3. Phylogenetic Analysis

Phylogenetic analysis was performed based on the polyprotein sequences of 56 strains of SBV. The phylogenetic tree clearly diverged into two main branches according to the host (Figure 2). The first branch was composed of the SBV strains from *A. mellifera*; within this branch, AmSBV-TW was closely related to AmSBV-UK and AmSBV-Kor19. The second branch was composed of SBV strains from either *A. cerana* or *A. mellifera*; moreover, a branch contained two groups, one composed of AcSBV from India, while the other consisted of AcSBV and AmSBV from Asian areas, which included AcSBV-TW from Taiwan, CSBV strains from China, AmSBV/AcSBV from Korea, and AmSBV/AcSBV from Vietnam. Especially according to the phylogenetic tree, AcSBV-TW is closely related to AmCSBV-SDLY and CSBV-FZ (Figure 2).

2.4. Variation of VP1 Region in AcSBV and AmSBV in Taiwan

As aforementioned, the deletion of 51 base pairs (17 amino acids) in the VP1 region were found in most of AcSBV and, thereby, named as AC-genotype and vice versa (AM-genotype without any deletion in the VP1 region). To better understand whether the AcSBV AM-genotype and AmSBV AC-genotype exist in the populations of *A. cerana* and *A. mellifera*, the partial VP1 sequence of three AmSBV and four AcSBV from Taiwan were further compared to those of AmSBV-TWand AcSBV-TW (Figure 3A). The results showed that the 17-amino-acid deletion (AC genotype) was only detected in one AmSBV sample in Taichung; besides, one AcSBV with AM genotype was also detected in the sample from Hsinchu (Figure 3A) The phylogenetic analysis was also performed based on the partial VP1 region of 63 strains of SBV. It revealed similar result to those of polyprotein phylogeny. Moreover, the AcSBV-AC genotype and AmSBV-AC genotype in Taiwan were grouped in the same clade, which was closed to CSBV-FZ and CSBV-JL, and the AmSBV-AM genotype and AcSBV-AM genotype in Taiwan were grouped in the same clade, which was closed to AmSBV-UK (Figure 3B). These results supported that the cross-infection between AcSBV and AmSBV in *A. cerana* and *A. mellifera*.

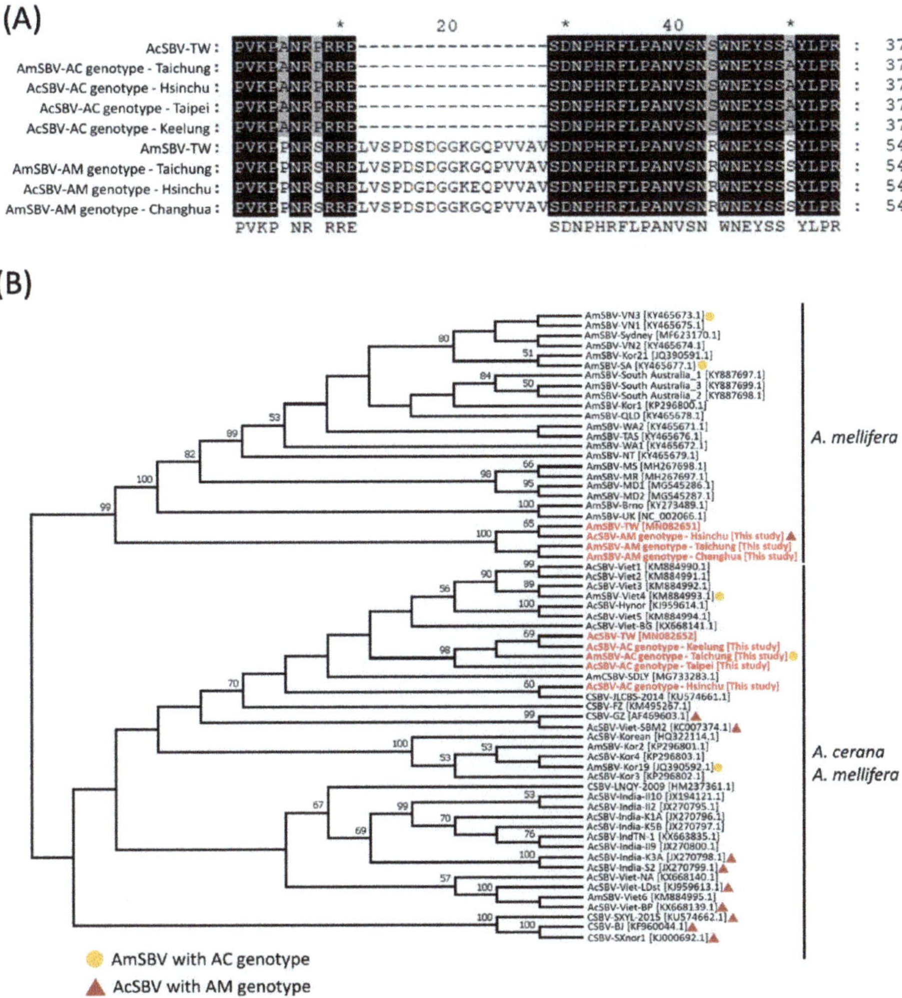

Figure 3. Comparison and phylogenetic analysis of partial VP1 region of AcSBV and AmSBV from Taiwan. (**A**) The pairwise alignment indicated the 17-anmion deletion presented not in AcSBV but also AmSBV vice versa. (**B**) Phylogenetic tree construct based on the partial VP-1 amino acid sequences of Taiwan AmSBV and AcSBV, and other 54 SBV strains from NCBI. The phylogenetic tree was constructed using the neighbor-joining (NJ) method and 1000 bootstrap replications. red font: The Taiwan sequences from this study. *: A note for every 10 bases.

3. Discussion

AcSBV has recently been recorded in Taiwan and caused serious losses of *A. cerana* from 2015 to 2019 [33–35]. In Taiwan, most *A. cerana* populations are reared in Northern Taiwan, and according to our observations, some of these apiaries are crossbreeding with *A. mellifera* populations. The detection of AcSBV prevalence in *A. mellifera* populations from the sampling sites where *A. cerana* and *A. mellifera* were crossbreeding confirmed that AcSBV prevalence rates gradually developed a similar trend in the *A. cerana* and *A. mellifera* crossbreeding apiaries, and the existence of AcSBV cross-infection between *A. cerana* and *A. mellifera* was also confirmed by phylogenetic analysis based on partial VP1 sequences [33]. Similar to our case, some SBV strains from *A. mellifera* included in this study were found to be distinct from other AmSBV strains in terms of genomic features and were clustered with AcSBVs based on whole genome comparisons and phylogenetic analysis [24,30,36,40]. Strains from the same or closer geography distance showed higher similarity, and the phylogenetic analysis also indicated the same result [24,30,36,40,44]. Indeed, it was shown that the cross-infection of SBV strains occurs between two honeybee species in other countries, including China, Vietnam, and Korea, leading to the high genetic divergence among SBV strains [24,30,36,40].

As mentioned above, the comparison of different genome sequences could provide precise and reliable information for detecting variations within closely related species. In this study, complete SBV genome sequences from *A. cerana* and *A. mellifera* in Taiwan were determined and were designated AcSBV-TW and AmSBV-TW, respectively. Our comparisons revealed greater divergence in 5′ and 3′ UTRs than in ORF region not only between AmSBV-TW and AcSBV-TW but also compared with those of SBVs from other countries (Table 3). The structures of 5′ UTR play many functions in RNA viruses, including viral replication, translation, virus–host protein interactions, and virulence [45,46]. It has been also reported that the 5′ UTR of RNA viruses in *Iflaviridae* functions as an internal ribosome entry site (IRES) [47]. Similar result was also described from [30], that 5′ UTRs of VN-SBVs (including AcSBV and AmSBV in Vietnam) showed greater divergence from SBV strains from other countries [30]. The structure of 5′UTR of SBV might have a critical function for virus replication, therefore the sequence divergence may reveal different viral activities among different viruses.

It should be noted that deletions in the 712–730 amino acid (VP1) region of the ORF were found in most AcSBVs [24,30,40]. Since it has been mentioned that VP1 has the highest sequence variation among SBVs [24,30,40,48]. Based on our comparisons, there are three types of deletion patterns: 17-amino-acid deletions, 10–13-amino-acid deletions, and deletions of less than 10 amino acids (Figure 2). Most of the examined AcSBVs, including AcSBV-TW, exhibit a 17- or 10–13-amino-acid deletion in VP1 region, while there were nine AcSBVs from China, India, and Vietnam exhibiting no deletions, and a deletion of less than 10 amino acids was found in SBVs from *A. mellifera* in Australia, including AmSBV-VN3 and SA. AcSBVs from India all have 10-amino-acid deletion, which were clustered in same branch, suggesting that the occurrence of the 17-amino-acid difference in the VP1 region tends to be host-preference. Interestingly, some AmSBV from Asia countries, where have *A. cerana* population, including AmSBV-Viet4 and AmSBV-Kor19, also harbor the same 17-amino-acid deletion in their VP1 region, indicating the cross-infection of SBV at different geographic origins [49].

Further investigation of the VP1 variations of AcSBV and AmSBV in Taiwan indicated that most AcSBVs have 17-amino-acid deletion in their VP1 region compared to AmSBV, while the AmSBV-AC genotype and AcSBV-AM genotype were also detectable in AmSBV and AcSBV, respectively. It has been reported that high variability exists among SBV genomes, especially between AC-genotype SBV and AM-genotype SBV, and this genetic diversity is supported by the geographic distances or viral cross-infections between different honeybee species [30,40]. These characteristics might also provide clues regarding SBV adaption in different hosts [24,30]. In conclusion, the genomic differences in AmSBV-TW and AcSBV-TW compared with other SBVs could be further applied to identify genetic markers for host-specific and geographic distance evaluations.

The phylogenetic analysis based on the polyproteins and partial VP1 region of AmSBV-TW and AcSBV-TW and other SBV strains revealed that the SBV strains diverged into two distinct branches, which could represent host affiliation and geographic origin. According to comparisons with the current 54 strains of SBV available in NCBI, AmSBV-TW and AcSBV-TW were grouped onto different branches. AcSBV-TW is closely related to AmCSBV-SDLY and CSBV-FZ and was clustered into the AcSBV group with the AC genotype; therefore, it was assumed that the AcSBV in Taiwan may have originated from China and currently be experiencing host adaption and evolution. In contrast, AmSBV-TW was grouped into the AM-genotype SBVs, which originated from *A. mellifera*; however, AmSBV-

TW was separated from other AmSBVs in this group, suggesting that geographic distance might be involved in the process of genomic divergence.

The comparison and phylogenetic analysis of partial VP1 region in another seven SBVs in Taiwan showed that most of AcSBV and AmSBV were grouped into AcSBV-TW and AmSBV-TW, respectively, except one AmSBV-AC genotype (grouped with the AcSBV-TW) and one AcSBV-AM genotype (grouped with the AmSBV-TW). These results suggested that AcSBVs in Taiwan presented the closely geographical relationship to those of China, while AmSBVs in Taiwan revealed the geographic distance-based evolution. Additionally, the AcSBV-AM genotype and AmSBV-AC genotype clearly showed the viral cross-infection between these two species.

4. Materials and Methods

4.1. Sample Collection

For vial genomic sequencing, *A. cerana* and *A. mellifera* were collected from two apiaries located in Taipei City and Yilan City, respectively, in 2018 (Figure 4; Supplementary Table S1). Besides, 3 samples of *A. mellifera* and 4 samples of *A. cerana* were selected for the investigation of variations in VP1 region (Figure 4; Supplementary Table S1). The midguts of 10 randomly selected adult bees were collected as a single sample in each apiary. The collected samples were preserved in 0.5 mL of RNA Keeper™ Tissue Sample Storage Reagent (Protech, Taipei, Taiwan) in a 1.5 mL microtube and stored at −20 °C for the following experiment of RNA extraction.

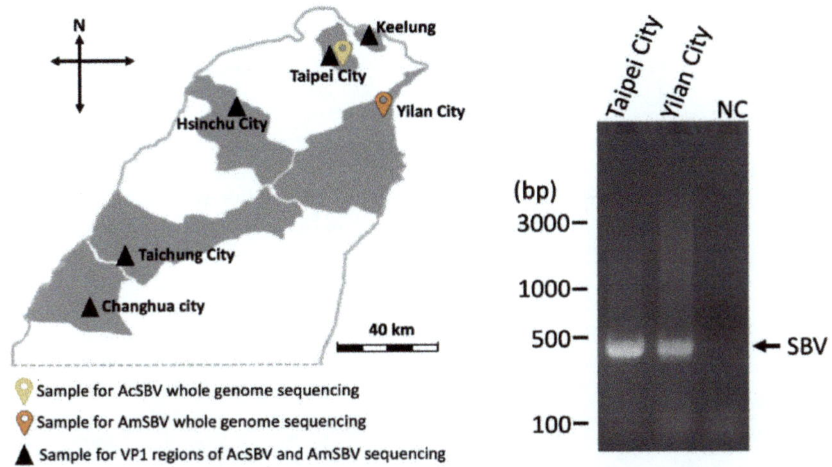

Figure 4. Locations of sample collection and the electrophoretic screening of SBV infection-positive samples by RT-PCR with primer set of VP1-F/VP1-R (Supplementary Table S2). The two sample sites for genomic sequencing were located at northern Taiwan; the AcSBV and AmSBV samples were collected in Taipei City and Yilan City, respectively. The black triangle represents the sampling site for detection of partial VP1 region (335 bp) variations in AcSBV and AmSBV in Taiwan. bp = base pair; NC = negative control. The black arrow indicated the signals of SBV positive.

4.2. RNA Extraction and RT-PCR Screening

Each sample was homogenized with a sterile plastic pestle. Total RNA was extracted from the midgut tissues using TRIzol reagent (Invitrogen, Waltham, MA, USA) following the manufacturer's instructions. The quantity and purity of the RNA were measured using a ScanDrop² Nanovolume spectrophotometer (Analytik Jena, Jena, Germany). For copy DNA (cDNA) synthesis, total RNA (1 µg) samples were treated with DNase I (Roche Molecular Biochemicals, Basel, Switzerland) and then primed with random hexamer primers and reverse-transcribed with Super Script III (Invitrogen, Waltham, MA, USA) at 42 °C for 3 h, after which the reaction was stopped at 70 °C.

All of the samples were first screened with the VP1-F/VP1-R specific primer set (Supplementary Table S2) via PCR with cycling at 95 °C initial denaturation for 45 s and then followed by 35 cycles of 95 °C denaturation for 45 s, 50 °C primer annealing for 45 s, and 72 °C extension for 1 min, followed by a 10 min final extension at 72 °C and storage at 20 °C. The RT-PCR products were analyzed by electrophoresis on a 2% agarose gel in 1× TAE buffer to check the SBV infection-positive samples for the following experiments. For the investigation of variations in VP1 region, the infection-positive samples were amplified by using VP1-F/SBV_R4 primer set (Supplementary Table S2), and the PCR products were subjected to commercial DNA sequencing.

4.3. Whole Genome Sequencing and Assembly of AcSBV and AmSBV

The AcSBV infection-positive samples from *A. cerana* in Taipei City and the AmSBV infection-positive samples from *A. mellifera* in Yilan City were subjected to whole-genome sequencing by RT-PCR with 15 primer sets (Supplementary Table S2). PCR amplification was performed as described above. The RT-PCR products were analyzed by electrophoresis on a 2% agarose gel in 1× TAE buffer [35]. The PCR products with positive signals were purified (Geneaid, New Taipei City, Taiwan) and subjected to commercial DNA sequencing. The obtained sequences were subjected to the genome assembly using SeqMan (DNASTAR, Madison, WI, USA).

4.4. Viral Genomic 5′ and 3′ End Sequencing

The 5′ and 3′ untranslated regions of the AcSBV and AmSBV genomes were obtained by the rapid amplification of cDNA ends (RACE) method, which was slightly modified from [50]. For the 3′ end of the viral genome, 1 μL of an anchor-dTv primer at 50 μM was used to prime 1 μg of total RNA in a 20 μL reaction containing 10 mM dNTPs at 70 °C for 5 min, after which the reaction mixture was placed on ice for 1 min. RNA was reverse transcribed by using Super-Script III (Invitrogen, Waltham, MA, USA) at 42 °C for 1 h, and the reaction was stopped by heating at 70 °C for 15 min. The viral 3′ end sequences were amplified with genome-specific forward primers (GSP-F) and an anchor primer (Supplementary Table S2) using PCR Master Mix (Thermal, Riverside County, CA, USA).

The sequence of the viral 5′ end was decoded as described by [51], with slight modifications [51]. A total of 5 μg RNA was used for 5′ RACE, and the RN was primed with 0.5 μL of the GSP-RT primer (100 ng/μL) in a 20 μL reaction at 80 °C for 3 min, after which the mixture was rapidly transferred to ice. The iScript™ cDNA Synthesis Kit (BIO-RAD, Hercules, CA, USA) was used for reverse transcription at 42 °C for 1 h, and the reaction was inactivated at 95 °C for 5 min. Then, the RNA templates of the cDNA samples were digested with 1.5 U of RNase H (Thermo Fisher Scientific, Waltham, MA, USA) at 37 °C for 20 min. The sample was subsequently cleaned using a GenepHlow™ Gel/PCR Kit (Geneaid, New Taipei City, Taiwan). The 15 μL eluted cDNA sample was treated with transferase (Tdt) in the following reaction mixture: 5 μL of 10 × terminal deoxynucleotidyl transferase (Tdt) buffer (NEB, Ipswich, MA, USA), 5 μL CoCl$_2$ (2.5 mM), 0.5 μL dATP (10 mM), and 0.5 μL Tdt for 5′ end tailing, performed at 37 °C for 25 min, and the reaction was then stopped by heating at 70 °C for 10 min. The sample was next subjected to two rounds of PCR amplification using PCR Master Mix (Thermal, Riverside County, CA, USA). For the first round of PCR, 1 μL of cDNA template was used for amplification by three primers: GSP-R1, QO, and QT, at 25 pmols each (Supplementary Table S2). The PCR program was as follows: 98 °C initial denaturation for 5 min, 48 °C annealing for 2 min, 72 °C extension for 40 min, followed by 30 cycles of 94 °C denaturation for 10 s, 50 °C primer annealing for 30 s, 72 °C extension for 2 min, and a final extension at 72 °C for 15 min. The first-round PCR product was diluted 20-fold in ddH$_2$O for the second round of amplification. A total of 1 μL of the diluted PCR product and 25 pmols of each of the

GSP-R2 and QI primers (Supplementary Table S2) were mixed for PCR amplification via the following program: 98 °C for 5 min, 30 cycles at 94 °C for 10 s, 50 °C for 30 s, 72 °C for 1 min, and a final extension at 72 °C for 15 min. The amplified PCR products were checked by electrophoresis on a 4% agarose gel in 1× TAE buffer. The amplified DNA fragments were purified using a GenepHlow™ Gel/PCR Kit (Geneaid, New Taipei City, Taiwan) and cloned into the TA vector (RBC Bioscience, New Taipei City, Taiwan); the ligated plasmid DNAs were transformed into *Escherichia coli* DH5α (RBC Bioscience, New Taipei City, Taiwan) following the user manual. The plasmids were extracted from cultured bacterial colonies with a Presto™ Mini Plasmid Kit (Geneaid, New Taipei City, Taiwan) and were sequenced bidirectionally with the M13F and M13R primers (Supplementary Table S2).

4.5. Nucleotide Sequence Analysis and Comparison

The genomes of AcSBV-TW and AmSBV-TW were annotated by using NCBI ORFfinder, and proteins were predicted by using NCBI BLASTp [51]. The nucleotide sequences and the amino acid sequences of these two viruses were further compared to each other or to those of other SBVs from other countries.

For the nucleotide sequences, the whole genome sequence, 5′ UTR, ORF region, and 3′ UTR were compared; for the amino acid sequences, the polyproteins, structural proteins (rhv_like_1 and rhv_like_2), and nonstructural proteins (helicase, protease, and RNA-dependent RNA polymerase) of AcSBV-TW and AmSBV-TW were compared with other SBV sequence data from NCBI databases [52] (Table 1). Besides, the partial VP1 sequence of 3 AmSBV and 4 AcSBV from Taiwan were further compared to those of AmSBV-TW, AcSBV-TW, and other 54 SBVs. Multiple alignments of the sequences were obtained using ClustalX and edited in GeneDoc.

4.6. Phylogenetic Analysis

Phylogenetic analysis was performed based on the polyprotein sequences and partial VP1 region of the SBVs as follows. For the polyprotein phylogenetic analysis, the sequences of 54 SBV strains were obtained from the GenBank database and aligned, compared with AcSBV-TW and AmSBV-TW by using ClustalX and GeneDoc. For the partial VP1 phylogenetic analysis, 3 AmSBV and 4 AcSBV from Taiwan were included. Molecular Evolutionary Genetics Analysis Version 7.0 (MEGA7) was used for phylogenetic analyses of these two conserved domains with the neighbor-joining method. The nodes were determined via bootstrap analysis with 1000 replicates [53].

5. Conclusions

The whole genomes of SBV strains from *A. mellifera* and *A. cerana* were determined, and the origin of AcSBV-TW was indicated to be close to China, while AmSBV-TW presented novel genomic features. The cross-infection of *A. mellifera* with AcSBV was demonstrated in the apiaries, where *A. mellifera* and *A. cerana* were crossbreeding in Northern Taiwan in our previous report [33], suggesting that the variations identified in the genomes of AcSBV-TW and AmSBV-TW. According to the whole genome data, the sequences of 5′ and 3′ UTR revealed divergence compared to the polyprotein coding sequences either between AcSBV-TW and AmSBV-TW or among those of SBV from other countries, assuming there is less evolutionary pressure on the untranslated regions of the viral genomes. A comparison of partial VP1 region in Taiwan SBVs and phylogenetic analysis showed a deletion feature in VP1 region. The deletion feature in VP1 region, also mainly observed in most of AcSBV in other regions, suggested the host-preference phenomenon. However, it should be noted that some AmSBV also have a deletion in the VP1 region, it might be a consequence of cross-infection and viral–host adaptions. Therefore, cross-infection might be a high-risk factor for SBV resurgence [18,30,37,40]. For long-term surveillance, the features of VP1 in the genome sequences of SBV strains might provide molecular markers for the detection of

SBV adaption in different honeybee hosts. More detailed investigations of this issue will be needed in the future.

Supplementary Materials: The following are available online at https://www.mdpi.com/2076-0817/10/1/14/s1, Supplementary Table S1: Information of sampling sites; Supplementary Table S2: Primers used in this study.

Author Contributions: Conceptualization, Y.-W.C. and Y.-S.N.; methodology, J.-C.C. and Y.-S.N.; validation, J.-C.C. and Y.-S.N.; formal analysis, J.-C.C., Z.-T.C. and C.-Y.K.; supervision, Y.-W.C. and Y.-S.N.; project administration, Y.-W.C. and Y.-S.N.; funding acquisition, Y.-W.C. and Y.-S.N.; writing—review and editing, J.-C.C., Z.-T.C., C.-Y.K., Y.-W.C. and Y.-S.N.; All authors have read and agreed to the published version of the manuscript.

Funding: This research was funded by Ministry of Science and Technology, Taiwan, Grant MOST 107-2313-B-197-004-MY3 and 109-2313-B-005-048-MY3.

Institutional Review Board Statement: Ethical review and approval were waived for this study, due to there is no any animal experiment involved the ethical issue.

Data Availability Statement: The sequences generated in this study were submitted to NCBI Gen-Bank and also are available from the corresponding author (Yu-Shin Nai) on reasonable request.

Acknowledgments: This research was supported by the Bureau of Animal and Plant Health Inspection and Quarantine, the Council of Agriculture and the Grant MOST 107-2313-B-197-004-MY3 from the Ministry of Science and Technology, Taiwan.

Conflicts of Interest: The authors declare there is no conflict of interest involved in this work.

References

1.	Chen, Y.P.; Pettis, J.S.; Collins, A.; Feldlaufer, M.F. Prevalence and transmission of honeybee viruses. *Appl. Environ. Microbiol.* **2006**, *72*, 606–611. [CrossRef] [PubMed]
2.	Bailey, L.; Gibbs, A.J.; Woods, R.D. Sacbrood Virus of the Larval Honey Bee (*Apis mellifera* Linnaeus). *Virology* **1964**, *23*, 425–429. [CrossRef]
3.	Baker, C.A.; Schroeder, D.C. The use of RNA-dependent RNA polymerase for the taxonomic assignment of Picorna-like viruses (order Picornavirales) infecting *Apis mellifera* L. populations. *Virol. J.* **2008**, *5*, 10. [CrossRef] [PubMed]
4.	Brcak, J.; Kralik, O. On the Structure of the Virus Causing Sacbrood of the Honey Bee. *J. Invertebr. Pathol.* **1965**, *20*, 110–111. [CrossRef]
5.	Zhang, J.; Feng, J.; Liang, Y.; Chen, D.; Zhou, Z.H.; Zhang, Q.; Lu, X. Three-dimensional structure of the Chinese Sacbrood bee virus. *Sci. China C Life Sci.* **2001**, *44*, 443–448. [CrossRef]
6.	Ghosh, R.C.; Ball, B.V.; Willcocks, M.M.; Carter, M.J. The nucleotide sequence of sacbrood virus of the honey bee: An insect picorna-like virus. *J. Gen. Virol.* **1999**, *80 Pt 6*, 1541–1549. [CrossRef]
7.	Bailey, L. The multiplication and spread of sacbrood virus of bees. *Ann. Appl. Biol.* **1969**, *63*, 483–491. [CrossRef]
8.	Chen, Y.P.; Siede, R. Honey bee viruses. *Adv. Virus Res.* **2007**, *70*, 33–80. [CrossRef]
9.	White, G.F. *Sacbrood, a Disease of Bees*; Department of Agriculture, Bureau of Entomology: Washington, DC, USA, 1917; p. 431.
10.	Bailey, L. Recent Research on Honeybee Viruses. *Bee World* **1975**, *56*, 55–64. [CrossRef]
11.	Drescher, N.; Klein, A.M.; Neumann, P.; Yanez, O.; Leonhardt, S.D. Inside Honeybee Hives: Impact of Natural Propolis on the Ectoparasitic Mite Varroa destructor and Viruses. *Insects* **2017**, *8*, 15. [CrossRef]
12.	Dall, D.J. Inapparent infection of honey bee pupae by Kashmir and sacbrood bee viruses in Australia. *Ann. Appl. Biol.* **1985**, *106*, 461–468. [CrossRef]
13.	Tentcheva, D.; Gauthier, L.; Zappulla, N.; Dainat, B.; Cousserans, F.; Colin, M.E.; Bergoin, M. Prevalence and seasonal variations of six bee viruses in *Apis mellifera* L. and *Varroa destructor* mite populations in France. *Appl. Environ. Microbiol.* **2004**, *70*, 7185–7191. [CrossRef] [PubMed]
14.	Antunez, K.; D'Alessandro, B.; Corbella, E.; Ramallo, G.; Zunino, P. Honeybee viruses in Uruguay. *J. Invertebr. Pathol.* **2006**, *93*, 67–70. [CrossRef]
15.	Nielsen, S.; Nicolaisen, M.; Kryger, P.I. Incidence of acute bee paralysis virus, black queen cell virus, chronic bee paralysis virus, deformed wing virus, Kashmir bee virus and sacbrood virus in honey bees (*Apis mellifera*) in Denmark. *Apidologie* **2008**, *39*, 310–314. [CrossRef]
16.	Freiberg, M.; De Jong, D.; Message, D.; Cox-Foster, D. First report of sacbrood virus in honey bee (*Apis mellifera*) colonies in Brazil. *Genet. Mol. Res.* **2012**, *11*, 3310–3314. [CrossRef] [PubMed]
17.	Choi, Y.S.; Lee, M.Y.; Hong, I.P.; Kim, N.S.; Kim, H.K.; Lee, K.G.; Lee, M.L. Occurrence of sacbrood virus in Korean apiaries from *Apis cerana* (Hymenoptera: Apidae). *J. Apicult.* **2010**, *25*, 187–191.

18. Gong, H.R.; Chen, X.X.; Chen, Y.P.; Hu, F.L.; Zhang, J.L.; Lin, Z.G.; Yu, J.W.; Zheng, H.Q. Evidence of *Apis cerana* Sacbrood virus Infection in *Apis mellifera*. *Appl. Environ. Microbiol.* **2016**, *82*, 2256–2262. [CrossRef]

19. Kim, H.K.; Choi, Y.S.; Lee, M.L.; Lee, M.Y.; Lee, K.G.; Ahn, N.H. Detection of sacbrood virus (SBV) from the honeybee in Korea. *J. Apic.* **2008**, *23*, 103–109.

20. Rana, B.S.; Garg, I.D.; Khurana, S.P.; Verma, L.; Agrawal, H. Thai sacbrood virus of honeybees (*Apis cerana indica* F) in north-west Himalayas. *Indian J. Virol.* **1986**, *2*, 127–131.

21. Verma, L.R.; Rana, B.S.; Verma, S. Observations on *Apis cerana* colonies surviving from Thai Sacbrood Virus infestation. *Apidologie* **1990**, *21*, 169–174. [CrossRef]

22. Zhang, G.; Han, R. Advances on sacbrood of honeybees. *China J. Invertebr. Pathol.* **2012**, *109*, 160–164.

23. Ai, H.; Yan, X.; Han, R. Occurrence and prevalence of seven bee viruses in *Apis mellifera* and *Apis cerana apiaries* in China. *J. Invertebr. Pathol.* **2012**, *109*, 160–164. [CrossRef] [PubMed]

24. Choe, S.E.; Nguyen, L.T.; Noh, J.H.; Kweon, C.H.; Reddy, K.E.; Koh, H.B.; Chang, K.Y.; Kang, S.W. Analysis of the complete genome sequence of two Korean sacbrood viruses in the *Honey bee*, *Apis mellifera*. *Virology* **2012**, *432*, 155–161. [CrossRef] [PubMed]

25. Forsgren, E.; Wei, S.; Guiling, D.; Zhiguang, L.; Tran, T.V.; Tang, P.T.; Truong, T.A.; Dinh, T.Q.; Fries, I. Preliminary observations on possible pathogen spill-over from *Apis mellifera* to *Apis cerana*. *Apidologie* **2014**, *46*, 265–275. [CrossRef]

26. Wang, M.; Bi, J. Prevalence of Four Common Bee RNA Viruses in Eastern Bee Populations in Yunnan Province, China. *J. Vet. Sci. Technol.* **2015**, *7*, 284. [CrossRef]

27. Wu, Y.Y.; Jia, R.H.; Dai, P.L.; Diao, Q.Y.; Xu, S.F.; Wang, X.; Zhou, T. Multiple Virus Infections and the Characteristics of Chronic Bee Paralysis Virus in Diseased Honey Bees (*Apis mellifera* L.) in China. *J. Apic. Sci.* **2015**, *59*, 95–106. [CrossRef]

28. Thu, H.T.; Lien, N.T.K.; Linh, M.T.; Le, T.; Hoa, N.T.T.; Thai, P.H.; Reddy, K.E.; Yoo, M.S.; Kim, Y.; Cho, Y.S.; et al. Prevalence of bee viruses among *Apis cerana* populations in Vietnam. *J. Apic. Res.* **2016**, *55*, 379–385. [CrossRef]

29. Yañez, O.; Zheng, H.Q.; Su, X.L.; Hu, F.L.; Neumann, P.; Dietemann, V. Potential for virus transfer between the honey bees *Apis mellifera* and *A. cerana*. *J. Apic. Res.* **2016**, *54*, 179–191. [CrossRef]

30. Reddy, K.E.; Thu, H.T.; Yoo, M.S.; Ramya, M.; Reddy, B.A.; Lien, N.T.K.; Trang, N.T.P.; Duong, B.T.T.; Lee, H.J.; Kang, S.W.; et al. Comparative Genomic Analysis for Genetic Variation in Sacbrood Virus of *Apis cerana* and *Apis mellifera* Honeybees from Different Regions of Vietnam. *J. Insect. Sci.* **2017**, *17*. [CrossRef]

31. Anderson, D.L. Viruses of Apis cerana and Apis mellifera. In *The Asiatic Hive Bee: Apiculture, Biology, and Role in Sustainable Development in Tropical and Subtropical Asia*; Kevan, P., Ed.; Enviroquest: Cambridge, ON, Canada, 1995.

32. Blanchard, P.; Guillot, S.; Antunez, K.; Koglberger, H.; Kryger, P.; de Miranda, J.R.; Franco, S.; Chauzat, M.P.; Thiery, R.; Ribiere, M. Development and validation of a real-time two-step RT-qPCR TaqMan((R)) assay for quantitation of Sacbrood virus (SBV) and its application to a field survey of symptomatic honey bee colonies. *J. Virol. Methods* **2014**, *197*, 7–13. [CrossRef]

33. Ko, C.Y.; Chiang, Z.L.; Liao, R.J.; Chang, Z.T.; Chang, J.C.; Kuo, T.Y.; Chen, Y.W.; Nai, Y.S. Dynamics of *Apis cerana* Sacbrood Virus (AcSBV) Prevalence in *Apis cerana* (Hymenoptera: Apidae) in Northern Taiwan and Demonstration of its Infection in *Apis mellifera* (Hymenoptera: Apidae). *J. Econ. Entomol.* **2019**, *112*, 2055–2066. [CrossRef] [PubMed]

34. Huang, W.F.; Mehmood, S.; Huang, S.; Chen, Y.W.; Ko, C.Y.; Su, S. Phylogenetic analysis and survey of *Apis cerana* strain of Sacbrood virus (AcSBV) in Taiwan suggests a recent introduction. *J. Invertebr. Pathol.* **2017**, *146*, 36–40. [CrossRef]

35. Nai, Y.S.; Ko, C.Y.; Hsu, P.S.; Tsai, W.S.; Chen, Y.W.; Hsu, M.H.; Sung, I.H. The seasonal detection of AcSBV (*Apis cerana sacbrood virus*) prevalence in Taiwan. *J. Asia-Pacific Entomol.* **2018**, *21*, 417–422. [CrossRef]

36. Reddy, K.E.; Yoo, M.S.; Kim, Y.H.; Kim, N.H.; Ramya, M.; Jung, H.N.; Thao, L.T.B.; Lee, H.S.; Kang, S.W. Homology differences between complete Sacbrood virus genomes from infected *Apis mellifera* and *Apis cerana* honeybees in Korea. *Virus Genes* **2016**, *52*, 281–289. [CrossRef]

37. Aruna, R.; Srinivasan, M.R.; Balasubramanian, V.; Selvarajan, R. Complete genome sequence of sacbrood virus isolated from Asiatic honey bee *Apis cerana* indica in India. *Virusdisease* **2018**, *29*, 453–460. [CrossRef] [PubMed]

38. Fung, E.; Hill, K.; Hogendoorn, K.; Glatz, R.V.; Napier, K.R.; Bellgard, M.I.; Barrero, R.A. De novo assembly of honey bee RNA viral genomes by tapping into the innate insect antiviral response pathway. *J. Invertebr. Pathol.* **2018**, *152*, 38–47. [CrossRef]

39. Hu, Y.; Fei, D.; Jiang, L.; Wei, D.; Li, F.; Diao, Q.; Ma, M. A comparison of biological characteristics of three strains of Chinese sacbrood virus in *Apis cerana*. *Sci. Rep.* **2016**, *6*, 37424. [CrossRef]

40. Li, M.; Fei, D.; Sun, L.; Ma, M. Genetic and phylogenetic analysis of Chinese sacbrood virus isolates from *Apis mellifera*. *PeerJ* **2019**, *7*, e8003. [CrossRef]

41. Nguyen, N.T.; Le, T.H. Complete Genome Sequence of Sacbrood Virus Strain SBM2, Isolated from the Honeybee *Apis cerana* in Vietnam. *Genome Announc* **2013**, *1*, e00076-12. [CrossRef]

42. Roberts, J.M.K.; Anderson, D.L.; Durr, P.A. Absence of deformed wing virus and Varroa destructor in Australia provides unique perspectives on honeybee viral landscapes and colony losses. *Sci. Rep.* **2017**, *7*, 6925. [CrossRef]

43. Xia, X.; Zhou, B.; Wei, T. Complete genome of Chinese sacbrood virus from *Apis cerana* and analysis of the 3C-like cysteine protease. *Virus Genes* **2015**, *50*, 277–285. [CrossRef] [PubMed]

44. Yildirim, Y.; Cagirgan, A.A.; Usta, A. Phylogenetic analysis of sacbrood virus structural polyprotein and non-structural RNA dependent RNA polymerase gene: Differences in Turkish strains. *J. Invertebr. Pathol.* **2020**, *176*, 107459. [CrossRef] [PubMed]

45. Barton, D.J.; O'Donnell, B.J.; Flanegan, J.B. 5′ cloverleaf in poliovirus RNA is a cis-acting replication element required for negative-strand synthesis. *EMBO J.* **2001**, *20*, 1439–1448. [CrossRef] [PubMed]

46. Kloc, A.; Rai, D.K.; Rieder, E. The Roles of Picornavirus Untranslated Regions in Infection and Innate Immunity. *Front. Microbiol.* **2018**, *9*, 485. [CrossRef]

47. Wu, T.Y.; Wu, C.Y.; Chen, Y.J.; Chen, C.Y.; Wang, C.H. The 5′ untranslated region of Perina nuda virus (PnV) possesses a strong internal translation activity in baculovirus-infected insect cells. *FEBS Lett.* **2007**, *581*, 3120–3126. [CrossRef]

48. Mingxiao, M.; Yanna, Y.; Xiaoli, X.; Lin, Z.; Yongfei, L.; Zhidong, L. Genetic characterization of VP1 gene of seven Sacbrood virus isolated from three provinces in northern China during the years 2008–2012. *Virus Res.* **2013**, *176*, 78–82. [CrossRef]

49. Mingxiao, M.; Ming, L.; Jian, C.; Song, Y.; Shude, W.; Pengfei, L. Molecular and Biological Characterization of Chinese Sacbrood Virus LN Isolate. *Comp. Funct. Genomics* **2011**, *2011*, 409386. [CrossRef]

50. Yang, Y.T.; Nai, Y.S.; Lee, S.J.; Lee, M.R.; Kim, S.; Kim, J.S. A novel picorna-like virus, Riptortus pedestris virus-1 (RiPV-1), found in the bean bug, R. pedestris, after fungal infection. *J. Invertebr. Pathol.* **2016**, *141*, 57–65. [CrossRef]

51. Scotto-Lavino, E.; Du, G.; Frohman, M.A. Amplification of 5′ end cDNA with 'new RACE'. *Nat. Protoc.* **2006**, *1*, 3056–3061. [CrossRef]

52. Pearson, W.R.; Lipman, D.J. Improved tools for biological sequence comparison. *Proc. Natl. Acad. Sci. USA* **1988**, *85*, 2444–2448. [CrossRef]

53. Kumar, S.; Stecher, G.; Tamura, K. MEGA7: Molecular Evolutionary Genetics Analysis Version 7.0 for Bigger Datasets. *Mol. Biol. Evol.* **2016**, *33*, 1870–1874. [CrossRef] [PubMed]

Communication

Histopathological Features of Symptomatic and Asymptomatic Honeybees Naturally Infected by Deformed Wing Virus

Karen Power *, Manuela Martano, Gennaro Altamura, Nadia Piscopo and Paola Maiolino

Department of Veterinary Medicine and Animal Productions, University of Naples "Federico II", Via F. Delpino 1, 80137 Naples, Italy; manuela.martano@unina.it (M.M.); gennaro.altamura@unina.it (G.A.); nadia.piscopo@unina.it (N.P.); paola.maiolino@unina.it (P.M.)
* Correspondence: karen.power@unina.it

Abstract: Deformed wing virus (DWV) is capable of infecting honeybees at every stage of development causing symptomatic and asymptomatic infections. To date, very little is known about the histopathological lesions caused by the virus. Therefore, 40 honeybee samples were randomly collected from a naturally DWV infected hive and subjected to anatomopathological examination to discriminate between symptomatic (29) and asymptomatic (11) honeybees. Subsequently, 15 honeybee samples were frozen at $-80°$ and analyzed by PCR and RTqPCR to determinate the presence/absence of the virus and the relative viral load, while 25 honeybee samples were analyzed by histopathological techniques. Biomolecular results showed a fragment of the expected size (69bp) of DWV in all samples and the viral load was higher in symptomatic honeybees compared to the asymptomatic group. Histopathological results showed degenerative alterations of the hypopharyngeal glands (19/25) and flight muscles (6/25) in symptomatic samples while 4/25 asymptomatic samples showed an inflammatory response in the midgut and the hemocele. Results suggest a possible pathogenic action of DWV in both symptomatic and asymptomatic honeybees, and a role of the immune response in keeping under control the virus in asymptomatic individuals.

Keywords: deformed wing virus; hypopharyngeal glands; flight muscles; honeybee immunity; honeybee pathology

Citation: Power, K.; Martano, M.; Altamura, G.; Piscopo, N.; Maiolino, P. Histopathological Features of Symptomatic and Asymptomatic Honeybees Naturally Infected by Deformed Wing Virus. *Pathogens* **2021**, *10*, 874. https://doi.org/10.3390/pathogens10070874

Academic Editor: Giovanni Cilia

Received: 28 May 2021
Accepted: 8 July 2021
Published: 10 July 2021

Publisher's Note: MDPI stays neutral with regard to jurisdictional claims in published maps and institutional affiliations.

1. Introduction

Among the factors that threaten the health and wellbeing of honeybees, a noteworthy variety of pathogens such as bacteria, viruses, fungi and parasites are mentioned. In recent decades, honeybee viruses have been studied for their potential impact on beekeeping productions, acquiring more and more importance in the research world. Viruses in honeybees were first described in 1913 [1] when an American researcher attributed to a virus the "sac" appearance showed by some diseased larvae, although the causative agent (Sacbrood virus) was not characterized until 1964 [2]. To date, at least 22 viruses that can infect honeybees have been described. Most investigated hives are found to be infected by at least one virus, but often multiple viruses are detected in one hive [3–5].

In addition to causing high economic losses, viruses negatively affect the morphology, physiology, and behavior of honeybees and, although individuals do not always show clinical signs, they are frequently associated with weakening and colony collapse [6]. Depending on the different pathways of infection and on the health status of the colonies, viruses can cause symptomatic infections, i.e., overt or clinical, and asymptomatic infections, i.e., covert or subclinical [7]. Symptomatic infections are characterized by clinical signs and by high levels of viral particle production, to which the insect either succumbs or survives according to the status of the immune system.

These symptomatic infections can be further divided into acute and chronic: the acute involve the active replication of the virus, with a high titer of viral particles in a short time and cause rapid death of the host with evident clinical signs. In extreme cases, when the

99

production of high viral titers occurs during a short time, sudden death can occur without previous clinical signs (hyperacute infections) [8]. Chronic infections, on the other hand, imply a slow but constant production of viral particles during the life of the host, or during the duration of the infected life stage, with subsequent appearance of clinical signs. On the contrary, asymptomatic infections are characterized by persistence of the virus beyond life stage, vertical transmission and the absence of obvious symptoms, although there could still be a hidden cost for the host [8–10]. Asymptomatic infections can be latent and persistent [6]. In the first case the viral genome may be present as an extrachromosomal episome or may be integrated into the host genome with incomplete replication or no replication at all. In the second case, there is a constant but low production of viral particles in the host cells, and either the infected cell survives, or the limited number of dead cells is counterbalanced by the production of new cells. Persistent infections, therefore, represent a balance between host and persistent viral replication, where despite the infection, the host does not die. Moreover, asymptomatic infections can become symptomatic when the host homeostasis is unbalanced by stressors such as other pathologies, food deficiencies, and other environmental factors [9].

Deformed wing virus (DWV) is positive single-stranded RNA virus belonging to the genus Iflavirus, family Iflaviridae of the order Picornavirales [11] and is the most prevalent virus in honeybees, with a minimum average of 55% of apiaries infected across 32 countries [12]. The virus was first isolated in the 1982 in the UK by Bill Baley and Brenda Ball from dead Japanese honeybees showing particular deformity of wings [13]. Soon after, honeybees from the UK, Belize and South Africa died showing DWV symptoms. Ten years after, in the UK the virus was found in Varroa destructor-infested colonies, and it was then found in every location where *V. destructor* was well established [12]. Due to the link with *V. destructor* and following the huge spread of it around the world between 1970 and 1980, DWV altered its epidemiology and has currently a global distribution [14]. Except for Australia, Uganda and the Canadian island of Newfoundland, where the *V. destructor* mite has not been found, the presence of this particular virus has been reported in Africa, Asia, Europe, North America and South America [15]. DWV appears in three master variants DWV-A, DWV-B and DWV-C, plus numerous recombinations, often more virulent than the masters [16]. DWV-A was the first variant to be detected and it is closely associated to colony collapse [17]; however, DWV-B, previously termed Varroa Destructor virus-1, was found to be equally or more virulent than DWV-A when injected in high viral loads [18]. DWV-C was first described in U.K. honeybee samples from 2007 and linked in combination with DWV-A to the death of overwintering colonies [19].

Recent studies have shown that DWV is present in more than 64 species of insects and highly prevalent not only in honeybees, but also in more than 29 arthropod species associated with honeybee hives [12,20,21]. Within insects, DWV was found in bumblebee species *Bombus terrestris* and *Bombus pascuorum*, wasp species *Vespula vulgaris* and *Vespa crabro* and *Lasius* spp. ants [12,22,23], besides *A. mellifera*. DWV is a major pathogen of honeybees and its prevalence, strongly connected with the ectoparasite *V. destructor*, strongly increases honeybee colony mortality [24]. DWV is a low pathogenic virus that is capable of infecting all stages of development of honeybees, from eggs to adults, although it shows a higher replication in pupae [25,26]. It takes its name from the characteristic symptom that manifests itself in newly hatched honeybees with deformed or underdeveloped wings; these honeybees, unable to fly, can die shortly after emerging from the cell. Initially, the deformity of the wings had been attributed to the action of the *V. destructor* mite, as the symptom was more evident in conjunction with a strong infestation by the parasite [27]. Subsequently, DWV was identified as the etiological agent of wing deformity, emphasizing the association between the viral titers and the symptom [25,28]. However, although DWV is one of the few honeybee viruses to have its own characteristic clinical manifestation, it is known also to be present in apparently healthy colonies [10].

Although the pathology and virulence of DWV remain linked to horizontal vectored transmission by *V. destructor*, the presence of DWV has been demonstrated also in the absence of *V. destructor* [29]. Varroa-mediated virus transmission from adult honeybees to developing pupae is responsible for the display of the symptoms [24,30] such as early pupal death, deformed wings, shortened and swollen abdomen and discoloration of the cuticle in adult bees, and learning deficiencies [14]. Symptomatic DWV infection occurs primarily during autumn and in highly mite-infested colonies, where it constitutes predictive marker for winter colony losses [30,31]. According to the epidemiological model proposed by Chen et al. [9] two distinct moments of viral presence and infection can be recognized: in healthy and viable colonies, the virus remains latent / persistent without determining evident symptoms. Vice versa, in weak "stressed" colonies, the virus can abandon the state of latency, considerably replicate, increasing its virulence and causing the death of single individuals and depopulation of the colony. Among the main stressors identified in DWV infection, temperature decline could increase severity of viral infection in newly emerged honeybees (probably explaining the high levels of winter losses), while pesticides and poor nutrition could trigger the honeybee immune system making them more susceptible to viral infection, leading to colony collapse [32,33]. However, the main trigger for DWV symptomatic infection remains the uncontrolled Varroa infestation.

There is no doubt that the impact of viral diseases, especially DWV, in apiaries is a global threat to beekeeping and it is associated to honeybee colony loss [34]. Possible treatments against viral infections in honeybees are not known and legally recognized to date. Currently, a suitable treatment against Varroa is the best approach to fight DWV, since, after treatment, there is a gradual reduction of viral titers in colonies [35,36].

A deep knowledge of the crucial aspects of the viral pathogenicity, is important for realizing an effective control program, therefore, the aim of this preliminary study was to analyze any anatomo-histopathological findings in symptomatic and asymptomatic honeybees collected from a hive infected from DWV to try to better understand the pathological events underlying the infection.

2. Results

2.1. Anatomopathological Results

Anatomopathological examination confirmed the presence of alterations in 29/40 (72.5%) honeybees, namely deformed and crippled wings, discolored and shortened abdomens while 11/40 (27.5%) honeybees showed no lesions. Samples displaying alterations were classified as symptomatic (S) honeybees, while samples not showing anatomopathological alterations were classified as asymptomatic (A).

2.2. Biomolecular Results

A fragment of the expected size (69bp) of DWV was successfully amplified from 10/10 (100%) S samples and 5/5 (100%) A samples by RT-PCR but not in negative control (NTC) (data not shown). Act β amplification (151bp product) confirmed the integrity of all analyzed cDNAs. To gain insights on the possible difference of viral load between S honeybees and A group, samples were further investigated by qPCR. A successful and reproducible Cq of reference and target genes was obtained in 10/10 (100%) S and 5/5 (100%) A samples. RQ analysis according to $2^{-\Delta\Delta Cq}$ method revealed that the viral load was higher in 9/10 S samples (90%) compared to the A group (Figure 1). Further variant specific PCR analysis for identification of the DWV variant has revealed the presence of DWV-A but not DWV-B in all the 15/40 analyzed samples (data not shown).

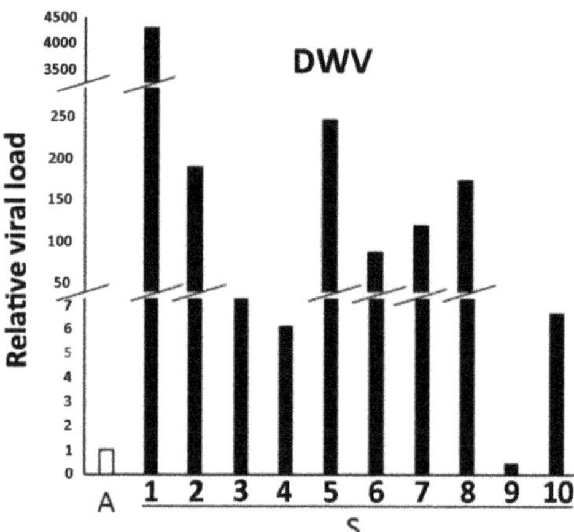

Figure 1. Analysis of relative viral load in honeybee samples showing clinical signs (S 1–10) compared with apparently healthy honeybee samples (A). Relative quantization data obtained by Real-time qPCR are expressed as fold change of each S sample with respect to the A samples considered as group (n = 5), which were set equal to 1, according to the $2^{-\Delta\Delta Cq}$ method.

Moreover, multiplex PCR for six honeybee viruses (Acute Bee Paralysis Virus-ABPV, Chronic Bee Paralysis Virus-CBPV, Sacbrood Virus-SBV, Black Queen Cell Virus-BQCV, Kashmir Bee Virus-KBV, Israeli Acute Paralysis Virus-IAPV) revealed the presence in all the 15/40 previously analyzed samples (symptomatic and asymptomatic honeybees) of ABPV (data not shown).

2.3. Histopathological Results

The histopathological analysis of symptomatic honeybees (19/25; 76%) revealed alterations of the hypopharyngeal glands in 19/19 (100%) honeybees and of flight muscles in 6/19 (31%) honeybees. The hypopharyngeal glands were characterized by small irregularly shaped acini, consisting of cells showing hyperchromic often fragmented nuclei and more or less abundant cytoplasm filled with few small vacuoles and numerous eosinophilic granules. Moreover, in the gland lumen and in the hemocele, it was noticed the presence of small cells with strongly basophilic nuclei and eosinophilic cytoplasm (Figure 2a).

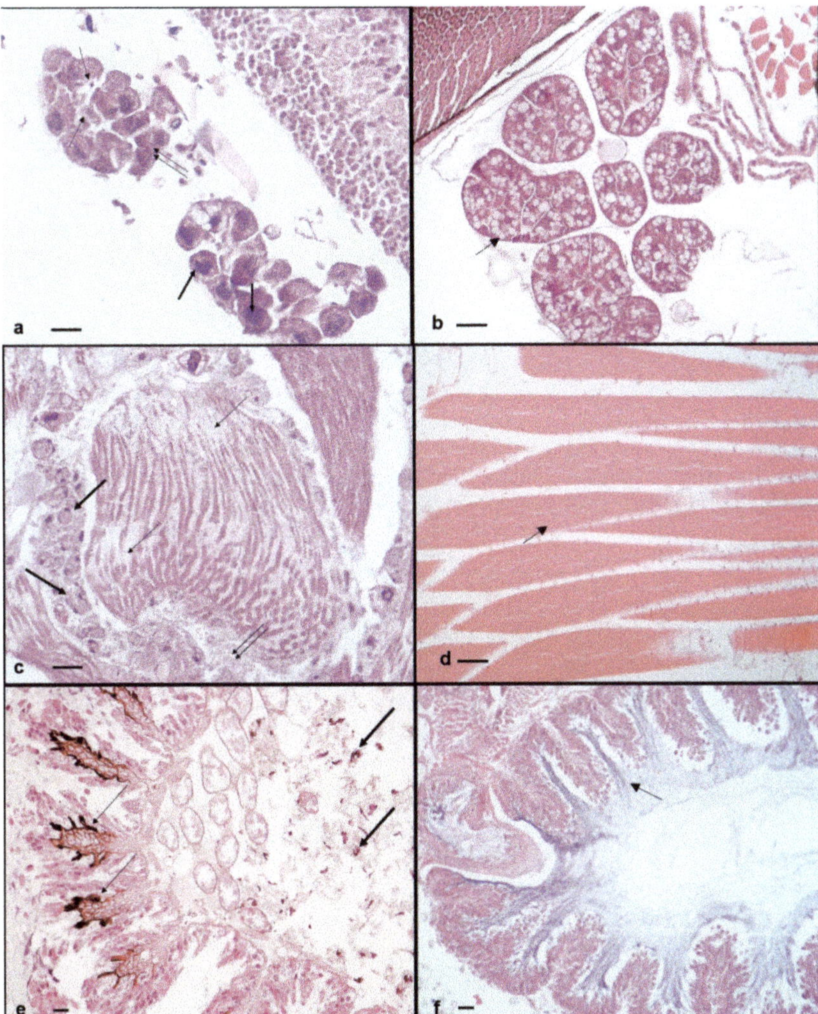

Figure 2. *A. mellifera* with symptomatic and asymptomatic infection of DWV and confirmed presence of ABPV. (**a**) Symptomatic honeybee. Hypopharyngeal glands. Small irregular acini showing cells with hyperchromic nuclei (thick arrows), cytoplasm filled with few small vacuoles and eosinophilic granules (double arrow), plasmatocytes in the gland lumen (thin arrows). H-E. 400× (40× objective and 10× ocular). (**b**) Asymptomatic honeybee. Hypopharyngeal glands. Large acini showing cells with cytoplasm filled with numerous large vacuoles with clear foamy material (thin arrow). H-E. 400×. (40× objective and 10× ocular). (**c**) Symptomatic honeybee. Flight muscles. Fibers with few not completely formed myofibrils (thin arrows), numerous muscle-forming nuclei (thick arrows), trophocytes with nuclear fragmentation and eosinophilic material between the muscle fibers (double arrow). H-E. 400× (40× objective and 10× ocular). (**d**) Asymptomatic honeybee. Flight muscles. Numerous fibers with many well-formed myofibrils (thin arrow). No trophocytes are present. H-E. 400× (40× objective and 10×ocular). (**e**) Asymptomatic honeybee. Midgut. Melanin accumulation between the fold of the villi (thin arrows) and in the hemocele (thick arrows). H-E. 200× (20× objective and 10× ocular). (**f**) Symptomatic honeybee. Midgut. Absence of melanization and hemocytes. The midgut epithelium appears intact and the peritrophic membrane appears well lined (thin arrow). H-E. 200× (20× objective and 10× ocular). Scale bar: 50 µm.

In contrast, hypopharyngeal glands of asymptomatic honeybees appeared composed of larger acini, consisting of cells showing numerous large vacuoles with clear foamy material in the cytoplasm (Figure 2b). The flight muscles showed absence of tonofibrils, few myofibrils often not completely formed and many new muscle-forming nuclei indicative of an ongoing myogenesis and incomplete maturation. Moreover, trophocytes with nuclear fragmentation or absence of nuclei, intermingled with eosinophilic material were evident between the muscle fibers. (Figure 2c). In contrast, the flight muscles of asymptomatic honeybees consisted of numerous fibers showing many well-formed myofibrils. No trophocytes were observed in asymptomatic samples (Figure 2d). The histopathological evaluation of asymptomatic honeybees (6/25) highlighted the presence in 4/6 (66%) honeybee samples of a great amount of melanin between the folds of the villi and in the lumen of the midgut, and scattered in the hemocele (Figure 2e). Moreover, the presence of two cell populations (hemocytes) was observed: one population characterized by small cells, showing small and hyperchromic nuclei, often localized at the periphery, and clear, bright eosinophilic cytoplasm, identified as plasmatocytes; the second population characterized by bigger cells with dark nuclei and granular light eosinophilic cytoplasm, identified as granulocytes. Plasmatocytes were localized in the epithelium of the midgut and in the hemocele; granulocytes were mainly present near the abdominal fat body (Figure 3).

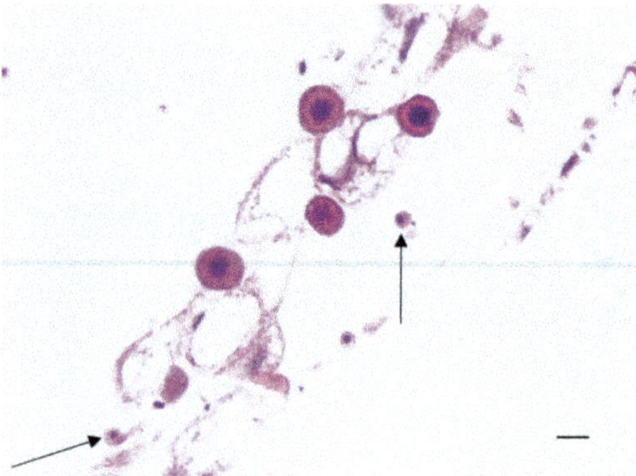

Figure 3. *A. mellifera* with asymptomatic infection of DWV and confirmed presence of ABPV. Fat Body. Granulocytes near the abdominal fat body (thin arrows). H-E. 400× (40× objective and 10× ocular). Scale bar: 40 μm.

Where melanin deposition occurred at the basal lamina level of the midgut villi and high infiltration of plasmatocytes was present at this level, high level of midgut epithelial cell exfoliation and only few regenerative cell nests were observed; in the severest cases, epithelial cells showed pyknotic nuclei, and disruption of whole villi was noticed (Figure 4). On the contrary, in symptomatic honeybees melanization was not present and hemocytes were not observed in the midgut neither in the hemocele. The midgut epithelium appeared intact and the peritrophic membrane appeared well lined (Figure 2f).

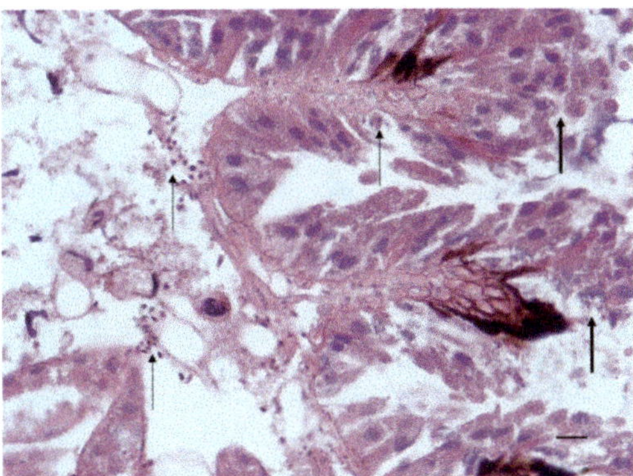

Figure 4. *A. mellifera* with asymptomatic infection of DWV and confirmed presence of ABPV. Midgut. Plasmatocytes in the epithelium and in the hemocele (thin arrows), epithelium exfoliation and disruption of the villi (thick arrows). H-E. 400× (40× objective and 10× ocular). Scale bar: 40 μm.

Moreover, no spores of Nosema spp. were observed in any of the 25/40 analyzed samples.

3. Discussion

DWV is recognized, in association with *V. destructor*, as one of the main causes of colony collapse.

Unlike many other viruses, it is characterized by typical symptomatic infections showing high pupal mortality, wing deformities, shortened, bloated and discolored abdomens; however, the virus is also capable of infecting the entire colony silently [10,37]. Therefore, the scarce presence or absence of clinical signs may not reflect the actual state of health of the colony.

In this study, symptomatic and asymptomatic honeybee samples were collected and subjected to biomolecular analysis to highlight the presence of viral genome and to anatomo-histopathological analysis to evaluate the presence of any alterations of organs and tissues. Biomolecular results showed elevated DWV viral titers in S samples compared to A samples. Despite the limited number of samples analyzed, the results obtained agree with previous studies [10,35]. We can therefore imply that also the samples used for histopathological had high viral titers in symptomatic honeybees and lower viral titers in asymptomatic honeybees.

Honeybees exhibiting anatomopathological alterations showed also histopathological alterations of the hypopharyngeal glands and of the flight muscles.

In *A. mellifera* the hypopharyngeal glands are part of the digestive system and, according to the role played in the colony, they are responsible for the production of royal jelly, storage of glycogen for the flight muscles, synthesis of enzymes important for the transformation of nectar into honey and for social immunity [38,39]. Moreover, the presence of vitellogenin, a glycoprotein necessary to produce immune system components and for longevity, has been demonstrated in the hypopharyngeal glands [40]. In this study the hypopharyngeal glands of symptomatic honeybees appeared hypotrophic, containing few small vacuoles (mucous origin) and numerous eosinophilic granules (serous origin). This seems to suggest an alteration of the secretory activity, particularly a shift towards an increase production of serous secretion, typical of foragers [41,42]. A possible early passage of honeybees to their role as foragers could be responsible for an unbalance in the castes and premature aging of the colony. Considering the role of hypopharyngeal glands in producing components of worker and royal jelly, essential for the efficient development of larvae [43,44], a modification in secretion, could lead to an altered production of the components of this substance and a consequent altered development of the larvae, which could be weaker and more susceptible to the action of the virus and of other pathogens [45,46]. Moreover, it can be hypothesized that alterations of the hypopharyngeal glands could also lead to a reduced secretion of vitellogenin, and a consequent, at least partial, impairment of the immune system [47,48]. These effects, in the long run could compromise colony fitness and survival. Alterations of hypopharyngeal and mandibular glands of honeybees infected with DWV, have already been described by Koziy et al. [49] and our observations match what previously found, corroborating the theory of an action of the virus on these organs.

At the thoracic level, symptomatic honeybees showed incomplete development of the flight muscles. In healthy honeybees, the mature muscles begin to form during pupal development by replacement of the larval muscles with mature muscles, starting from new muscle nuclei with an end-to-end trend. At the same time there is a gradual reduction of the fat body due to the degeneration of the trophocytes. The myogenesis process ends 70 h after cell capping with the attachment of the muscles to the epidermis of the cuticle using tonofibrils [50]. The histopathological study of the flight muscles of symptomatic honeybees has highlighted the presence of eosinophilic material and trophocytic nuclear debris between the muscle fibers, most of which appeared immature and detached from the cuticle, consequent to the absence of tonofibrils. These aspects, found in adult honeybees, could be indicative of incomplete myogenesis and could be responsible of an altered development of honeybees and of a reduction of their size and inability to hatch and fly. Localization of DWV in the flight muscles of symptomatic honeybees was described by Lamp et al. [51], using immunohistochemical techniques, and we here describe for the first time the presence of lesions at this level. Additionally, in this study not all samples showed incomplete myogenesis, and the reason could be found in the different developmental moment in which the virus infects the honeybee or in the titer of the virus.

Interestingly, honeybees showing no anatomopathological alterations, despite being infected by the virus, did not show the same tissue alterations as the symptomatic ones, but revealed the presence of a high number of inflammatory cells (plasmatocytes and granulocytes) and melanin accumulation between the midgut villi and in the hemocele. These findings suggest a strong activation of the immune system, particularly of the cellular response. Honeybees can try to keep the virus under control thanks to an efficient individual immune system, which is mainly composed by a first line defense and a second line defense. Honeybee venom is present on the cuticle of adult honeybee and can be considered as a chemical barrier and a first line defense against pathogens in the individual [52]. The exoskeleton cuticle and the peritrophic membranes of the digestive tract, also are considered as a first line defense as they prevent pathogens from entering the body and have access to the cells [53]. If unfortunately, a pathogen manages to surpass these physical barrier, cellular and humoral immune responses will be activated as a second line of defense [54]. The cellular response consists in activation of hemocytes function including phagocytosis, nodulation, encapsulation of the pathogen, what in pathology is defined as "granulomatosis reaction", and melanization [55]. The humoral response involves secretion of antimicrobial peptides (AMP), and other effectors, melanization, and the enzymatic degradation of pathogens by different pathways [54]. Richardson et al. [56] have identified and described the presence of two predominant cell types involved in the cellular response: granulocytes and plasmatocytes. Granulocytes exhibit a strong propensity for phagocytosis while plasmatocytes are involved in the encapsulation activity [57]. A strong and efficient immune system is the key for honeybee health and colony fitness.

In our study, the midgut epithelium of asymptomatic honeybees showed slugged epithelial cells, and as only few regenerative cell nests were present the adequate turnover that could restore the non-functional epithelium was not guaranteed. It is intuitive that a midgut showing these alterations cannot be functional both in absorption of nutrients and secretion of substances useful for the wellbeing of the peritrophic membrane, and consequently of the honeybees. It seems evident that, although no symptoms are evident, the virus is still acting on cells and tissues and that the activation of the immune response comes with a cost for the host.

This study has highlighted the presence of significant morphological alterations in symptomatic and asymptomatic honeybees infected with DWV and the results could suggest a possible pathological action of the virus in both groups of honeybees, and a possible role of the immune system, particularly of the cellular response, in keeping under control the virus in asymptomatic infections. It could be discussed that the alterations found could be linked to the action of other pathogens such other viruses or Nosema spp. However, histopathological examination of the midgut has been proven to be an efficient diagnostic tool for identifying the parasite in honeybees [58,59] and, as no spores have been observed in our samples, we can exclude the role of the parasite in generating the lesions observed. Regarding the possible action of other viruses, in this study we have screened for the presence of six different viruses and ABPV was found in all samples. ABPV is often associated to DWV in honeybee colonies [60], yet ABPV alone does not trigger humoral or cellular immune response in honeybees and therefore should not be considered as directly responsible for generating the immune response and melanization observed in the midgut and in the hemocele [61]. However, we cannot exclude a co-participation of ABPV to the generation of the alterations here found.

Therefore, further studies using other techniques such as FISH and immunohisto-chemistry, are necessary to deepen this preliminary study and better understand the etiopathogenesis of the findings here described.

4. Materials and Methods

During a regular visit to a beehive at an apiary located in Naples, Campania Region, it was possible to observe the presence of numerous small honeybees with deformed wings and shortened and discolored abdomens, suggesting the presence of a DWV infection.

The infected hive was clinically inspected, and the levels of *V. destructor* infestation were evaluated using the icing sugar technique [62] and assessed at 6% (18 mites/ 300 honeybees). A total of 40 adult honeybee samples were randomly captured from the frames and transported in 50 mL tubes to the laboratory of Veterinary General Pathology and Anatomical Pathology of the Department of Veterinary Medicine and Animal Productions, University of Naples "Federico II".

4.1. Anatomopathological Analysis

After immobilization with chilling for 3 min at −20 °C [63], all collected samples (40) were observed at the stereo microscope (Microscope Axioskop HBO50, Zeiss, Milan, Italy) to better identify any anatomopathological lesions and classify individuals in symptomatic and asymptomatic according to the presence/absence of typical clinical signs of the disease.

4.2. RNA Extraction, Reverse Transcription (RT) and PCR

A total of 15/40 honeybees were subjected to biomolecular investigation to verify and, in case of positive results, quantify the presence of viral RNA.

Samples were individually chopped up with a sterile blade to facilitate subsequent homogenization with the TissueLyser mechanical homogenizer (Qiagen, Hilden, Germany). Each sample was put in 2 mL tubes along with a grinding metal bead and subjected to lysis by two steps of five minutes at 50 Hz, interspersed with a cycle of ice cooling of 2 min to avoid overheating and preserve the integrity of the biological molecules.

RNA was extracted and purified from genomic DNA using the RNeasy Plus Mini Kit (Qiagen, Hilden, Germany), according to the protocol provided by the manufacturer, and RNA concentration was measured by spectrophotometric reading.

For each sample, 250 ng of RNA were subjected to RT using the commercial iScript cDNA Synthesis Kit (Bio-Rad Laboratories, Hercules, CA, USA), according to the manufacturer's recommendations.

Subsequently, 12.5 ng of cDNA for each sample were subjected to PCR to amplify a segment of DWV genetic material and verify the presence/absence of the virus in the samples using the AmpliTaq Gold DNA Polymerase kit (Applied Biosystems, ThermoFisher

Scientific, Waltham, MA, USA) according to manufacturer's instructions. The housekeeping gene β-actin (Act β) of *A. mellifera* was also amplified to ensure the presence of amplifiable cDNA in each sample. One no template control (NTC) was included in each PCR reaction as negative control.

Subsequently, a new PCR was performed on the same samples to discriminate between the two different variants DWV-A and DWV-B according to the protocols found in the literature [64,65].

Moreover, a multiplex PCR was executed on the previous samples (15/40) to screen for the presence of six other relevant honeybee viruses (ABPV, CBPV, SBV, BQCV, KBV, IAPV) according to the protocol proposed and validated by Cagirgan and Yazici [66]. The set of primers used for amplification of the genetic material of viruses and Act β used in this study were found in literature and a complete list, together with the product size, annealing temperature and application is reported in Supplementary Materials.

Amplification products were migrated by electrophoresis on 2.5% agarose gel in TBE buffer (Tris-Borate-EDTA) along with a 50 bp molecular marker (Bioline), stained with ethidium bromide and observed under UV with the ChemiDoc gel scanner (Bio-Rad).

4.3. Real-Time PCR (qPCR) for Detection of Relative Viral Load

In order to determine a relative quantization (RQ) of viral load of the samples, a Real-Time PCR (qPCR) was carried out using the primers described above.

For each sample tested positive for DWV in PCR, 12.5 ng of cDNA were subjected to qPCR using iTaq Universal SYBR Green Supermix kit (Bio-Rad), according to the manufacturer's instructions.

Amplification of honeybee Act β as reference gene was also performed in parallel to allow normalization of the results and an NTC was included in the reaction as negative control.

Relative quantization of DWV viral load was calculated by using the $2^{-\Delta\Delta Cq}$ method as previously described [67,68]. Briefly, fold change in viral load was estimated for each individual S sample against A samples considered as control group.

4.4. Histopathological Analysis

Samples were processed as previously described [69]. Briefly, honeybees were individually injected with 10 μL of 10% buffered formalin and then stored for 24 h in 50 mL tubes containing the same fixative.

Subsequently, each sample was placed in an embedding cassette and processed. 3 μm sections were cut, stained with hematoxylin and eosin, and observed by light microscopy (Microscope Nikon Eclipse E-600, Tokyo, Japan). All tissues were observed to identify possible alterations and analyzed for the presence of visible pathogens, i.e., Nosema spp.

Supplementary Materials: The following are available online at https://www.mdpi.com/article/10.3390/pathogens10070874/s1, File S1: Oligonucleotides used for amplification of viruses and Act β in this study. Sequences, products size, annealing temperature and applications are indicated.

Author Contributions: Conceptualization, K.P. and P.M.; Methodology, K.P. and G.A.; Validation, M.M., N.P. and P.M.; Formal Analysis, N.P.; Investigation, K.P. and G.A.; Writing—Original Draft Preparation, K.P.; Writing—Review and Editing, M.M. and P.M.; Supervision, P.M. All authors have read and agreed to the published version of the manuscript.

Funding: This research was funded by PSR 14/20 Campania. Tipologia intervento 16.1.1 "sostegno per costituzione e funzionamento dei GO del PEI in materia di produttività e sostenibilità dell'agricoltura". Azione 2 "Sostegno ai POI". "Uso tecnologico e muove pratiche a carattere innovativo per la gestione, il controllo e la valorizzazione economica del cinghiale (Sus scrofa) in maniera sostenibile in Regione Campania". S.U.S Campania (CUP B58H19004460009).

Institutional Review Board Statement: Ethical review and approval were waived for this study, as according to the D.L. 4 March 2014 n.26, and national implementing decree following the European regulation 2010/63/UE, ethical approval is not necessary for insects with the except of cephalopoda.

Informed Consent Statement: Not applicable.

Data Availability Statement: Data are available on reasonable request to the corresponding author.

Conflicts of Interest: The authors declare no conflict of interest.

References

1. White, G.F. *Sacbrood, a Disease of Bees*; US Department of Agriculture, Bureau of Entomology: Washington, DC, USA, 1913.
2. Bailey, L.; Gibbs, A.J.; Woods, R.D. Sacbrood virus of the larval honey bee (*Apis mellifera* linnaeus). *Virology* **1964**, *23*, 425–429. [CrossRef]
3. Tantillo, G.; Bottaro, M.; Di Pinto, A.; Martella, V.; Di Pinto, P.; Terio, V. Virus Infections of Honeybees *Apis Mellifera*. *Ital. J. Food Saf.* **2015**, *4*, 5364. [CrossRef]
4. Tentcheva, D.; Gauthier, L.; Zappulla, N.; Dainat, B.; Cousserans, F.; Colin, M.E.; Bergoin, M. Prevalence and seasonal variations of six bee viruses in *Apis mellifera* L. and Varroa destructor mite populations in France. *Appl. Environ. Microbiol.* **2004**, *70*, 7185–7191. [CrossRef]
5. Matthijs, S.; De Waele, V.; Vandenberge, V.; Verhoeven, B.; Evers, J.; Brunain, M.; Saegerman, C.; De Winter, P.J.J.; Roels, S.; de Graaf, D.C.; et al. Nationwide Screening for Bee Viruses and Parasites in Belgian Honey Bees. *Viruses* **2020**, *12*, 890. [CrossRef]
6. Tantillo, G.; Ullah, A.; Tlak Gajger, I.; Majoros, A.; Dar, S.A.; Khan, S.; Kalimullah Haleem Shah, A.; Nasir Khabir, M.; Hussain, R.; Khan, H.U.; et al. Viral impacts on honey bee populations: A. review. *Saudi J. Biol. Sci.* **2021**, *28*, 523–530. [CrossRef]
7. De Miranda, J.R.; Genersch, E. Deformed wing virus. *J. Invertebr. Pathol.* **2010**, *103* (Suppl. 1), S48–S61. [CrossRef] [PubMed]
8. Dimmock, N.J.; Primrose, S.B. *Introduction to Modern Virology*, 3rd ed.; Blackwell Scientific Publications: Oxford, UK, 1987.
9. Chen, Y.; Evans, J.; Feldlaufer, M. Horizontal and vertical transmission of viruses in the honey bee, *Apis mellifera*. *J. Invertebr. Pathol.* **2006**, *92*, 152–159. [CrossRef]
10. De Jong, D.; De Jong, P.H.; Goncalves, L.S. Weight loss and other damage to developing worker honeybees from infestation with Varroa jacobsoni. *J. Apic. Res.* **1982**, *21*, 165–167. [CrossRef]
11. Lanzi, G.; de Miranda, J.R.; Boniotti, M.B.; Cameron, C.E.; Lavazza, A.; Capucci, L.; Camazine, S.M.; Rossi, C. Molecular and biological characterization of deformed wing virus of honeybees (*Apis mellifera* L.). *J. Virol.* **2006**, *80*, 4998–5009. [CrossRef] [PubMed]
12. Martin, S.J.; Brettell, L.E. Deformed Wing Virus in Honeybees and Other Insects. *Annu. Rev. Virol.* **2019**, *6*, 49–69. [CrossRef]
13. Bailey, L.; Ball, B.V. *Honey Bee Pathology*; Academic Press: London, UK, 1991.
14. Yue, C.; Genersch, E. RT-PCR analysis of Deformed wing virus in honeybees (*Apis mellifera*) and mites (*Varroa destructor*). *J. Gen. Virol.* **2005**, *86*, 3419–3424. [CrossRef]
15. Allen, M.F.; Ball, B.V. The incidence and world distribution of honey bee viruses. *Bee World.* **1996**, *77*, 141–162. [CrossRef]
16. Gisder, S.; Möckel, N.; Eisenhardt, D.; Genersch, E. In vivo evolution of viral virulence: Switching of deformed wing virus between hosts results in virulence changes and sequence shifts. *Environ. Microbiol.* **2018**, *20*, 4612–4628. [CrossRef]
17. Kevill, J.L.; de Souza, F.S.; Sharples, C.; Oliver, R.; Schroeder, D.C.; Martin, S.J. DWV-A Lethal to Honey Bees (*Apis mellifera*): A Colony Level Survey of DWV Variants (A, B, and C) in England, Wales, and 32 States across the US. *Viruses* **2019**, *11*, 426. [CrossRef] [PubMed]
18. McMahon, D.P.; Natsopoulou, M.E.; Doublet, V.; Fürst, M.; Weging, S.; Brown, M.J.F.; Gogol-Döring, A.; Paxton, R.J. Elevated virulence of an emerging viral genotype as a driver of honeybee loss. *Proc. R. Soc. Lond. B Biol. Sci.* **2016**, *283*, 20160811. [CrossRef] [PubMed]
19. Kevill, J.L.; Highfield, A.; Mordecai, G.J.; Martin, S.J.; Schroeder, D.C. ABC assay: Method development and application to quantify the role of three DWV master variants in overwinter colony losses of European honey bees. *Viruses* **2017**, *9*, 314. [CrossRef]
20. Dalmon, A.; Gayral, P.; Decante, D.; Klopp, C.; Bigot, D.; Thomasson, M.; Herniou, E.A.; Alaux, C.; Le Conte, Y. Viruses in the Invasive Hornet Vespa velutina. *Viruses* **2019**, *11*, 1041. [CrossRef]
21. Eyer, M.; Chen, Y.P.; Schäfer, M.O.; Pettis, J.; Neumann, P. Small hive beetle, Aethina tumida, as a potential biological vector of honeybee viruses. *Apidologie* **2009**, *40*, 419–428. [CrossRef]
22. Forzan, M.; Felicioli, A.; Sagona, S.; Bandecchi, P.; Mazzei, M. Complete Genome Sequence of Deformed Wing Virus Isolated from Vespa crabro in Italy. *Genome Announc.* **2017**, *5*, e00961-17. [CrossRef]
23. Schläppi, D.; Chejanovsky, N.; Yañez, O.; Neumann, P. Foodborne Transmission and Clinical Symptoms of Honey Bee Viruses in Ants Lasius spp. *Viruses* **2020**, *12*, 321. [CrossRef]
24. Dainat, B.; Evans, J.D.; Chen, Y.P.; Gauthier, L.; Neumann, P. Dead or alive: Deformed wing virus and Varroa destructor reduce the life span of winter honeybees. *Appl. Environ. Microbiol.* **2012**, *78*, 981–987. [CrossRef]
25. Chen, Y.P.; Higgins, J.A.; Feldlaufer, M.F. Quantitative real-time reverse transcription-PCR analysis of deformed wing virus infection in the honeybee (*Apis mellifera* L.). *Appl. Environ. Microbiol.* **2005**, *71*, 436–441. [CrossRef]
26. Sanpa, S.; Chantawannakul, P. Survey of six bee viruses using RT-PCR in Northern Thailand. *J. Invertebr. Pathol.* **2009**, *100*, 116–119. [CrossRef]

27. Tentcheva, D.; Gauthier, L.; Jouve, S.; Canabady-Rochelle, L.; Dainat, B.; Cousserans, F.; Colin, M.E.; Ball, B.V.; Bergoin, M. Polymerase Chain Reaction detection of deformed wing virus (DWV) in *Apis mellifera* and Varroa destructor. *Apidologie* **2004**, *35*, 431–439. [CrossRef]

28. Brettell, L.E.; Mordecai, G.J.; Schroeder, D.C.; Jones, I.M.; da Silva, J.R.; Vincente-Rubiano, M.; Martin, S.J. A Comparison of Deformed Wing Virus in Deformed and Asymptomatic Honey Bees. *Insects* **2017**, *8*, 28. [CrossRef] [PubMed]

29. Forsgren, E.; Friesm, I.; de Miranda, J.R. Adult honey bees (*Apis mellifera*) with deformed wings discovered in confirmed varroa-free colonies. *J. Apic. Res.* **2012**, *51*, 136–138. [CrossRef]

30. Bowen-Walker, P.L.; Martin, S.J.; Gunn, A. The transmission of deformed wing virus between honeybees (*Apis mellifera* L.) by the ectoparasitic mite varroa jacobsoni Oud. *J. Invertebr. Pathol.* **1999**, *73*, 101–106. [CrossRef] [PubMed]

31. Dainat, B.; Neumann, P. Clinical signs of deformed wing virus infection are predictive markers for honey bee colony losses. *J. Invertebr. Pathol.* **2013**, *112*, 278–280. [CrossRef] [PubMed]

32. Nazzi, F.; Pennacchio, F. Honey Bee Antiviral Immune Barriers as Affected by Multiple Stress Factors: A Novel Paradigm to Interpret Colony Health Decline and Collapse. *Viruses* **2018**, *10*, 159. [CrossRef] [PubMed]

33. Di Prisco, G.; Pennacchio, F.; Caprio, E.; Boncristiani, H.F., Jr.; Evans, J.D.; Chen, Y. Varroa destructor is an effective vector of Israeli acute paralysis virus in the honeybee, *Apis mellifera*. *J. Gen. Virol.* **2011**, *92*, 151–155. [CrossRef]

34. Tehel, A.; Vu, Q.; Bigot, D.; Gogol-Döring, A.; Koch, P.; Jenkind, C.; Doublet, V.; Theodouru, P.; Paxton, R. The Two Prevalent Genotypes of an Emerging Infectious Disease, Deformed Wing Virus, Cause Equally Low Pupal Mortality and Equally High Wing Deformities in Host Honey Bees. *Viruses* **2019**, *11*, 114. [CrossRef]

35. Locke, B.; Semberg, E.; Forsgren, E.; de Miranda, J.R. Persistence of subclinical deformed wing virus infections in honeybees following Varroa mite removal and a bee population turnover. *PLoS ONE* **2017**, *12*, e0180910. [CrossRef]

36. Martin, S.J.; Ball, B.V.; Carreck, N.L. Prevalence and persistance of deformed wing virus (DWV) in untreated or acaricide-treated Varroa destructor infested honey bees (*Apis mellifera*) colonies. *J. Apic. Res.* **2010**, *49*, 72–79. [CrossRef]

37. Amiri, E.; Kryger, P.; Meixner, M.D.; Strand, M.K.; Tarpy, D.R.; Rueppell, O. Quantitative patterns of vertical transmission of deformed wing virus in honey bees. *PLoS ONE* **2018**, *13*, e0195283. [CrossRef] [PubMed]

38. Chan, Q.W.T.; Melathopoulos, A.P.; Pernal, S.F.; Foster, L.J. The innate immune and systemic response in honey bees to a bacterial pathogen, Paenibacillus larvae. *BMC Genom.* **2009**, *10*, 387. [CrossRef] [PubMed]

39. Costa, R.A.C.; Cruz-Landim, C. Occurrence and morphometry of the hypopharyngeal glands in Scaptotrigona postica Lat. (Hymenoptera, Apidae, Meliponinae). *J. Biosci.* **1999**, *24*, 97–102. [CrossRef]

40. Corona, M.; Velarde, R.A.; Remolina, S.; Moran-Lauter, A.; Wang, Y.; Hughes, K.A.; Robinson, G.E. Vitellogenin, juvenile hormone, insulin signaling, and queen honey bee longevity. *Proc. Natl. Acad. Sci. USA* **2007**, *104*, 7128–7133. [CrossRef]

41. Suwannapong, G.; Chaiwongwattanakul, S.; Benbow, M.E. Histochemical Comparison of the Hypopharingeal Gland in Apis cerana Fabricius, 1793 Workers and *Apis mellifera* Linnaues, 1758 Workers. *Psyche A J. Entomol.* **2010**, 181025. [CrossRef]

42. Ahmad, S.; Khan, S.A.; Khan, K.A.; Li, J. Novel Insight into the Development and Function of Hypopharyngeal Glands in Honey Bees. *Front. Physiol.* **2020**, *11*, 615830. [CrossRef]

43. Albert, S.; Bhattàcharya, D.; Klaudiny, J.; Schmitzova, J.; Simuth, J. The family of major royal jelly proteins and its evolution. *J. Mol. Evol.* **1999**, *49*, 290–297. [CrossRef]

44. Schmitzová, J.; Klaudiny, J.; Albert, S.; Schröder, W.; Schreckengost, W.; Hanes, J.; Júdová, J.; Simúth, J. A family of major royal jelly proteins of the honeybee *Apis mellifera* L. *Cell Mol. Life Sci.* **1998**, *54*, 1020–1030. [CrossRef]

45. Crailsheim, K. The protein balance of the honey bee worker. *Apidologie* **1990**, *21*, 417–429. [CrossRef]

46. Milone, J.P.; Chakrabarti, P.; Sagili, R.R.; Tarpy, D.R. Colony-level pesticide exposure affects honey bee (*Apis mellifera* L.) royal jelly production and nutritional composition. *Chemosphere* **2021**, *263*, 128183. [CrossRef] [PubMed]

47. Amdam, G.V.; Fennern, E.; Havukainen, H. Vitellogenin in Honey Bee Behavior and Lifespan. In *Honeybee Neurobiology and Behavior*; Galizia, C., Eisenhardt, D., Giurfa, M., Eds.; Springer: Dordrecht, The Netherlands, 2012. [CrossRef]

48. Amdam, G.V.; Simões, Z.L.P.; Hagen, A.; Norberg, K.; Schrøder, K.; Mikkelsen, Ø.; Kirkwood, T.B.; Omholt, S.W. Hormonal control of the yolk precursor vitellogenin regulates immune function and longevity in honey-bees. *Exp. Gerontol.* **2004**, *39*, 767–773. [CrossRef] [PubMed]

49. Koziy, R.V.; Wood, S.C.; Kozii, I.V.; van Rensburg, C.J.; Moshynskyy, I.; Dvylyuk, I.; Simko, E. Deformed Wing Virus Infection in Honey Bees (*Apis mellifera* L.). *Vet. Pathol.* **2019**, *56*, 636–641. [CrossRef]

50. Snodgrass, R.E. *The Anatomy of the Honey Bee*; U.S. Government Printing Office: Washington, DC, USA, 1910.

51. Lamp, B.; Url, A.; Seitz, K.; Eichhorn, J.; Riedel, C.; Sinn, L.J.; Indik, S.; Köglberger, H.; Rümenapf, T. Construction and Rescue of a Molecular Clone of Deformed Wing Virus (DWV). *PLoS ONE* **2016**, *11*, e0164639. [CrossRef]

52. Baracchi, D.; Francese, S.; Turillazzi, S. Beyond the antipredatory defence: Honeybee venom function as a component of social immunity. *Toxicon* **2011**, *58*, 550–557. [CrossRef]

53. Evans, J.D.; Spivak, M. Socialized medicine: Individual and communal disease barriers in honey bees. *J. Invertebr. Pathol.* **2010**, *103* (Suppl. 1), S62–S72. [CrossRef]

54. Evans, J.D.; Aronstein, K.; Chen, Y.P.; Hetru, C.; Imler, J.L.; Jiang, H.; Kanost, M.; Thompson, G.J.; Zou, Z.; Hultmark, D. Immune pathways and defence mechanisms in honey bees *Apis mellifera*. *Insect. Mol. Biol.* **2006**, *15*, 645–656. [CrossRef] [PubMed]

55. DeGrandi-Hoffman, G.; Chen, Y. Nutrition, immunity and viral infections in honey bees. *Curr. Opin. Insect. Sci.* **2015**, *10*, 170–176. [CrossRef]

56. Richardson, R.T.; Ballinger, M.N.; Qian, F.; Christman, J.W.; Johnson, R.M. Morphological and functional characterization of honey bee *Apis mellifera*, hemocyte cell communities. *Apidologie* **2018**, *49*, 397–410. [CrossRef]
57. Ribeiro, C.; Brehélin, M. Insect haemocytes: What type of cell is that? *J. Insect. Physiol.* **2006**, *52*, 417–429. [CrossRef]
58. Maiolino, P.; Iafigliola, L.; Rinaldi, L.; De Leva, G.; Restucci, B.; Martano, M. Histopathological findings of the midgut in European honey bee (*Apis mellifera*, L.) naturally infected by Nosema spp. *Vet. Med. Anim. Sci.* **2014**, *2*, 4. [CrossRef]
59. Higes, M.; García-Palencia, P.; Urbieta, A.; Nanetti, A.; Martín-Hernández, R. Nosema apis and Nosema ceranae Tissue Tropism in Worker Honey Bees (*Apis mellifera*). *Vet. Pathol.* **2020**, *57*, 132–138. [CrossRef]
60. Brutscher, L.M.; McMenamin, A.J.; Flenniken, M.L. The Buzz about Honey Bee Viruses. *PLoS Pathog.* **2016**, *12*, e1005757. [CrossRef]
61. Azzami, K.; Ritter, W.; Tautz, J.; Beier, H. Infection of honey bees with acute bee paralysis virus does not trigger humoral or cellular immune responses. *Arch Virol.* **2012**, *157*, 689–702. [CrossRef]
62. Lee, K.; Moon, R.D.; Burknes, E.C.; Hutchinson, W.D.; Spivak, M. Practical sampling plans for Varroa destructor (Acari Varroidae) in *Apis mellifera* (Hymenoptera: Apidae) colonies and apiaries. *J. Econ. Entomol.* **2010**, *103*, 1039–1050. [CrossRef] [PubMed]
63. Human, H.; Brodschneider, R.; Dietemann, V. Miscellaneous Standard Methods for *Apis mellifera* Research. *J. Apic. Res.* **2013**, *52*, 1–53. [CrossRef]
64. Bradford, E.L.; Christie, C.R.; Campbell, E.M.; Bowman, A.S. A real-time PCR method for quantification of the total and major variant strains of the deformed wing virus. *PLoS ONE* **2017**, *12*, e0190017. [CrossRef] [PubMed]
65. Moore, J.; Jironkin, A.; Chandler, D.; Burroughs, N.; Evans, D.J.; Ryabov, E.V. Recombinants between Deformed wing virus and Varroa destructor virus-1 may prevail in Varroa destructor-infested honeybee colonies. *J Gen Virol.* **2011**, *92 Pt 1*, 156–161. [CrossRef]
66. Cagirgan, A.A.; Yazici, Z. Development of a multiplex RT-PCR assay for the routine detection of seven RNA viruses in *Apis mellifera*. *J. Virol. Methods* **2020**, *281*, 113858. [CrossRef] [PubMed]
67. Mazzei, M.; Forzan, M.; Carlucci, V.; Anfossi, A.G.; Alberti, A.; Albanese, F.; Binati, D.; Millanta, F.; Baroncini, L.; Pirone, A.; et al. A study of multiple Felis catus papillomavirus types (1, 2, 3, 4) in cat skin lesions in Italy by quantitative PCR. *J. Feline Med. Surg.* **2018**, *20*, 772–779. [CrossRef]
68. Altamura, G.; Cardeti, G.; Cersini, A.; Eleni, C.; Cocumelli, C.; Bartolomé Del Pino, L.E.; Razzuoli, E.; Martano, M.; Maiolino, P.; Borzacchiello, G. Detection of Felis catus papillomavirus type-2 DNA and viral gene expression suggest active infection in feline oral squamous cell carcinoma. *Vet. Comp. Oncol.* **2020**, *18*, 494–501. [CrossRef] [PubMed]
69. Power, K.; Martano, M.; Altamura, G.; Maiolino, P. Histopathological Findings in Testes from Apparently Healthy Drones of *Apis mellifera* ligustica. *Vet. Sci.* **2020**, *7*, 124. [CrossRef] [PubMed]

Article

Propolis Extract and Chitosan Improve Health of *Nosema ceranae* Infected Giant Honey Bees, *Apis dorsata* Fabricius, 1793

Sanchai Naree [1], Rujira Ponkit [1], Evada Chotiaroonrat [1], Christopher L. Mayack [2] and Guntima Suwannapong [1,*]

1. Biological Science Program, Faculty of Science, Burapha University, Chon Buri 20131, Thailand; 60810001@go.buu.ac.th (S.N.); 59810030@go.buu.ac.th (R.P.); 62810007@go.buu.ac.th (E.C.)
2. Molecular Biology, Genetics, and Bioengineering, Faculty of Engineering and Natural Sciences, Sabanci University, 34956 Istanbul, Turkey; christopher.mayack@sabanciuniv.edu
* Correspondence: guntima@buu.ac.th; Tel.: +66-3810-3088

Abstract: *Nosema ceranae* is a large contributing factor to the most recent decline in honey bee health worldwide. Developing new alternative treatments against *N. ceranae* is particularly pressing because there are few treatment options available and therefore the risk of increased antibiotic resistance is quite high. Recently, natural products have demonstrated to be a promising avenue for finding new effective treatments against *N. ceranae*. We evaluated the effects of propolis extract of stingless bee, *Tetrigona apicalis* and chito-oligosaccharide (COS) on giant honey bees, *Apis dorsata*, experimentally infected with *N. ceranae* to determine if these treatments could improve the health of the infected individuals. Newly emerged *Nosema*-free bees were individually inoculated with 10^6 *N. ceranae* spores per bee. We fed infected and control bees the following treatments consisting of 0%, 50%, propolis extracts, 0 ppm and 0.5 ppm COS in honey solution (*w/v*). Propolis extracts and COS caused a significant increase in trehalose levels in hemolymph, protein contents, survival rates and acini diameters of the hypopharyngeal glands in infected bees. Our results suggest that propolis and COS could improve the health of infected bees. Further research is needed to determine the underlying mechanisms responsible for the improved health of the infected bees.

Keywords: *Apis dorsata*; chito-oligosaccharide; *Nosema ceranae*; propolis

Citation: Naree, S.; Ponkit, R.; Chotiaroonrat, E.; Mayack, C.L.; Suwannapong, G. Propolis Extract and Chitosan Improve Health of *Nosema ceranae* Infected Giant Honey Bees, *Apis dorsata* Fabricius, 1793. *Pathogens* **2021**, *10*, 785. https://doi.org/10.3390/pathogens10070785

Academic Editor: Giovanni Cilia

Received: 25 May 2021
Accepted: 20 June 2021
Published: 22 June 2021

Publisher's Note: MDPI stays neutral with regard to jurisdictional claims in published maps and institutional affiliations.

1. Introduction

Honey bees play a vital role in agricultural crop production and ecosystem stability due to their pollination services [1–6]. Despite their importance there has been a global decline in bee health around the world at unsustainable rates [7–9]. The health decline can be attributed to a number of health factors such as pesticide exposure and poor nutrition, with parasitic infections as one of the major contributors [10,11]. *Nosema* disease or nosemosis is one of the most widespread parasitic infections of adult honey bees and has been implicated to play a major role in the most recent global bee health decline [12–14]. Nosemosis is caused by three species of microsporidia, *Nosema apis*, *N. ceranae* and *N. neumanni*, but the most prevalent strain found in honey bees that has emerged is *N. ceranae* displacing much of the *N. apis* infections worldwide [15,16]. Nosemosis is considered to be a chronic infection that does not exhibit obvious external disease symptoms, but can cause a poor nutrient and energy absorption leading to a suppressed immune function and ultimately a shortened life span [17–21]. Infected bees have evidence of lower trehalose and lipid levels, and a reduced hypopharyngeal gland resulting from the poor nutrient absorption across the gut lining [22–24]. *N. ceranae* primarily lives and reproduces in the gut lining which is likely the cause for the poor nutrient absorption in infected bees [25,26]. Consequently, infected individuals suffer from energetic stress, which results in increased bee mortality on the individual and colony level [24,27–30].

There are only a few treatment options on the market for controlling Nosemosis. The antibiotic Fumagillin has been on the market for a long time, but it is unable to kill the mature spore form of the parasite [31], so reinfections can occur [32,33]. Moreover, its use has been banned in the European Union because it has been shown to contaminate honey and could possibly lead to the buildup of antibiotic resistance in humans [34]. A natural product that is completely safe and environmentally friendly is desirable, especially for organic beekeepers. There have been a number of recent developments in this area which include using phytochemicals, Bee Cleanse, zeolite clinoptilolite, plant extracts, and propolis extract that have been documented to be effective alternative treatments [30,35–38]. These studies generally show lowered parasite loads and improved survival of the treated bees, but very few assess the health of the bee to determine how the survival of the treated bees are being increased. Among the various promising substances to control *N. ceranae*- infection, ApiHerb® and Api-Bioxal®, commercial dietary supplements were used as treatments against *N. ceranae*-infection effectively in both laboratory and colony level [39]. The commercial probiotics, Vetafarm Probotic, Protexin Concentrate single-strain (*Enterococcus faecium*), and Protexin Concentrate multi-strain (*Lactobacillus acidophilus, L. plantarum, L. rhamnosus, L. delbrueckii, Bifidobacterium bifidum, Streptococcus salivarius*, and *E. faecium* [40], and *Parasaccharibacter apium* (PC1 sp.) and *Bacillus* sp. (PC2 sp.) [41] also were used to reduce spore loads and mortality in *N. ceranae* infected-honey bees. Currently, the control of *N. ceranae*-infections involves the use of the natural compounds to stimulate the immunity of honey bee, *A. mellifera* by inducing resistance against pathogens. Chitosan and peptidoglycans were used to reduce *N. ceranae*-infection, and increase survivorship of *N. ceranae* infected-bees. In addition, peptidoglycan and chitosan promoted the gene expression of hymenoptaecin and defensin2 [42]. Another example of natural compounds being effective at reducing *N. ceranae* loads involves using Brassicaceae defatted seed meals (DSMs) containing antimicrobial and antioxidant properties [43]. Besides fumagillin, sulforaphane was used to control *N. ceranae*-infection in laboratory. It was reported that 1.25 mg/mL of sulforaphane showed 100% reduction of spore counts, but also caused 100% bee mortality. The antimicrobial properties of this new alternative treatment may be promising, however reducing its toxicity is required before it can be considered as an alternative treatment for controlling *N. ceranae* [44].

Propolis extract of stingless bees is emerging to be an effective treatment to control *N. ceranae* across three of the four honey bee species, *A. cerana, A. mellifera*, and *A. florea*, [30,45–47]. Propolis is collected by bees and contains a number of plant resins, which are considered to be a natural product. In general, the plant resins are known to have antimicrobial effects and are used by bees to aid in sanitizing their hives. These plant resins also have recently been found to have a potential inhibitory effect on microsporidian development [30,45–48]. However, the propolis has to be fed to the honey bee in order to observe a reduction in the proliferation of *N. ceranae* in the midgut cells as the bees do not preferentially consume food containing propolis when infected. When fed, propolis extract treatment significantly enhances bee survival [30,45–47]. Another natural product, chito-oligosaccharides (COS) promotes antimicrobial activity and has been shown to stimulate the immune system thereby reducing *N. apis* infection in *A. mellifera* [48–51]. COS is a derivative of chitosan which is known as a biopolymer and polysaccharide found in the exoskeleton of insects and crustaceans. This water-soluble glycoprotein molecule has been used as a pre-biotic for gastrointestinal infections and diarrhea. COS is also known to aid in increased amino acid absorption across the gut lining, and also promote gut health including anti-inflammation activity through activation of 5′ AMP-activated protein kinase (AMPK) [52–55]. We, therefore, hypothesize that this treatment can aid in treating the symptoms of a *N. ceranae* infection and consequently improve the health of the honey bee.

Whether the pathological effects from a *N. ceranae* infection is of the same magnitude across the honey bee species and can be generalized to the giant honey bee, *A. dorsata*, remains unknown. *A. dorsata* serves as a main pollinator for crop plants in Thailand and provides a substantial amount of honey for a number of Asian countries [6]. Thus, the first aim of this study is to investigate the pathological effects of a *N. ceranae* infection in *A. dorsata*. Secondly, we aim to determine the efficacy of propolis extract and COS, as alternative treatment options for *N. ceranae* infections, by measuring hemolymph trehalose levels, protein contents in the hypopharyngeal gland, survival rates and acini diameters of the hypopharyngeal glands as health status indicators.

2. Results

2.1. Hemolymph Trehalose Levels

N. ceranae-infected bees without any treatment had the lowest hemolymph trehalose levels on day 14 p.i. compared to all other treatment groups (χ^2 = 34.52, df = 3, p < 0.0001, Figure 1). The highest levels of hemolymph trehalose were found in uninfected bees treated with propolis extract, CO-50P (273.2 ± 6.69 µg/bee) followed by the control group, CO-0P (250.8 ± 2.26 µg/bee). However, *N. ceranae*-infected bees treated with 50% propolis extract (NO-50P) showed higher levels of trehalose (204.2 ± 5.13 µg/bee) than that of *N. ceranae*-infected bees without propolis extract treatment, NO-0P (148.0 ± 5.79 µg/bee). Interestingly, similar trend was found in bees treated with COS where the highest hemolymph trehalose levels were found in the control group with 0.5 COS (CO-0.5COS) and without COS(CO-0COS) treatment 250.8 ± 2.26 µg/bee and 254.2 ± 1.73 µg/bee, respectively. The lowest hemolymph trehalose levels were found in the *Nosema* infected bees without treatment (NO-0COS) 148.0 ± 5.79 µg/bee, while there was a significant increase in the infected bees that received a COS treatment (NO-0.5COS) 184.2 ± 5.14 µg/bee (χ^2 = 33.21, df = 3, p < 0.0001, Figure 2).

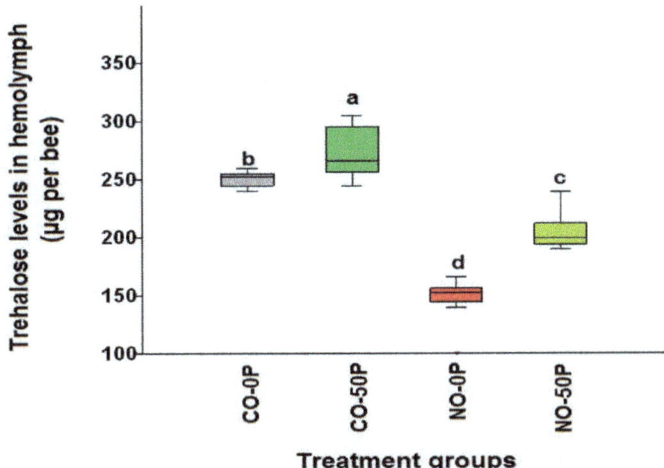

Figure 1. A box plot with the lines representing the median levels of hemolymph trehalose across the treatment groups of the propolis extract experiment. The control bees (CO-0P) (grey), propolis control bees (CO-50P) (green), *N. ceranae*-infected bees not treated with propolis extract (NO-0P) (red) and infected bees treated with 50% propolis extract (NO-50P) (light green) are represented by each box plot. The hemolymph trehalose levels are measured on 14 days p. i. The box indicates the inter-quartile range while the vertical bars indicate the range of the data. The different letters above each box represent significant differences (Kruskal–Wallis test: χ^2 = 34.52, df = 3, p < 0.0001).

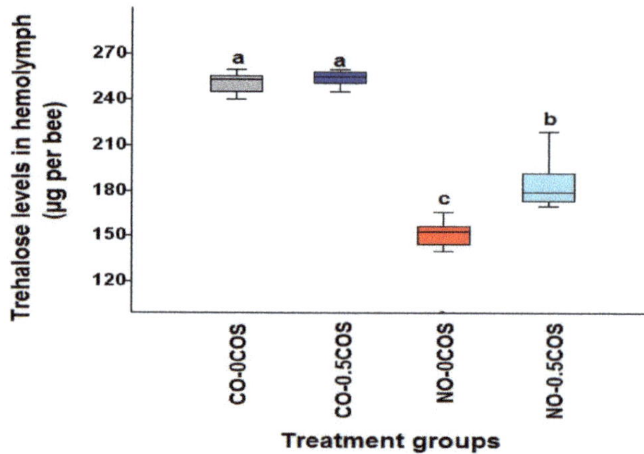

Figure 2. A box plot showing the median hemolymph trehalose levels data across treatments from the COS experiment. The control bees not treated (CO-0COS) (grey), the treated COS control bees (CO-0.5COS) (purple), *N. ceranae*-infected bees not treated (NO-0COS) (red) and the infected bees treated with 0.5 ppm (NO-0.5COS) (light blue) are each indicated by a box plot. The hemolymph trehalose levels were measured on day 14 p.i.. The boxes indicate the interquartile range, while the vertical bars represent the range of the data. The different letters above each box plot represent significant differences (Kruskal–Wallis test: χ^2 = 33.21, df = 3, p < 0.0001).

2.2. Hypopharyngeal Gland Protein Content

The hypopharyngeal gland protein contents of control bees treated with propolis extract (CO-50P) and not treated with 50% propolis extract (CO-0P) were significantly higher than the infected bees (χ^2 = 31.75, df = 3, p < 0.0001, Figure 3), they were 1470.0 ± 65.06 µg/bee and 1326.23 ± 103.4 µg/bee, respectively. The *N. ceranae*-infected bees treated with propolis extract (NO-50P) had significantly higher protein levels, 963.0 ± 52.77 µg/bee, in comparison to the infected bees not treated with propolis extract (NO-0P), which had 486.0 ± 32.5 µg/bee of protein, respectively (χ^2 = 31.75, df = 1, p = 0.0002). The similar trend was found in both groups of bees treated with 0.5COS (χ^2 = 31.39, df= 3, p < 0.0001, Figure 4). The highest protein content was found in the control bees treated with COS (CO-0.5COS) (1500.0 ± 76.01 µg/bee), and it was not significantly different from that of control bees not treated with COS (CO-0COS) (χ^2 = 31.39, df= 1, p = 0.5657). However, infected bees treated with 0.5 ppm COS had significantly increased protein content of the hypopharyngeal glands (1006.0 ± 44.4 µg/bee), in comparison with *N. ceranae*-infected bees that were not treated with COS (486.0 ± 32.5 µg/bee).

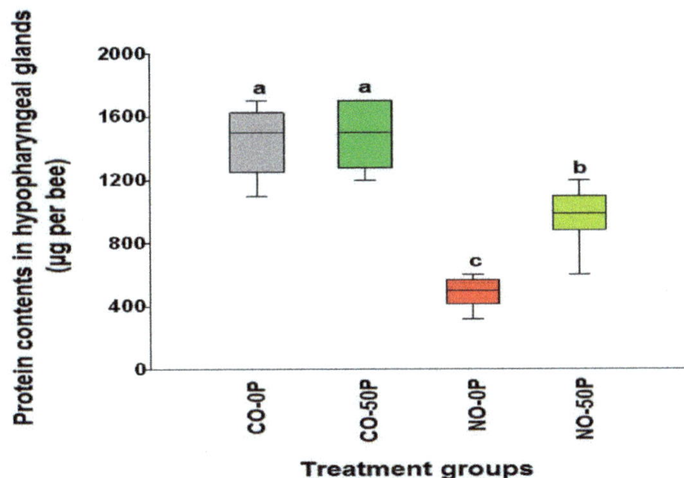

Figure 3. A box plot showing the median levels of hypopharyngeal gland protein contents across the treatments from the propolis extract experiment: control bees (CO-0P) (grey), control bees treated with propolis extract (CO-50P) (green), *N. ceranae*-infected bees not treated with propolis (NO-0P) (red) and infected bees treated with 50% propolis extract (NO-50P) (light green). The hypopharyngeal gland protein content was measured 14 days p.i. The boxes indicate the interquartile range, while the vertical bars indicate the range of the data. The different letters above each box plot represent significant differences (Kruskal–Wallis test: $\chi^2 = 31.75$, df = 3, $p < 0.0001$).

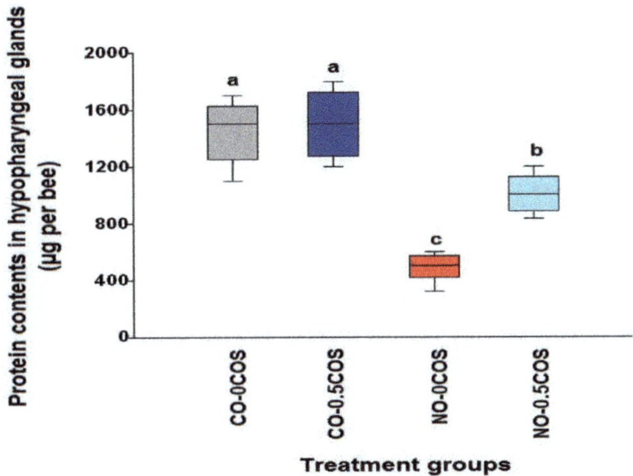

Figure 4. A box plot showing the median hypopharyngeal gland protein content levels from the COS experiment. The following treatments are shown with each box plot: control bees not treated with COS (CO-0COS) (grey), control bees treated with 0.5 ppm COS (CO-0.5COS) (purple), *N. ceranae*-infected bees not treated with COS (NO-0COS) (red) and infected bees treated with 0.5 ppm of COS (NO-0.5COS) (light blue). The hypopharyngeal gland was measured on day 14 p.i. The boxes represent interquartile ranges, while the vertical bars indicate the range of the data. The different letters above the box plots represent significant differences (Kruskal–Wallis test: $\chi^2 = 31.39$, df = 3, $p < 0.0001$).

2.3. Acini Diameters of Hypopharyngeal Glands

The smallest acini diameter on average was found in the *Nosema*-infected workers (NO) without any treatment, with a distance of 111.05 ± 0.4 μm. The mean diameters of acini of the hypopharyngeal glands were largest in the untreated control bees (CO) (134.55 ± 5.22 μm) and the COS treated control bees (CO-0.5COS) (137.13 ± 8.73 μm), followed by the control bees treated with propolis extract (CO-50P) (128.75 ± 2.9 μm), the infected bees treated with the propolis extract (NO-50P) (125.34 ± 2.9 μm), and the infected bees treated with COS (NO-0.5COS) (120.44 ± 6.8 μm). When we compare between *Nosema*-infected bees and the ones treated with propolis extract and COS, we see a significant increase in the acini diameter on average (χ^2 = 33.09, df = 5, p < 0.0001, Figure 5). However, the COS treated bees have significantly lower acini distances than the control bees (χ^2 = 33.09, df = 1, p = 0.0022), but the propolis extract treated bees do not have a significant difference in acini distance in comparison to the control bees treated with propolis (χ^2 = 33.09, df = 1, p = 0.0553).

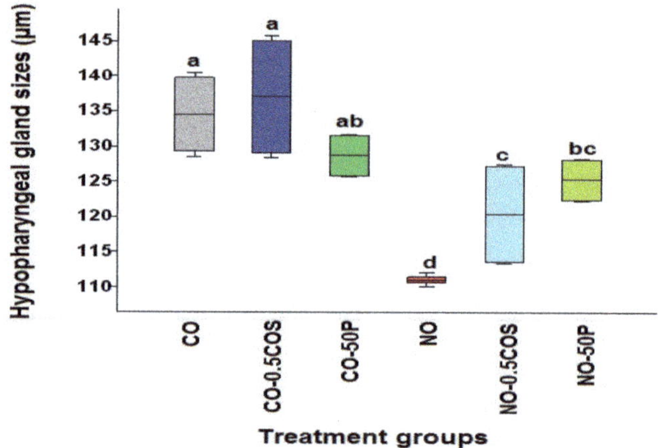

Figure 5. A box plot showing the median acini diameters across the treatments for the propolis extract and COS experiments. Each box plot represents a treatment: *A. dorsata* infected with *N. ceranae* dosages 10⁶ spores per bee without any treatment (NO) (red), *N. ceranae*-infected bees treated with propolis extract (NO-50P) (light green), *N. ceranae*-infected bees treated with 0.5 ppm COS (NO-0.5COS) (light blue), control bees without any treatment (CO) (grey), control bees treated with 0.5 ppm COS (CO-0.5COS) (purple), and control bees treated with 50% propolis extract (CO-50P) (green). The boxes indicate interquartile ranges, while the vertical bars represent the range of the data. The different letters above each box plot represents significant differences (Kruskal–Wallis test: χ^2 = 33.09, df = 5, p < 0.0001).

The histological structure of the hypopharyngeal glands of CO bees showed fully developed and contained with several secretory units or acini (oval to rounded shape), each unit composed of 5–8 secretory cells surrounded a central secretory duct. The secretory cells contained with numerous secretory granules stained red-pink with PAS that surround the large cell nuclei, stained greenish with light green (Figure 6), while the secretory units of the hypopharyngeal glands of *Nosema*-infected bees (NO) were incomplete developed in structure indicated by different irregular in shaped and sizes. Therefore, each cell cytoplasm consisted of numerous small secretory vesicles stained pink with PAS (Figure 7). Interestingly, the glands of NO-50P and NO-0.5COS showed fully developed acini. The secretory cell contains several vesicles giving both positive and negative staining with PAS, this indicated the cell storage both carbohydrate and non-carbohydrate molecules. In addition, the large extracellular space between adjacent acinar cells were found indicated by white gap between adjacent cells separating them from each other (Figures 8 and 9).

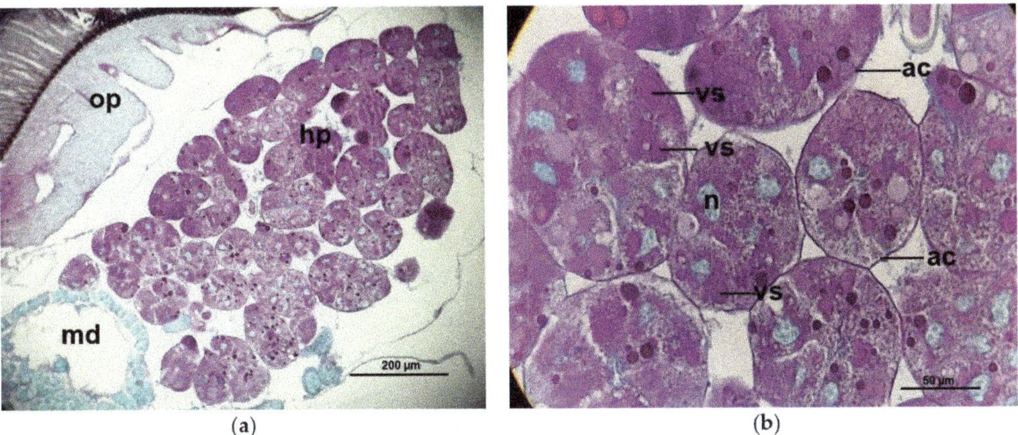

(a)   (b)

Figure 6. Histology cross sections of the hypopharyngeal gland of *A. dorsata* worker: (**a**) the completely developed secretory units of the glands on 14 dpi of control bees (CO). The secretory cells contain secretory granules surrounded the large nuclei of the secretory cells; (**b**) a section of the hypopharyngeal gland of CO bees with the high magnification of light microscope, the cytoplasm of the secretory cell is seen to contain variable numbers of secretory vesicles (stained red-pink with PAS). The oval nuclei are stained greenish with light green. Abbreviations: ac, acinus; hp, hypopharyngeal gland; md, mandibular gland; n, nucleus; op, optic lobe; vs, secretory vesicle.

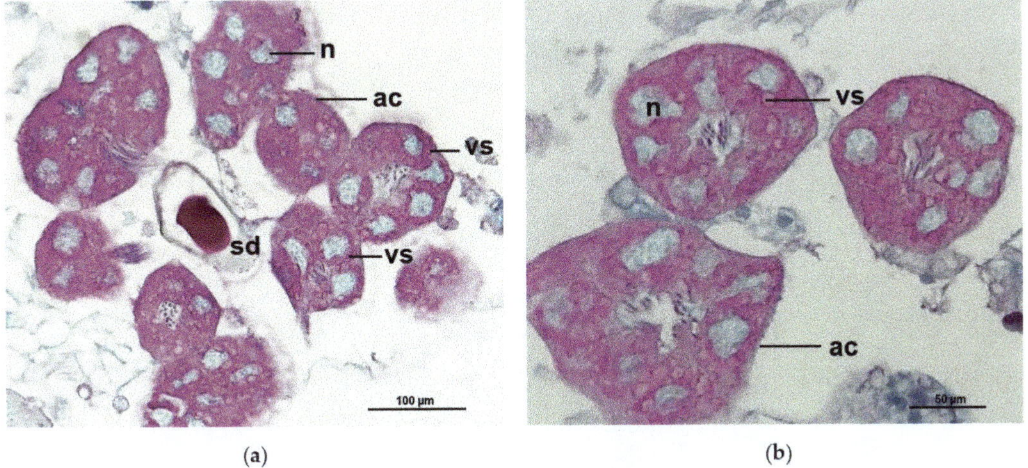

(a)   (b)

Figure 7. A section of the hypopharyngeal gland of 10^6 *N. ceranae*-infected *A. dorsata* worker (**a**) on 14 dpi of 10^6 *N. ceranae*-infected bees (NO), the cell cytoplasm contains variable numbers of secretory granules stained red-pink with PAS. The large oval loose nuclei are stained greenish from a light green dye used as a counterstain; (**b**) A medial section of NO bees on 14 dpi, the secretory cell contains secretory granules surround the large nuclei of the secretory cells. Abbreviations: ac, acinus; n, nucleus; sd, secretory duct; vs, secretory vesicle.

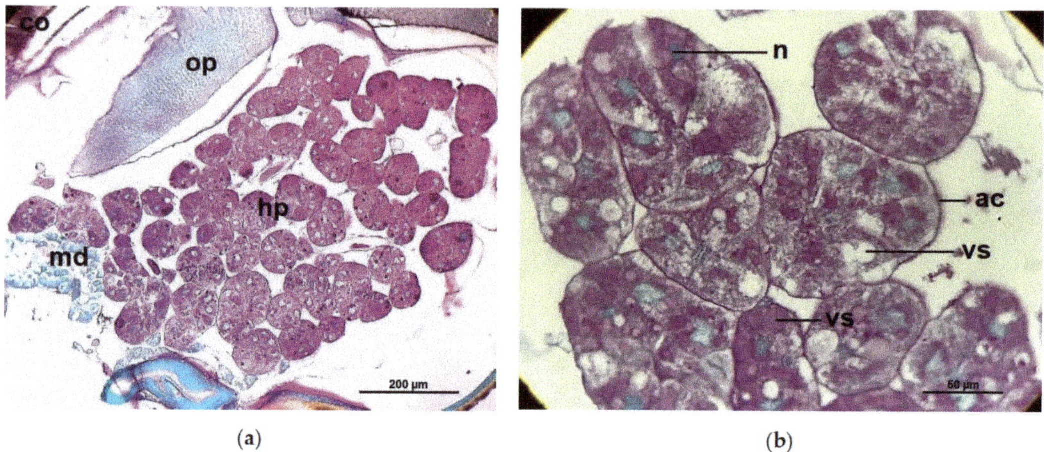

(**a**) (**b**)

Figure 8. The light micrographs of: (**a**) A section of the hypopharyngeal glands of 10^6 *N. ceranae*-infected bees on14 dpi, treated with 50% propolis (NO-50P). The cytoplasm of the secretory cell contains variable numbers of secretory granules stained red-pink with PAS. The oval nuclei are stained a greenish color from light green; (**b**) with higher magnification of NO-50P shows the secretory cell contains several secretory vesicles with negative staining using PAS, and also contains secretory vesicles with smaller amounts of carbohydrate, which are characterized by a red-pink color from PAS staining. Abbreviations: ac, acinus; co, compound eyes; hp, hypopharyngeal gland; md, mandibular gland; n, nucleus; op, optic lobe; vs, secretory vesicle.

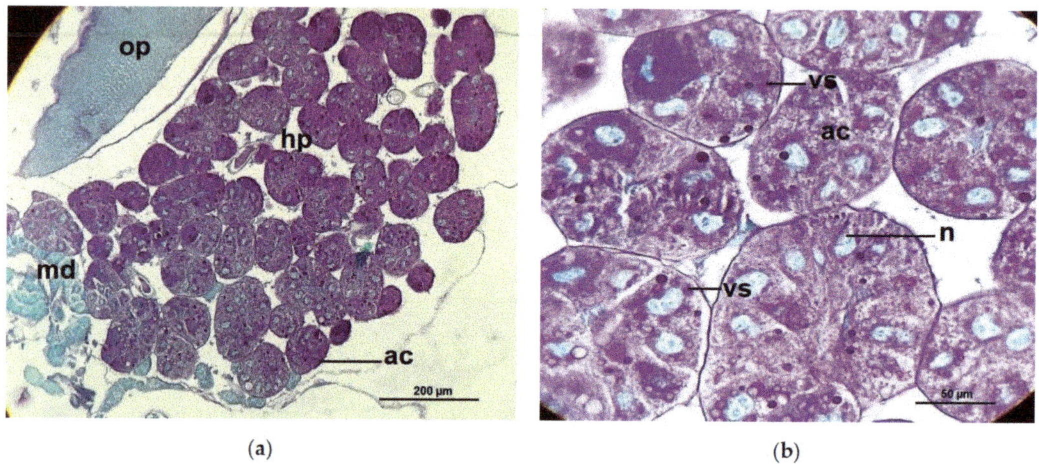

(**a**) (**b**)

Figure 9. (**a**) A cross section of the hypopharyngeal gland from 10^6 *N. ceranae*-infected *A. dorsata* bees on 14 dpi that were treated with 0.5 ppm COS (NO-0.5COS). The cytoplasm of the secretory cells contains variable numbers of secretory granules stained red-pink with PAS. The oval nuclei are stained greenish from a light green; (**b**) a medial cross section of the hypopharyngeal gland from 10^6 *N. ceranae*-infected *A. dorsata* bees on 14 dpi that were treated with 0.5 ppm COS (NO-0.5COS). Abbreviations: ac, acinus; hp, hypopharyngeal gland; md, mandibular gland; n, nucleus; op, optic lobe; vs, secretory vesicle.

2.4. Honey Bee Survival Rates

Kaplan–Meier curves showed that *A. dorsata* workers infected with *N. ceranae* dosed with 10^6 spores per bee (NO-0P) had significantly lower survival in comparison to the infected bees that received propolis treatment ($\chi^2 = 17.33$, df = 3, $p = 0.0005$, Figure 10). The control bees treated with propolis extract (CO-50P) had the highest survival, followed by CO-0P and NO-50P, respectively. A similar trend was found in bees treated with 0.5 ppm COS, except in this case there was no significant difference between the control bees and the control bees treated with COS (χ^2 = 16.08, df = 3, p = 0.0010, Figure 11).

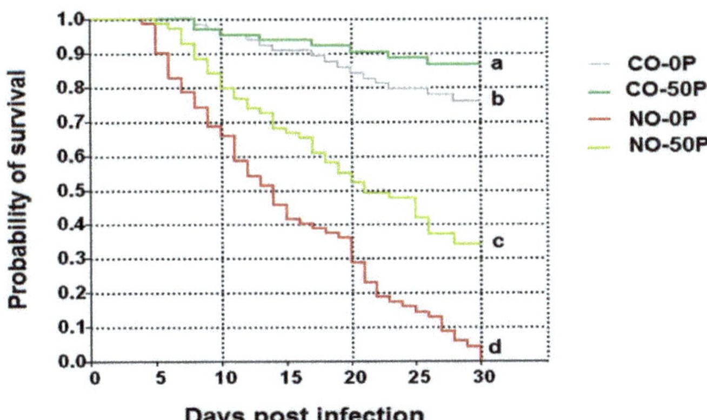

Figure 10. Kaplan–Meier survivorship curves of *A. dorsata* workers after being infected with 10^6 *N. ceranae* spores (NO-0P), versus infected bees that received a propolis treatment (NO-50P) or no infection and a propolis treatment (control: CO-0P and CO-50P). Survivorship curves with different letters within treatments are significantly different (Kruskal–Wallis test: χ^2 = 17.33, df = 3, p = 0.0005).

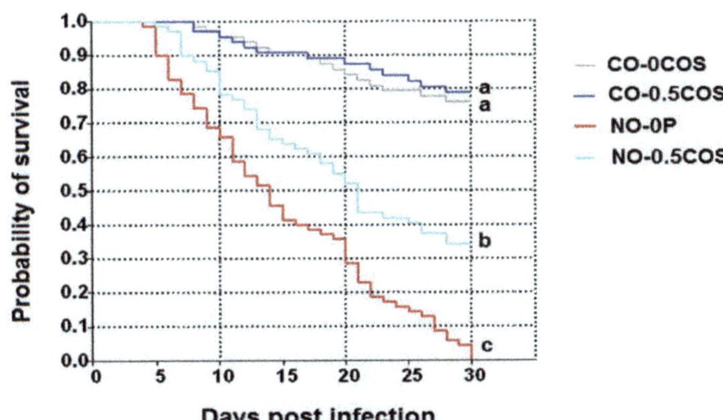

Figure 11. Kaplan–Meier survivorship curves of *A. dorsata* workers after *N. ceranae* infection at 10^6 spores (NO-0COS), NO-0.5COS or no infection (control: CO-0COS and CO-0.5COS). Survivorship curves with different letters within treatments are significantly different (Kruskal–Wallis test: χ^2 = 16.08, df = 3, p = 0.0010).

3. Discussion

The increase of hemolymph trehalose levels, protein content of the hypopharyngeal glands, and enhanced acini diameters of hypopharyngeal glands of the infected giant honey bee, *A. dorsata*, all indicate improved health after 50% propolis extract and 0.5 ppm COS treatment. Although the levels after treatment were not to the same level of the uninfected control bees, except for the acini diameter from the propolis treatment, there was still a significant increase for all health measures in comparison to the infected bees without any treatment. Based on our results, both stingless bee propolis extract and COS are effective treatments in improving the health of the honey bee. Propolis extract however may be a slightly better treatment as indicated by the full recovery of the acini diameter distance of the hypopharyngeal gland, which was not the case for the COS treatment. For both treatments there were no detrimental effects in the uninfected control bees for hemolymph trehalose levels, protein content in the hypopharyngeal gland, and the acini diameters in the hypopharyngeal gland, which suggests that these treatments are not having any damaging side-effects for these health parameters measured. Whether the increased trehalose levels and protein content of the hypopharyngeal gland is due to a lower parasite load or the improved nutrient absorption across the gut lining in treated bees—which results in higher tolerance of the parasite—remains to be investigated.

The higher trehalose levels in the uninfected bees treated with propolis extract in comparison to the uninfected bees without any treatment suggests that propolis might be affecting the sugar metabolism of the honey bee. This is interesting to note because hemolymph trehalose levels are central to buffering against the energetic stress suffered from the infected bees [24]. In addition, increased trehalose levels were found to be a key difference in honey bees selected to better tolerate *N. ceranae* infections [56]. Therefore, the increased hemolymph trehalose levels from propolis consumption could be one way in which increased survival results from infected bees treated with this [30]. Propolis extract also positively increased the protein contents of hypopharyngeal glands, and the acini diameters of hypopharyngeal glands of honey bee. This suggests that the increased health measures may also be due to the lowering of the *N. ceranae* load in the treated bees. This corresponds to the results of the previous study which showed the potential of propolis extracted from stingless bee to control *N. ceranae* infection in the red dwarf honey bee, *A. florea* [30]. Recently, it has also been demonstrated that propolis extract can increase the survival of *A. mellifera* [45,46]. Taken together, these results suggest that propolis extract may have a general positive impact on bee health, across all honey bee species. Further supporting this notion is the fact that stingless bee propolis extracts are known to specifically have antifungal properties [57]. Moreover, a previous study demonstrated abnormal structure of *N. ceranae* spores, inside *A. cerana* bees, after being treated with propolis extract, which corresponded with the interference of spore growth and development [47].

Bees treated with COS after infection with *N. ceranae* also had significantly higher trehalose levels, protein contents of hypopharyngeal glands, and increased acini diameters. However, it is more likely that these effects are resulting from indirect mechanisms such as enhanced bee immunity or increased nutrient absorption across the gut lining as opposed to directly reducing the reproduction and growth of the *N. ceranae* infection. COS is known to improve nutrient digestibility, gut functions and gut modifications in animals [50]. COS may affect *Nosema* development, but is more likely to achieve this through enhancing bee immunity, that will eventually result in higher hypopharyngeal gland protein contents, trehalose levels, and the increasing hypopharyngeal gland acini from decreased *N. ceranae* loads [58]. This is plausible because *N. ceranae* typically suppresses the immune system in infected bees in order for increased growth and reproduction inside the host [17,59].

It is important to note that *N. ceranae* can infect *A. dorsata* and develop well in this host. As previously shown, *A. mellifera*, *A. cerana* and *A. florea* can also be infected by *N. ceranae*. To date, all of the honey bee species have now been shown that not only can become infected with this parasite, but they are also suffering from the pathological effects of the infections as well. Based on our results *A. dorsata* is no exception, which raises concerns as the honey bee species may be suffering from some of the same behavioral and physiological changes that have been documented in *A. mellifera* from a *N. ceranae* infection [27]. Previous results show that the parasite develops well in each of the four honey bee species and that the intracellular life cycle is completed within three days p.i. [60–62]. Due to the successful reproduction in all four of the honey bee species there are opportunities for cross transmission between the species on a community level as they have overlapping foraging ranges and are known to share the same floral species when foraging [62–64].

The reduction in acini diameter of the hypopharyngeal glands of *N. ceranae*-infected bees might due to deficiency of amino acids used for secretory cell development. This is not surprising because previously it has been shown that the amino acid profiles in the hemolymph of infected bees is altered and feeding pollen can increase the survival of infected bees [65,66]. The recovery in the hypopharyngeal gland protein is important because it has been noted to play a role in protein synthesis of royal jelly production [6]. Metabolite dysregulation of royal jelly secretions has been documented in *N. ceranae* infected bee hives, which has implications for the antibacterial effectiveness of the secretions when feeding the brood [18]. Our findings suggest that COS and propolis extract treatment is likely to contribute to the increase of royal jelly productivity as well at the colony level, due to the increase of protein contents and the acini diameters of the hypopharyngeal glands. The lowering of the immune system is likely to be a result in the lack of protein nutrition resulting from the force feeding of spores that geminate and proliferate within the midgut epithelial cells where they disrupt host nutrient absorption [67,68]. Although infected bees do not exhibit obvious external disease symptoms, some of the main pathophysiological effects from an infection identified from omics studies have pointed to metabolic dysregulation [18,21,25,69]. Resulting from this metabolic dysregulation are the key symptoms of infection, which are lowered trehalose levels and reduced hypopharyngeal glands [22,24,62]. Therefore, we find that the measures used in this study to be accurate predictors of bee health and recovery from a *N. ceranae* infection. The effects of a *N. ceranae* infection on the colony level include lower colony population and the reduction of honey production [61,70]. Thus, we are interested if this treatment at the colony level might show an improvement in these colony level symptoms of infection.

The use of natural product such as propolis from stingless bees and COS will facilitate new strategies that can be used to control *Nosema* and improve honey bee health and beekeeping production. However, further experiments could be performed to determine the optimal doses to maximize the effect for each of the treatments. In addition, long term treatments could be investigated on a colony level to determine if they are great enough to effectively reduce the *N. ceranae* parasitic loads in a more natural setting. Perhaps using both treatments at the same time will synergistically improve the overall effectiveness in improving the health and survival of the honey bee. On one hand, *N. bombi* might also be another suitable target for this treatment which could lead to the health improvement of bumble bees as well, but on the other hand, the non-lethal side effects of these treatments should be investigated to determine if they pose any detriment to the health of bees. All of these would be interesting avenues to pursue in the future to further understand the practical use of stingless bee propolis extract and COS in terms of managing *N. ceranae* infections around the world.

4. Materials and Methods

4.1. Propolis Extraction

We collected propolis from three different stingless bee, *Tetrigona apicalis*, colonies from an apiary located in Chanthaburi Province, Thailand. We then dried the propolis in a hot air oven (Binder ED 53, BINDER GmbH, Tuttlingen, Germany) at 80 °C for 72 h, this was then frozen at −21 °C (Sharp SJ-X43T, Sharp Thai Co., Ltd. (STCL), Bangkok, Thailand) for 3 h and grinded using a motor and pestle. We extracted 60 g of propolis powder with 100 mL of 70% ethanol for 72 h, this was then followed by gravity filtration using a Whatman No. 4 filter paper [30]. After filtration a crude ethanol extract was formed that we defined as 100% propolis stock solution. For the experiments a 50% propolis solution was prepared by diluting the stock solution with water (*v/v*).

4.2. Chito-Oligosaccharide Solution Preparation

We made a 10^4 ppm stock of COS, by taking 0.25 g of COS (6081 Da) and dissolving it in 5 mL of pure *A. dorsata* honey (pH = 3.45). We then adjusted the final volume of this to 20 mL with 50% sucrose solution (*v/v*). We then diluted the 10^4 ppm stock solution of

COS to a 50% honey solution using water (*v/v*) to make a final concentration of 10^2 ppm. Afterwards we then prepared a 0.5 ppm COS solution using the same methods that had a pH of 3.77 (pH meter, Mettler Toledo Gmbh, Greifensee, Zurich, Switzerland).

4.3. Spore Preparation

Nosema ceranae spores were propagated from heavily infected *A. florea* colonies located in the Chon Buri Province of Thailand. We fed isolated spores to *A. mellifera* workers (5×10^7 spores for 50 bees) that were kept at $34 \pm 2\,°C$ (Memmert IPP 260, Schwabach, Germany) with relative humidity (Barigo-8861, Schwenningen, Germany) (RH) between 50–55% for 14 days in order to propagate more spores for the experimental infections. To propagate more spores, midguts were removed and transferred to a 1.5 mL microcentrifuge tube containing 100 µL distilled water. The midguts were then homogenized using a sterile pestle and centrifuged at $6000 \times g$ (Benchmark Scientific Z206-A, Sayreville, NJ, USA) for 10 min, this was repeated for 3 times [40]. We discarded the supernatant each time and the white sediment at the bottom was collected to be counted using a hemocytometer (Hausser Scientific, Horsham, PA, USA) under a light microscope (Olympus CX50, Shinjuku, Tokyo, Japan) [71]. After one more centrifugation, we re-suspended the spores in 50% (*w/v*) sucrose solution to make a final concentration of 5×10^5 spores per µL. We stored this syrup at room temperature overnight until further use.

4.4. Propolis Extract and COS Treatment Experiments

We obtained 3 frames of sealed brood from three *Nosema* free colonies of *A. dorsata* located in Samut Songkhram Province, Thailand. Colonies were confirmed to be *Nosema* free following standard procedures [72,73]. To obtain newly emerged bees, the brood frames were kept in an incubator (Memmert IPP 260, Schwabach, Germany) at $34 \pm 2\,°C$ with RH (Barigo-8861, Schwenningen, Germany) between 50–55%. The newly emerged bees, between 24–48 h of age, were confined to cages, in groups of 50, and divided into 8 groups. The first 4 groups, were individually force-fed with 2 µL 50% sucrose solution (*v/v*) containing 10^6 *N. ceranae* spores per bee. We then provided 2 groups with 2 mL of either 0% or 50% stingless bee propolis extracts, daily, and these groups were defined as NO-0P and NO-50P, respectively. For the other 2 groups, we provided 2 mL of 0 ppm or 0.5 ppm of COS defined as NO-0COS and NO-0.5COS, respectively.

The control groups were individually force-fed with only 50% sucrose solution (*v/v*), and were defined as the negative control bees CO-0P, CO-0COS, CO-0.5COS, and CO-50P, respectively. In the CO-50P control group was each bee was also treated daily with 2 mL of 50% stingless bee propolis extract, while in the CO-0.5COS control group the bees were treated daily with 2 mL of 0.5 ppm COS. For the duration of the experiment, each cage was fitted with two gravity feeders, one containing distilled water, and the other sugar syrup (50% *w/v* sucrose solution). We also supplied 60 g of pollen mixed with 17 mL of 50% sucrose solution (*w/v*), each was replenished as necessary throughout the experiment. All cages were placed in an incubator at $34 \pm 2\,°C$ (Memmert IPP 260, Schwabach, Germany), with a RH ranging from 50–55%. The 50% stingless bee propolis extract was provided in 2 mL at a time in a 1.5 mL micro-centrifuge tube from the start of the experiment (0 Day p.i.), until the end (30 Days p.i.), and was replaced as necessary. For the COS treatment, we provided 2 mL of 0.5 ppm COS in 50% honey solution in a 1.5 mL micro-centrifuge at the start of the experiment (0 Day p.i.) and this was replaced as necessary until the end of the experiment (30 Days p.i.).

4.5. Hemolymph Trehalose Measurements

On Day 14 p.i., 10 honey bees were removed from each cage and were anaesthetized at $-21\,°C$ for 5 min. Before we collected their hemolymph, honey bees were mounted on a wax plate by a pair of insect pins crossing over the waist. Using a glass microcapillary (Hirschmann® Laborgerate, Eberstadt, Germany), 5 µL per bee was collected by puncturing abdomen segments between tergites 3 and 4, and the hemolymph was transferred to a microcentrifuge tube (Eppendorf, Hamburg, Germany) containing 45 µL of 0.85% NaCl. For each sample, 2.9 mL of anthrone reagent was added and then vortexed for 30 s before we quickly put them into a boiling water bath for 15 min. After this they were placed into cold water (4 °C) for 20 min and read at 620 nm absorbance using a Shimadzu UV-visible spectrophotometer (UV-1610). Quantification of the hemolymph trehalose amounts were based on a standard curve.

4.6. Hypopharyngeal Gland Protein Content Measurements

Another 10 bees were randomly removed from each cage at 14 days post infection (p.i.). These bees were decapitated so that their hypopharyngeal glands could be removed under a stereomicroscope (Olympus CH30, Shinjuku, Tokyo, Japan). Glands of each bee were stored in 50 µL of phosphate buffer solution (pH 7.8) in a 1.5 mL microcentrifuge tube. These were then homogenized and centrifuged at $1000 \times g$ for 2 min. Supernatant from each tube was used in the Bradford protein assay [74]. Quantification of protein content was based on standard curves that were prepared using bovine serum albumin (BSA). Protein absorbance was measured at 595 nm absorbance against a blank reagent using a Shimadzu UV-visible spectrophotometer (UV-1610).

4.7. Measurements of Acinar Sizes of the Hypopharyngeal Glands and Histological Structure

Another 10 bees from each group were collected on 14 days p.i. and the heads were dissected in insect saline (NaCl 7.5 g/L, Na_2HPO_4 2.38 g/L, KH_2PO_4 2.72 g/L) and then fixed in Bouin's solution for 24 h. Samples were dehydrated using a series of increasing ethyl alcohol concentrations: 70%, 90%, 95%, and 100% for 10 min per concentration. Samples were then soaked in xylene for 1 h and then embedded in paraffin wax. The tissues were sectioned into 6 µm thickness using a rotary microtome (Leica, Wetzlar, Germany), and then stained with Periodic acid Schiff's reagent (PAS) followed by a counter staining of light green dye [75,76]. Measurement of acinar sizes of the hypopharyngeal glands were made under a light microscopy (Olympus CX 50, Shinjuku, Tokyo, Japan) using a micrometer (ERMA: ESM-11, Japan); $n = 10$ per bee each treatment.

4.8. Survival Analysis

Survivorship curves of all treatment groups were generated using the Kaplan–Meier approach by plotting number of surviving bees against days from initiation of the experiment [30]. Honey bee survival rates were compared across the treatment groups using a non-parametric, univariate analysis of variance and a corresponding post hoc test (Kruskal–Wallis test and the Mann–Whitney U test).

4.9. Statistical Analyses

Hemolymph trehalose levels, protein contents of the hypopharyngeal gland, the diameter of the hypopharyngeal gland acini of *N. ceranae*-infected bees on day 14 p.i., and the survival rates were normally distributed (Jarque–Bera JB test: $p > 0.05$) but had unequal variances (Levene's test: $p < 0.05$). We, therefore, used a non-parametric Kruskal–Wallis test and a Mann–Whitney U test to compare across the treatment groups. Multiple comparisons were accounted for using a Bonferroni correction.

Author Contributions: Conceptualization, G.S. and S.N.; methodology, R.P. and E.C.; software, S.N.; validation, G.S., S.N. and C.L.M.; formal analysis, C.L.M. and S.N.; investigation, R.P. and E.C.; resources, R.P.; data curation, R.P.; E.C., S.N.; writing—original draft preparation, G.S.; writing—review and editing, G.S. and C.L.M.; visualization, S.N. and E.C.; supervision, G.S. and C.L.M.; project administration, G.S.; funding acquisition, G.S., S.N. and R.P. All authors have read and agreed to the published version of the manuscript.

Funding: This research was funded by The Royal Golden Jubilee Ph.D. Scholarship (RGJ) no. PHD58K0155; PHD/0078/2559, Thailand Research Funds.

Institutional Review Board Statement: The study was conducted according to the guidelines of the Declaration of Helsinki, and approved by Thai IACUC (Animal care and uses Committee of Thailand) (Protocol Code: U1-03211-2559 and date of approval: 20 March 2016).

Informed Consent Statement: Informed consent was sought from the owner of honey bee colonies. Honey bee samples were collected from that owner who has agreed to have his honey bee sampled.

Data Availability Statement: The data presented in this study are available on requested from the corresponding author.

Acknowledgments: We would like to thank Department of Biology, Faculty of Science, Burapha University for providing research facilities.

Conflicts of Interest: The authors declare no conflict of interest.

References

1. Breeze, T.D.; Vaissière, B.E.; Bommarco, R.; Petanidou, T.; Seraphides, N.; Kozák, L.; Scheper, J.; Biesmeijer, J.C.; Kleijn, D.; Gyldenkaerne, S.; et al. Agricultural Policies Exacerbate Honeybee Pollination Service Supply-Demand Mismatches Across Europe. *PLoS ONE* **2014**, *9*, e82996. [CrossRef] [PubMed]
2. Genersch, E. Honey bee pathology: Current threats to honey bees and beekeeping. *Appl. Microbiol. Biotechnol.* **2010**, *87*, 87–97. [CrossRef]
3. Klein, A.-M.; Vaissière, B.E.; Cane, J.H.; Steffan-Dewenter, I.; Cunningham, S.A.; Kremen, C.; Tscharntke, T. Importance of pollinators in changing landscapes for world crops. *Proc. R. Soc. B Biol. Sci.* **2007**, *274*, 303–313. [CrossRef] [PubMed]
4. Kremen, C.; Williams, N.M.; Aizen, M.A.; Gemmill-Herren, B.; Lebuhn, G.; Minckley, R.; Packer, L.; Potts, S.G.; Roulston, T.; Steffan-Dewenter, I.; et al. Pollination and other ecosystem services produced by mobile organisms: A conceptual framework for the effects of land-use change. *Ecol. Lett.* **2007**, *10*, 299–314. [CrossRef]
5. Potts, S.G.; Petanidou, T.; Roberts, S.; O'Toole, C.; Hulbert, A.; Willmer, P. Plant-pollinator biodiversity and pollination services in a complex Mediterranean landscape. *Biol. Conserv.* **2006**, *129*, 519–529. [CrossRef]
6. Suwannapong, G. *Honeybees of Thailand*; Nova Science Publishers, Inc.: New York, NY, USA, 2019; pp. 1–378.
7. Goulson, D.; Nicholls, E.; Botías, C.; Rotheray, E.L. Bee declines driven by combined stress from parasites, pesticides, and lack of flowers. *Science* **2015**, *347*, 1255957. [CrossRef] [PubMed]
8. Kulhanek, K.; Steinhauer, N.; Rennich, K.; Caron, D.M.; Sagili, R.R.; Pettis, J.S.; Ellis, J.D.; Wilson, M.E.; Wilkes, J.T.; Tarpy, D.R.; et al. A national survey of managed honey bee 2015–2016 annual colony losses in the USA. *J. Apic. Res.* **2017**, *56*, 328–340. [CrossRef]
9. López-Uribe, M.M.; Ricigliano, V.A.; Simone-Finstrom, M. Defining Pollinator Health: A Holistic Approach Based on Ecological, Genetic, and Physiological Factors. *Annu. Rev. Anim. Biosci.* **2020**, *8*, 269–294. [CrossRef]
10. McMahon, D.P.; Fürst, M.; Caspar, J.; Theodorou, P.; Brown, M.J.F.; Paxton, R. A sting in the spit: Widespread cross-infection of multiple RNA viruses across wild and managed bees. *J. Anim. Ecol.* **2015**, *84*, 615–624. [CrossRef]
11. Paxton, R. Does infection by Nosema ceranae cause "Colony Collapse Disorder" in honey bees (*Apis mellifera*)? *J. Apic. Res.* **2010**, *49*, 80–84. [CrossRef]
12. Higes, M.; Hernández, R.M.; Botías, C.; Bailón, E.G.; González-Porto, A.V.; Barrios, L.; Nozal, M.J.; Bernal, J.L.; Jiménez, J.J.; Palencia, P.G.; et al. How natural infection by *Nosema ceranae* causes honeybee colony collapse. *Environ. Microbiol.* **2008**, *10*, 2659–2669. [CrossRef]
13. Klee, J.; Besana, A.M.; Genersch, E.; Gisder, S.; Nanetti, A.; Tam, D.Q.; Chinh, T.X.; Puerta, F.; Ruz, J.M.; Kryger, P.; et al. Widespread dispersal of the microsporidian *Nosema ceranae*, an emergent pathogen of the western honey bee, Apis mellifera. *J. Invertebr. Pathol.* **2007**, *96*, 1–10. [CrossRef] [PubMed]
14. Martín-Hernández, R.; Bartolomé, C.; Chejanovsky, N.; Le Conte, Y.; Dalmon, A.; Dussaubat, C.; García-Palencia, P.; Meana, A.; Pinto, M.A.; Soroker, V.; et al. *Nosema ceranaein* in *Apis mellifera*: A 12 years postdetectionperspective. *Environ. Microbiol.* **2018**, *20*, 1302–1329. [CrossRef]
15. Chemurot, M.; De Smet, L.; Brunain, M.; De Rycke, R.; de Graaf, D.C. *Nosema neumanni* n. sp. (Microsporidia, Nosematidae), a new microsporidian parasite of honeybees, *Apis mellifera* in Uganda. *Eur. J. Protistol.* **2017**, *61*, 13–19. [CrossRef]
16. Zander, E. Tierische Parasiten als Krankheitserreger bei der Biene. *Munch. Bienenztg.* **1909**, *21*, 196–204.
17. Antúnez, K.; Hernández, R.M.; Prieto, L.; Meana, A.; Zunino, P.; Higes, M. Immune suppression in the honey bee (*Apis mellifera*) following infection by *Nosema ceranae* (Microsporidia). *Environ. Microbiol.* **2009**, *11*, 2284–2290. [CrossRef]
18. Broadrup, R.L.; Mayack, C.; Schick, S.J.; Eppley, E.J.; White, H.K.; Macherone, A. Honey bee (*Apis mellifera*) exposomes and dysregulated metabolic pathways associated with *Nosema ceranae* infection. *PLoS ONE* **2019**, *14*, e0213249. [CrossRef]
19. Goblirsch, M.; Huang, Z.Y.; Spivak, M. Physiological and Behavioral Changes in Honey Bees (*Apis mellifera*) Induced by *Nosema ceranae* Infection. *PLoS ONE* **2013**, *8*, e58165. [CrossRef]
20. Lecocq, A.; Jensen, A.B.; Kryger, P.; Nieh, J.C. Parasite infection accelerates age polyethism in young honey bees. *Sci. Rep.* **2016**, *6*, 22042. [CrossRef] [PubMed]

21. Vidau, C.; Panek, J.; Texier, C.; Biron, D.G.; Belzunces, L.P.; Le Gall, M.; Broussard, C.; Delbac, F.; El Alaoui, H. Differential proteomic analysis of midguts from *Nosema ceranae*-infected honeybees reveals manipulation of key host functions. *J. Invertebr. Pathol.* **2014**, *121*, 89–96. [CrossRef] [PubMed]

22. Corby-Harris, V.; Deeter, M.E.; Snyder, L.; Meador, C.; Welchert, A.C.; Hoffman, A.; Obernesser, B.T. Octopamine mobilizes lipids from honey bee (*Apis mellifera*) hypopharyngeal glands. *J. Exp. Biol.* **2020**, *223*, 216135. [CrossRef] [PubMed]

23. Li, W.; Chen, Y.; Cook, S.C. Chronic *Nosema ceranae* infection inflicts comprehensive and persistent immunosuppression and accelerated lipid loss in host *Apis mellifera* honey bees. *Int. J. Parasitol.* **2018**, *48*, 433–444. [CrossRef]

24. Mayack, C.; Naug, D. Parasitic infection leads to decline in hemolymph sugar levels in honeybee foragers. *J. Insect Physiol.* **2010**, *56*, 1572–1575. [CrossRef]

25. Dussaubat, C.; Brunet, J.-L.; Higes, M.; Colbourne, J.K.; Lopez, J.; Choi, J.-H.; Hernández, R.M.; Botías, C.; Cousin, M.; McDonnell, C.; et al. Gut Pathology and Responses to the Microsporidium *Nosema ceranae* in the Honey Bee *Apis mellifera*. *PLoS ONE* **2012**, *7*, e37017. [CrossRef] [PubMed]

26. García-Palencia, P.; Hernández, R.M.; González-Porto, A.-V.; Marin, P.; Meana, A.; Higes, M. Natural infection by *Nosema ceranaecauses* similar lesions as in experimentally infected caged-worker honey bees (*Apis mellifera*). *J. Apic. Res.* **2010**, *49*, 278–283. [CrossRef]

27. Mayack, C.; Natsopoulou, M.E.; McMahon, D. *Nosema ceranaealters* a highly conserved hormonal stress pathway in honeybees. *Insect Mol. Biol.* **2015**, *24*, 662–670. [CrossRef]

28. Mayack, C.; Naug, D. Energetic stress in the honeybee *Apis mellifera* from *Nosema ceranae* infection. *J. Invertebr. Pathol.* **2009**, *100*, 185–188. [CrossRef]

29. Mayack, C.; Naug, D. Individual energetic state can prevail over social regulation of foraging in honeybees. *Behav. Ecol. Sociobiol.* **2013**, *67*, 929–936. [CrossRef]

30. Suwannapong, G.; Maksong, S.; Phainchajoen, M.; Benbow, M.; Mayack, C. Survival and health improvement of *Nosema* infected *Apis florea* (Hymenoptera: Apidae) bees after treatment with propolis extract. *J. Asia-Pacific Èntomol.* **2018**, *21*, 437–444. [CrossRef]

31. Shimanuki, H.; Knox, D.A.; Furgala, B.; Caron, D.M.; Williams, J.L. Diseases and pests of honey bee. In *The Hive and the Honey Bee*; Graham, J.M., Ed.; Dadant and Sons: Hamilton, IL, USA, 1992; pp. 1083–1152.

32. Giacobino, A.; Rivero, R.; Molineri, A.I.; Cagnolo, N.B.; Merke, J.; Orellano, E.; Salto, C.; Signorini, M. Fumagillin control of *Nosema ceranae* (Microsporidia:Nosematidae) infection in honey bee (Hymenoptera:Apidae) colonies in Argentina. *Vet. Ital.* **2016**, *52*, 145–151. [PubMed]

33. Mendoza, Y.; Diaz-Cetti, S.; Ramallo, G.; Santos, E.; Porrini, M.; Invernizzi, C. *Nosema ceranae* Winter Control: Study of the Effectiveness of Different Fumagillin Treatments and Consequences on the Strength of Honey Bee (Hymenoptera: Apidae) Colonies. *J. Econ. Èntomol.* **2016**, *110*, 1–5. [CrossRef]

34. Maistrello, L.; Lodesani, M.; Costa, C.; Leonardi, F.; Marani, G.; Caldon, M.; Mutinelli, F.; Granato, A. Screening of natural compounds for the control of nosema disease in honeybees (*Apis mellifera*). *Apidologie* **2008**, *39*, 436–445. [CrossRef]

35. Gajger, I.T.; Ribaric, J.; Matak, M.; Svecnjak, L.; Kozaric, Z.; Nejedli, S.; Skerl, I.S. Zeolite clinoptilolite as a dietary supplement and remedy for honeybee (*Apis mellifera* L.) colonies. *Veterinární Med.* **2017**, *60*, 696–705. [CrossRef]

36. Tlak-Gajger, I.; Tomljanovic, Z.; Stanisavljevic, L. An environmentally friendly approach to the control of *Varroa destructor* mite and *Nosema ceranae* disease in Carniolan honeybee (*Apis mellifera* Carnica) colonies. *Arch. Biol. Sci.* **2013**, *65*, 1585–1592. [CrossRef]

37. Higes, M.; Gómez-Moracho, T.; Rodríguez-García, C.; Botías, C.; Hernández, R.M. Preliminary effect of an experimental treatment with "Nozevit®", (a phyto-pharmacological preparation) for *Nosema ceranae* control. *J. Apic. Res.* **2014**, *53*, 472–474. [CrossRef]

38. Porrini, M.P.; Fernández, N.J.; Garrido, P.M.; Gende, L.B.; Medici, S.K.; Eguaras, M.J. In vivo evaluation of antiparasitic activity of plant extracts on *Nosema ceranae* (Microsporidia). *Apidologie* **2011**, *42*, 700–707. [CrossRef]

39. Cilia, G.; Garrido, C.; Bonetto, M.; Tesoriero, D.; Nanetti, A. Effect of Api-Bioxal® and ApiHerb® Treatments against *Nosema ceranae* Infection in *Apis mellifera* Investigated by Two qPCR Methods. *Veter-Sci.* **2020**, *7*, 125. [CrossRef] [PubMed]

40. Borges, D.; Guzman-Novoa, E.; Goodwin, P. Effects of Prebiotics and Probiotics on Honey Bees (*Apis mellifera*) Infected with the Microsporidian Parasite *Nosema ceranae*. *Microorganisms* **2021**, *9*, 481. [CrossRef]

41. El Khoury, S.; Rousseau, A.; Lecoeur, A.; Cheaib, B.; Bouslama, S.; Mercier, P.-L.; Demey, V.; Castex, M.; Giovenazzo, P.; Derome, N. Deleterious Interaction Between Honeybees (*Apis mellifera*) and its Microsporidian Intracellular Parasite *Nosema ceranae* Was Mitigated by Administrating Either Endogenous or Allochthonous Gut Microbiota Strains. *Front. Ecol. Evol.* **2018**, *6*, 58. [CrossRef]

42. Valizadeh, P.; Guzman-Novoa, E.; Goodwin, P.H. Effect of Immune Inducers on *Nosema ceranae* Multiplication and Their Impact on Honey Bee (*Apis mellifera* L.) Survivorship and Behaviors. *Insects* **2020**, *11*, 572. [CrossRef]

43. Nanetti, A.; Ugolini, L.; Cilia, G.; Pagnotta, E.; Malaguti, L.; Cardaio, I.; Matteo, R.; Lazzeri, L. Seed Meals from *Brassica nigra* and *Eruca sativa* Control Artificial *Nosema ceranae* Infections in *Apis mellifera*. *Microorganisms* **2021**, *9*, 949. [CrossRef]

44. Borges, D.; Guzman-Novoa, E.; Goodwin, P.H. Control of the microsporidian parasite *Nosema ceranae* in honey bees (*Apis mellifera*) using nutraceutical and immuno-stimulatory compounds. *PLoS ONE* **2020**, *15*, e0227484. [CrossRef]

45. Mura, A.; Pusceddu, M.; Theodorou, P.; Angioni, A.; Floris, I.; Paxton, R.J.; Satta, A. Propolis Consumption Reduces *Nosema ceranae* Infection of European Honey Bees (*Apis mellifera*). *Insects* **2020**, *11*, 124. [CrossRef]

46. Naree, S.; Ellis, J.D.; Benbow, M.E.; Suwannapong, G. The use of propolis for preventing and treating *Nosema ceranae* infection in western honey bee (*Apis mellifera* Linnaeus, 1787) workers. *J. Apic. Res.* **2021**, 1–11. [CrossRef]

47. Yemor, T.; Phiancharoen, M.; Benbow, M.E.; Suwannapong, G. Effects of stingless bee propolis on *Nosema ceranae* infected Asian honey bees, *Apis cerana*. *J. Apic. Res.* **2015**, *54*, 468–473. [CrossRef]
48. Arismendi, N.; Vargas, M.; López, M.D.; Barría, Y.; Zapata, N. Promising antimicrobial activity against the honey bee parasite *Nosema ceranae* by methanolic extracts from Chilean native plants and propolis. *J. Apic. Res.* **2018**, *57*, 522–535. [CrossRef]
49. Saltykova, E.S.; Karimova, A.A.; Gataullin, A.R.; Gaifullina, L.; Matniyazov, R.T.; Frolova, M.A.; Albulov, A.I.; Nikolenko, A.G. The effect of high-molecular weight chitosans on the antioxidant and immune systems of the honeybee. *Appl. Biochem. Microbiol.* **2016**, *52*, 553–557. [CrossRef]
50. Thongsong, B.; Suthongsa, S.; Pichyangkura, R.; Kalandakanond-Thongsong, S. Effects of chito-oligosaccharide supplementation with low or medium molecular weight and high degree of deacetylation on growth performance, nutrient digestibility and small intestinal morphology in weaned pigs. *Livest. Sci.* **2018**, *209*, 60–66. [CrossRef]
51. Yousef, M.; Pichyangkura, R.; Soodvilai, S.; Chatsudthipong, V.; Muanprasat, C. Chitosan oligosaccharide as potential therapy of inflammatory bowel disease: Therapeutic efficacy and possible mechanisms of action. *Pharmacol. Res.* **2012**, *66*, 66–79. [CrossRef]
52. Azagra-Boronat, I.; Rodríguez-Lagunas, M.J.; Castell, M.; Pérez-Cano, F.J. Chapter 14—Prebiotics for gastrointestinal infections and acute diarrhea. In *Dietary Interventions in Gastrointestinal Diseases*; Watson, R., Preedy, V., Eds.; Academic Press: Cambridge, MA, USA, 2019; pp. 179–191.
53. Guan, G.; Azad, A.K.; Lin, Y.; Kim, S.W.; Tian, Y.; Liu, G.; Wang, H. Biological Effects and Applications of Chitosan and Chito-Oligosaccharides. *Front. Physiol.* **2019**, *10*, 516. [CrossRef]
54. Kunanusornchai, W.; Witoonpanich, B.; Tawonsawatruk, T.; Pichyangkura, R.; Chatsudthipong, V.; Muanprasat, C. Chitosan oligosaccharide suppresses synovial inflammation via AMPK activation: An in vitro and in vivo study. *Pharmacol. Res.* **2016**, *113*, 458–467. [CrossRef]
55. Zhang, C.; Jiao, S.; Wang, Z.A.; Du, Y. Exploring Effects of Chitosan Oligosaccharides on Mice Gut Microbiota in in vitro Fermentation and Animal Model. *Front. Microbiol.* **2018**, *9*, 2388. [CrossRef] [PubMed]
56. Kurze, C.; Mayack, C.; Hirche, F.; Stangl, G.I.; Le Conte, Y.; Kryger, P.; Moritz, R.F.A. *Nosema* spp. infections cause no energetic stress in tolerant honeybees. *Parasitol. Res.* **2016**, *115*, 2381–2388. [CrossRef] [PubMed]
57. Drescher, N.; Klein, A.-M.; Neumann, P.; Yañez, O.; Leonhardt, S.D. Inside Honeybee Hives: Impact of Natural Propolis on the Ectoparasitic Mite *Varroa destructor* and Viruses. *Insects* **2017**, *8*, 15. [CrossRef] [PubMed]
58. Saltykova, E.S.; Gaifullina, L.R.; Kaskinova, M.; Gataullin, A.R.; Matniyazov, R.T.; Poskryakov, A.V.; Nikolenko, A.G. Effect of Chitosan on Development of *Nosema apis* Microsporidia in Honey Bees. *Microbiology* **2018**, *87*, 738–743. [CrossRef]
59. Paris, L.; El Alaoui, H.; Delbac, F.; Diogon, M. Effects of the gut parasite *Nosema ceranae* on honey bee physiology and behavior. *Curr. Opin. Insect Sci.* **2018**, *26*, 149–154. [CrossRef]
60. Higes, M.; García-Palencia, P.; Hernández, R.M.; Meana, A. Experimental infection of *Apis mellifera* honeybees with *Nosema ceranae* (Microsporidia). *J. Invertebr. Pathol.* **2007**, *94*, 211–217. [CrossRef] [PubMed]
61. Suwannapong, G.; Maksong, S.; Seanbualuang, P.; Benbow, M.E. Experimental infection of red dwarf honeybee, *Apis florea*, with *Nosema ceranae*. *J. Asia-Pacific Èntomol.* **2010**, *13*, 361–364. [CrossRef]
62. Suwannapong, G.; Yemor, T.; Boonpakdee, C.; Benbow, M.E. *Nosema ceranae*, a new parasite in Thai honeybees. *J. Invertebr. Pathol.* **2011**, *106*, 236–241. [CrossRef]
63. Suwannapong, G.; Benbow, M.E.; Nieh, J.C. Biology of Thai honeybees: Natural history and threats. In *Bees: Biology, Threats and Colonies*; Florio, R.M., Ed.; Nova Science Publishers, Inc.: New York, NY, USA, 2011; pp. 1–98.
64. Suwannapong, G.; Maksong, S.; Yemor, T.; Junsuri, N.; Benbow, M.E. Three species of native Thai honey bees exploit overlapping pollen resources: Identification of bee flora from pollen loads and midguts from *Apis cerana*, *A. dorsata* and *A. florea*. *J. Apic. Res.* **2013**, *52*, 196–201. [CrossRef]
65. Aliferis, K.A.; Copley, T.; Jabaji, S. Gas chromatography–mass spectrometry metabolite profiling of worker honey bee (*Apis mellifera* L.) hemolymph for the study of *Nosema ceranae* infection. *J. Insect Physiol.* **2012**, *58*, 1349–1359. [CrossRef]
66. Jack, C.J.; Uppala, S.S.; Lucas, H.M.; Sagili, R.R. Effects of pollen dilution on infection of *Nosema ceranae* in honey bees. *J. Insect Physiol.* **2016**, *87*, 12–19. [CrossRef] [PubMed]
67. Chen, Y.; Evans, J.D.; Smith, I.B.; Pettis, J.S. *Nosema ceranae* is a long-present and wide-spread microsporidian infection of the European honey bee (*Apis mellifera*) in the United States. *J. Invertebr. Pathol.* **2008**, *97*, 186–188. [CrossRef]
68. Chen, Y.P.; Evans, J.D.; Murphy, C.; Gutell, R.; Zuker, M.; Gundensen-Rindal, D.; Pettis, J.S. Morphological, Molecular, and Phylogenetic Characterization of *Nosema ceranae*, a Microsporidian Parasite Isolated from the European Honey Bee, Apis mellifera. *J. Eukaryot. Microbiol.* **2009**, *56*, 142–147. [CrossRef]
69. Kurze, C.; Dosselli, R.; Grassl, J.; Le Conte, Y.; Kryger, P.; Baer, B.; Moritz, R.F. Differential proteomics reveals novel insights into *Nosema*–honey bee interactions. *Insect Biochem. Mol. Biol.* **2016**, *79*, 42–49. [CrossRef]
70. Malone, L.; Giacon, H.A.; Newton, M.R. Comparison of the responses of some New Zealand and Australian honey bees (*Apis mellifera* L) to Nosema apis Z. *Apidologie* **1995**, *26*, 495–502. [CrossRef]
71. Cantwell, G.E. Standard method for counting *Nosema* spores. *Am. Bee J.* **1970**, *110*, 222–223.
72. Fries, I.; Chauzat, M.-P.; Chen, Y.-P.; Doublet, V.; Gensersch, E.; Gisder, S.; Higes, M.; McMahon, D.P.; Martín-Hernández, R.; Natsopoulou, M.; et al. Standard methods for *Nosema* research. *J. Apic. Res.* **2013**, *52*, 1–28. [CrossRef]
73. Higes, M.; Martín-Hernández, R.; Garrido-Bailón, E.; García-Palencia, P.; Meana, A. Detection of infective *Nosema ceranae* (Microsporidia) spores in corbicular pollen of forager honeybees. *J. Invertebr. Pathol.* **2008**, *97*, 76–78. [CrossRef] [PubMed]

74. Bradford, M.M. A rapid and sensitive method for the quantitation of microgram quantities of protein utilizing the principle of protein-Dye binding. *Anal. Biochem.* **1976**, *72*, 248–254. [CrossRef]
75. Suwannapong, G.; Chaiwongwattanakul, S.; Benbow, M.E. Histochemical Comparison of the Hypopharyngeal Gland in *Apis cerana* Fabricius, 1793 Workers and *Apis mellifera* Linnaeus, 1758 Workers. *Psyche A J. Èntomol.* **2010**, *2010*, 1–7. [CrossRef]
76. Suwannapong, G.; Seanbualuang, P.; Wongsiri, S. A histochemical study of the hypopharangeal glands of the dwarf honey bees *Apis andreniformis* and *Apis florea*. *J. Apic. Res.* **2007**, *46*, 260–263. [CrossRef]

Case Report

A Case Report of Chronic Stress in Honey Bee Colonies Induced by Pathogens and Acaricide Residues

Elena Alonso-Prados [1], Amelia-Virginia González-Porto [2], José Luis Bernal [3], José Bernal [3], Raquel Martín-Hernández [4,5] and Mariano Higes [5,*]

[1] Unidad de Productos Fitosanitarios, Instituto Nacional de Investigación y Tecnología Agraria y Alimentaria (INIA, CSIC), 28040 Madrid, Spain; aprados@inia.es

[2] Laboratorio de Mieles y Productos de las Colmenas Centro de Investigación Apícola y Agroambiental, IRIAF, Consejería de Agricultura de la Junta de Comunidades de Castilla-La Mancha, 19180 Marchamalo, Spain; avgonzalezp@jccm.es

[3] Analytical Chemistry Group, Instituto Universitario Centro de Innovación en Química y Materiales Avanzados (I.U.CINQUIMA), Universidad de Valladolid, 47011 Valladolid, Spain; joseluis.bernal@uva.es (J.L.B.); jose.bernal@uva.es (J.B.)

[4] Instituto de Recursos Humanos para la Ciencia y la Tecnología (INCRECYT-FEDER), Fundación Parque Científico y Tecnológico de Castilla—La Mancha, 02006 Albacete, Spain; rmhernandez@jccm.es

[5] Laboratorio de Patología Apícola, Centro de Investigación Apícola y Agroambiental, IRIAF, Consejería de Agricultura de la Junta de Comunidades de Castilla-La Mancha, 19180 Marchamalo, Spain

* Correspondence: mhiges@jccm.es; Tel.: +34-949-88-88-56

Abstract: In this case report, we analyze the possible causes of the poor health status of a professional *Apis mellifera iberiensis* apiary located in Gajanejos (Guadalajara, Spain). Several factors that potentially favor colony collapse were identified, including *Nosema ceranae* infection, alone or in combination with other factors (e.g., BQCV and DWV infection), and the accumulation of acaricides commonly used to control *Varroa destructor* in the beebread (coumaphos and tau-fluvalinate). Based on the levels of residues, the average toxic unit estimated for the apiary suggests a possible increase in vulnerability to infection by *N. ceranae* due to the presence of high levels of acaricides and the unusual climatic conditions of the year of the collapse event. These data highlight the importance of evaluating these factors in future monitoring programs, as well as the need to adopt adequate preventive measures as part of national and international welfare programs aimed at guaranteeing the health and fitness of bees.

Keywords: honey bees; *Apis mellifera*; acaricides; pesticides; toxic unit; *Varroa destructor*; *Nosema ceranae*; bee viruses; tau-fluvalinate; coumaphos

Citation: Alonso-Prados, E.; González-Porto, A.-V.; Bernal, J.L.; Bernal, J.; Martín-Hernández, R.; Higes, M. A Case Report of Chronic Stress in Honey Bee Colonies Induced by Pathogens and Acaricide Residues. *Pathogens* **2021**, *10*, 955. https://doi.org/10.3390/pathogens10080955

Academic Editor: Giovanni Cilia

Received: 10 June 2021
Accepted: 26 July 2021
Published: 29 July 2021

Publisher's Note: MDPI stays neutral with regard to jurisdictional claims in published maps and institutional affiliations.

1. Introduction

Bees, including honey bees, bumble bees, and solitary bees, are a prominent and economically important group of pollinators worldwide. In fact, 35% of the global food crop production depends on these pollinators [1], and in Europe, the production of 84% of crop species is, to some extent, dependent on animal pollination [2]. Bees also fulfill an important role in the pollination of wild plants. Thus, habitat fragmentation seems to have a negative effect on pollination and plant reproduction [3]. However, there is still a debate on how to approach the pollen limitation in plant dynamics [4,5]. Furthermore, honey bees provide additional economic inputs in temperate areas where honey production is a fundamental source of income to professional beekeepers.

The honey bee colony is a complex system in which thousands of individuals work together to ensure its sustainability. Multiple factors play an important role in colony viability, such as climate, environment, nutrition, and pathogens, and consequently, in colony pollination and production capabilities. Thus, any deterioration of honey bee colonies has a direct negative environmental impact and, in the case of honey bees, economic consequences in countries where there is a large proportion of professional beekeepers as in Mediterranean areas [6].

Many factors have been related to the decline in honey bee colonies over the past decades (revised in [7]). On one hand, the global spread of pathogens related to colony losses entails changes to the ecological and evolutionary dynamics of both pathogens and hosts, often leading to the selection of the most virulent variant of the pathogen and reducing the heterogeneity of the host. A range of treatments exist to combat pathogens and diseases during the last decades. Thus, chemicals that keep *Varroa* mite populations under control may accumulate in different hive matrices, chronically exposing honey bees to the residues of chemicals. Due to foraging activities, honey bees may also be exposed to a wider range of potentially toxic compounds (naturally produced or not), which may also accumulate inside the hive. The action of pathogens and xenobiotics or the combination of both provokes physiological changes, immunosuppression, and gut microbiota disruption on honey bees, which may finally produce the collapse of colonies. These changes may be also influenced by the nutrition quality of the collected pollen, which depends on the seasonal climatological conditions [8–10]. Overall, the impact of pathogens on this decline (mainly parasites and related viruses) may be particularly important [11–13], probably in conjunction with the accumulation of pesticide residues in hive matrices. Thus, the combined effect of two or more such stressors could drive the mortality of individuals, eventually leading to colony collapse [14–16].

Accordingly, we present here a screening study of a professional Spanish apiary that reported a problematic health situation. To investigate the factors that had possibly provoked this situation, we performed a comprehensive evaluation of multiple drivers, including pathogens and pesticides, also analyzing the foraging flora, in an attempt to determine whether new factors should be examined in future monitoring programs.

Severe *Nosema ceranae* infection, in conjunction with the accumulation of acaricides used to control *Varroa* mite infestations in honey bee hives, represents a real threat to colonies, indicating that appropriate preventative strategies should be adopted in honey bee health programs.

2. Results

2.1. Veterinary Inspection

A professional beekeeper who managed 400 honey bee colonies (*Apis mellifera iberiensis*) in Gajanejos (Guadalajara, Central Spain; lat. 40.8423, long. −2.8933) reported health problems in his colonies. He revised all his colonies in September 2015 while applying a compulsory treatment for *Varroa destructor* with CheckMite® strips (a.m.: coumaphos), according to the manufacturer's recommendations. No health problems had previously been noted in the colonies, which appeared to be in satisfactory health at the beginning of autumn, September 2015, with a normal population of worker honey bees. When the acaricide strips were removed 6 weeks after their application (November 2015), the beekeeper noticed a reduction in worker honeybee population in some of the colonies, and consequently, he began to inspect the apiary more often. The first dead colonies were detected that winter, and losses continued until the next spring, March 2016.

Upon veterinary inspection in early March 2016, around 50% of the bee colonies had died. The hives of the dead colonies were stored at the beekeeper's warehouse in the same conditions that he found them in the field, awaiting cleaning (Figure 1A).

Only a few dead bees were found in the brood chamber frames of these hives, (Figure 1B), with no anatomical deformities and no *Varroa* mites detected at the bottom of the hives or in worker bees and sealed brood. Moreover, there were no clinical signs of chalkbrood and American or European foulbrood. The beekeeper also conserved the acaricide strips used in the previous autumn in plastic bags (Figure 1C).

Figure 1. Pictures obtained during the inspection of colonies. (**A**) Hives stored in the beekeeper's warehouse; (**B**) brood combs from the dead colonies with a few honey bees and sealed brood in the frames; (**C**) CheckMite® strips kept by the beekeeper after their removal; (**D**) apiary with many empty spots and a few surviving colonies; (**E**) weak surviving colony with a small honey bee population; (**F**) details of some CheckMite® strips with low interaction with honey bees.

In the surviving colonies (Figure 1D), there were no more than two combs from the brood chamber that were covered with adult honey bees, much fewer than the five to seven combs that would be expected to be covered in this geographical area at that time (Figure 1E). The acaricide strips used in the autumn treatment were essentially untouched by the honey bees (Figure 1F), and there were no clinical signals of varroosis or other diseases in the brood or in adult worker honey bees. *Varroa* mites were not identified in brood cells or at the bottom of the hives.

The presence of accessible pollen and honey reserves in both the dead and surviving colonies ruled out death by starvation.

Most of the surviving colonies (approximately 200) did not have a large-enough adult population to ensure their future survival.

Samples of worker honey bees and stored pollen were collected from the brood chamber from the dead and surviving colonies by veterinarians from the Centro de Investigación Apícola y Agroambiental' (CIAPA) in Marchamalo (Guadalajara, Spain) to make a diagnosis of the causes of the collapsing event.

With the permission of the beekeeper, dead colonies (*n* = 5) and surviving colonies (*n* = 10) with sufficient honey bees in the frames were randomly selected to take around 300 worker honey bees per hive to study pathogens according to the methodology described in Section 4. In addition, four or five pieces of honey bee combs (10 × 15 cm each) that contained stored pollen were also randomly taken from different areas of the brood chamber from each colony surveyed in order to conduct a chemical and palynological analysis of the stored pollen (beebread). Finally, climatic parameters were also considered in the diagnostic analysis.

2.2. Pathogen Screening

In only one of the sampled dead colonies, there was a 5% parasitization by *V. destructor*, and the mite was not detected in any of the other hives. Of the surviving colonies sampled, only two were positive for *Varroa* mite with infestation rates of 25% and 1%.

All the samples were positive for *N. ceranae*, with severe percentages of parasitization that were statistically higher in dead colonies (*p*-value = 0.001312).

The presence of deformed wing virus (DWV) was confirmed in surviving colonies infected by *Varroa* and in six other samples. Moreover, black queen cell virus (BQCV) was present in six samples, one of which was positive for *V. destructor* (Table 1). Finally, *N. apis*, *Acarapis woodi*, Trypanosomatids, Neogregarines, Lake Sinai virus complex (LSV), and acute bee paralysis virus–Kashmir bee virus–Israeli acute paralysis virus complex (AKI) were not detected in any sample.

Table 1. Results of pathogen and pesticide residue screening and palynological analysis of beebread samples from dead (D) and weak yet surviving (S) colonies. Pathogen screening: *V. destructor* parasitization (% VD), prevalence of *N. ceranae* (% NC), and detection (+ or −) of deformed wing virus (DWV) and black queen cell virus (BQCV). Residues quantified in beebread (ppb): tau-fluvalinate (FVT) and coumaphos (CMF). Natural logarithm of the toxic unit of the mixture in each colony (Ln(TUm)). Palynological analysis: percentage of wild foraging plants (%WP). Grey cells in the table indicate parameters not analyzed.

Status of the Colonies	Colony Code	%VD	%NC *	DWV	BQCV	FVT (*)	CMF *	LN(TUm) *	%WP
Dead	D1	5	70			7	435	−6.38	96.3
	D2	0	80			7	415	−6.43	45.1
	D3	0	75			<LOQ	202	−7.154	89.3
	D4	0	92			13	350	−6.59	65.1
	D5	0	89			9	323	−6.68	79.4
Surviving	S1	25	20	+	+	7	283	−6.81	90.5
	S2	1	30	+	−	9	545	−6.16	30.7
	S3	0	25	+	−	10	2230	−4.75	15.2
	S4	0	60	+	+	15	465	−6.31	30.5
	S5	0	36	+	+	20	1165	−5.42	33.7
	S6	0	20	+	−	16	305	−6.73	77.7
	S7	0	30	−	+	18	775	−5.80	92.1
	S8	0	45	−	+	19	850	−5.71	55.6
	S9	0	35	+	−	13	936	−5.62	85.4
	S10	0	35	+	+	13	845	−5.72	93.7

* Statistically significant differences between dead and surviving colonies at α = 0.05. LOQ: level of quantification.

Varroa Mite Resistance to Acaricides

Only 1 of the 15 colonies sampled (S1) complied with the criteria for conducting a resistance test [17]. The number of alive mites was lower (0, 1) than in control (7, 9) after 6 or 24 h of incubation, respectively. Therefore, it is concluded that this colony did not show any evidence of *Varroa* mite resistant to acaricides in any batch exposed to different acaricides (Table 2).

Table 2. Results of the screening to identify *Varroa mites* from the S1 colony resistant to acaricides.

Incubation Time	Mites	Control	CheckMite® (Coumaphos)	Apistan® (Tau-Fluvalinate)	Apitraz® (Amitraz)
6 h	Dead	1	10	5	8
	Alive	7	0	0	0
24 h	Dead	2	11	7	8
	Alive	9	0	1	1

2.3. Stored Pollen Analysis

Of the 67 substances analyzed in beebread samples, only tau-fluvalinate and coumaphos were detected (Table 1), and there were statistically higher concentrations of tau-fluvalinate and coumaphos in the beebread samples from the surviving colonies (W = 43.5, *p*-value = 0.01312; W = 42.0, *p*-value = 0.02165, respectively; Table 1). The mean TUm value of the whole apiary was 0.00262 ± 0.00199. TUm values were <1 in both the dead and surviving colonies (Table S1), indicating that the residue levels did not, in principle, reach the threshold of acute toxicity. Tau-fluvalinate represented less than 1% of TUm of the colonies (Table S1).

All the compounds identified in a given mixture contribute to TUm in accordance with their potency and the levels of their residues. Thus, as the levels of residues were significantly higher in the surviving colonies, TUm was also significantly higher in these hives (Table S1).

No significant differences were found between the dead and surviving colonies regarding the presence of wild flora in the beebread samples (W = 18.0, *p*-value = 0.21299). In four of the five samples from the dead colonies, wild plants were majorly present in the beebread. The most frequent taxa of wild plants identified were: *Araliaceae, Labiatae, Asteraceae, Chenopodiaceae,* and *Diplotaxis* spp. The major cultivated taxa detected was sunflower (*Helianthus annuus,* L.). In the surviving colonies, the predominant pollen was from wild plants in 6 out of 10 samples belonging to eight taxa: *Araliaceae, Labiatae, Caryophyllaceae, Cichorioideae, Convolvulaceae, Asteraceae, Chenopodiaceae,* and *Diplotaxis* spp. In the remaining 4 samples, the predominant taxa identified in the beebread were *H. annuus, Prunus* spp., and *Brassicaceae* (Figure S1).

2.4. Meteorological Data

Figure 2a shows a Walter–Leith diagram (see Section 4.3 for details) of the meteorological station nearest the apiary for historical data (2013–2021). The mean annual temperature was 12.7 °C, and the yearly precipitation for this period of time was 485.3 mm. The seasonal distribution of the precipitation showed a maximum peak during the spring season and a secondary one during the fall–winter season. The dry season extends from June to September. The different bioclimatic indices estimated showed that the climate in the location corresponds to the eutemperate latitudinal belt, and it can be classified as oceanic-low continental. The bioclimate is classified as Mediterranean pluviseasonal-oceanic, within the low supramediterranean upper dry bioclimatic belt.

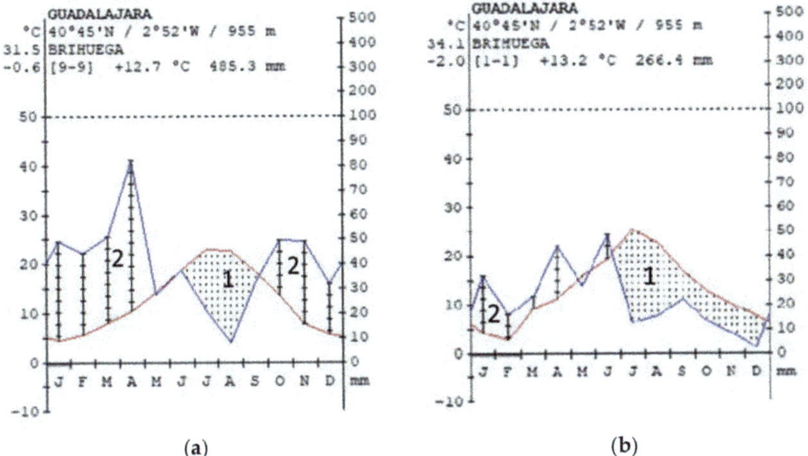

(a)   (b)

Figure 2. Climatic diagram drawn according to Walter and Leith (1960) for (**a**) the period 2013–2021 and (**b**) the studied year (2015) data of the Brihuega station. (——) monthly mean temperature (°C); (——) monthly precipitation. The plot shows 1 = dry season and 2 = rainy season.

In contrast, the year of the colony collapse event (2015) was characterized to be drier than the historical data (*p* = 266.4 mm), especially during the autumn–winter season, and it had a wider temperature amplitude between the coldest and hottest months (Figure 2b).

3. Discussion

Here we investigated a specific case study of the weakening and death of honey bee colonies in the field, a situation that has occurred quite frequently in Spain in recent years.

The first suspected cause of colony death in this apiary was the action of the *Varroa* mite due to a failure in acaricidal treatment, often proposed to cause these effects [18]. However, the visual examination and pathogen screening suggested a different origin of the collapse, and in fact, *V. destructor* was detected in only 20% of the colonies sampled (3 out of 15). There were no clinical signs consistent with generalized varroosis in the apiary or in the hives in which the honey bee colonies died. Moreover, 2 of the positive colonies had a parasitic mite load of only 1% and 5%, not apparently representing an immediate risk to bee health [18]. Only 1 of the surviving honey bee colonies sampled (10%) had a higher parasite load (25%) and was thus at risk of suffering the negative effects of this mite [18]. After decades of miticide use against varroosis, there is a general concern about the selection of *Varroa* mites tolerant to acaricides [19]. However, the preliminary results of the acaricide resistance test and the absence of clinical signs of varroosis upon inspection suggest that resistant *Varroa* mites do not affect treatment efficacy in this apiary. Hence, it was concluded that the symptoms observed were not due to clinical varroosis after therapeutic failure. Overall, these results indicate that the *Varroa* mite might have exerted a degree of pressure on some individual colonies, but it is insufficient to provoke the collapse/weakening evident across the entire apiary.

By contrast, severe *N. ceranae* infection was detected in all cases, especially in dead colonies, with clinical signs and symptoms in all cases (dead and surviving) in line with the infections observed previously in colonies that collapsed in winter due to such infestation [20]. The dynamics of nosemosis C in the colony provokes the mean spore count to fluctuate greatly from the start to the end of the disease in interior bees, and it is not a reliable measure of a colony's health when bees are infected with *N. ceranae* [20]. In fact, the proportion of foraging bees infected with *N. ceranae* was the strongest indicator of the spread of the disease in the colony. In this sense, forager bees are always more infected than house bees. The more foragers that are infected, the smaller size of the bee colony. However, percentages of parasitization by *N. ceranae* in house bees higher than 35%–40% indicate a serious risk for the colony of suffering from nosemosis C, which could reach collapse [21]. Moreover, as expected, a lower infestation was seen in early spring, as described in the phases 3 and 4 of the disease [20] and consistent with its evolution in different seasons [20,22,23]. The fact that the acaricide strips did not change color or underwent propolization probably reflects a change in honey bee behavior due to *N. ceranae* infection, which could produce a serious risk that the acaricide treatment would lose efficacy if the *V. destructor* mites were abundant [23,24]. Indeed, *N. ceranae* alters various physiological processes in individual honey bees involving immunomodulation [25–27] and energetic stress [28,29], inducing early foraging activities [30,31]. These alterations have a direct impact on the colony [20,21,32], especially in geographical areas with warmer climates where there is a large concentration of professional beekeeping [33–35], in contrast to colder climates ([36–39] reviewed in [23]).

Although there was a scattered presence of BQCV and DWV in the samples, their detection may be a consequence of a side effect of the presence of *Varroa* mites [18,40–42] and *Nosema* spp. [40,41,43,44]. The viral loads were not determined because workers did not display DWV clinical signs. The latent presence of viruses in *Apis mellifera* is well known in the literature, and while showing no signs of disease, they may destroy bee fitness and health during favorable conditions (e.g., *V. destructor* infestations). However, in the case of infection by *N. ceranae*, it has been demonstrated that these two pathogens are not acting synergistically [45–48]. Moreover, under laboratory conditions, it has been observed that the inoculation of DWV does not have an impact on *N. ceranae* infection. On the contrary, prior establishment of *N. ceranae* has a significant negative impact on the load of DWV [49].

Neonicotinoids and other agrochemicals were not detected in the beebread, consistent with the fact that honey bees mainly visited wild flora. Nevertheless, the high concentrations of tau-fluvalinate and coumaphos detected in beebread samples were assumed to have a beekeeping origin as these chemicals are registered in Spain to control *Varroa* mite. Moreover, tau-fluvalinate and coumaphos have octanol, water partitioning coefficient (logKow) > 3 [50,51], indicating high lipophilicity and potential to accumulate in wax [52,53] and other hive matrices [54], where they may remain relatively stable for long periods of time [55–57]. Indeed, both of these acaricides are estimated to need 5 years to completely disappear from bee matrices [58]. Moreover, their concentration in wax may increase due to the wax recycling processes [55,56], explaining why their residues are frequently found in wax and beebread worldwide [53,59–61]. If the acaricide residue levels in beebread reaches toxic levels, the health of the honey bee colony might be compromised. Thus, a synergetic toxic effect between these acaricides cannot be ruled out. Indeed, acute contact toxicity of coumaphos increased up to 3- to 4-fold when 4-day-old honey bees were pretreated with tau-fluvalinate at a dose of 1 or 3 μg/bee, and the contact toxicity of tau-fluvalinate increased up to 32-fold when the individuals were pretreated with coumaphos at a dose of 10 μg/bee [62].

In addition, high concentrations of acaricides may have made the honey bee colonies more sensitive to *N. ceranae* infection [63]. In this sense, following the toxic unit (TU) approach and based on acute toxicity, LC50, it has been proposed that Ln (TU) = −6.706 may represent as a preliminary break point regarding the increment of *N. ceranae* when assessed in the presence of a mixture of xenobiotics [14]. The mean Ln(TUm) in the present apiary was −5.95, suggesting that it may have been more vulnerable to *N. ceranae* infection [63]. However, Ln(TUm) values were higher in surviving colonies, which also had lower *N. ceranae* infection. This may lead to the erroneous conclusion that a high miticide concentration contributes to colony survival. However, this was not the case because their viability was compromised in early spring due to the small adult honey bee population. Thus, while more than 50% of the dead honey bees are expected to be infected in colonies that collapsed due to nosemosis C in the cold months, this percentage is lower when colonies collapse later in the year, probably due to an increment in the proportion of uninfected newborn honey bees [20].

Finally, the unusual climatic conditions of the year of the collapse event may have had an influence on the strength of the colonies. On one side, the warmer and drier conditions of this year may have provoked a change in the phenology and physiology of the vegetation of the zone, with a decrease in the length of the flowering period and, therefore, and quality of the collected pollen [64], which may indirectly have affected the strength of the colonies [65,66]. In this direction, the impact of weather conditions on the overwintering survival of colonies has recently been studied in Pennsylvania for 3 years [67]. Despite the short database, the authors found adverse effects of both too-cool and too-hot summer on overwintering colony survival [67]. These results are in line with the ones found in other countries [67]. In addition, these climatic conditions may have favored the infection with *N. ceranae*, whose spores resist high temperatures and desiccation, and they complete their life cycle more efficiently at high temperatures [68].

4. Material and Methods

4.1. Pathogen Screening

The number of bees present in each sample (around 300 bees) was counted, and the bees were examined individually to detect the presence of *Varroa* mites and collect them by means of sterile tweezers [14]. Each mite detected was analyzed macroscopically to confirm the species. A honey bee colony was considered infested with *V. destructor* when at least 1 *Varroa* mite was found in the sample. The rate of infestation of the bee colony was estimated by assessing the number of *Varroa* mites in relation to the number of adult bees in each sample, and it was expressed as the number of *Varroa* mites/100 bees/sample [69].

The presence of *Nosema* spp., Trypanosomatids, Neogregarines, and *Acarapis woodi* was evaluated in a sample (*n* = 60) from each colony. The remaining bees were kept frozen at −80 °C.

The presence of different viruses was only analyzed in the surviving colonies sampled because the viral RNA integrity could not be ensured in the dead colonies.

The subsample of each colony ($n = 60$) was macerated in 50% AL buffer (Qiagen GmbH, Hilden, Germany) before DNA and RNA were extracted, as detailed in [69–71].

Briefly, macerated bees were centrifuged at 3000 rpm for 10 min, and the resulting pellets were used for DNA extraction, and the supernatants for RNA extraction. Both the pellets and supernatants were stored at −80 °C prior to nucleic acid (DNA or RNA) extraction. For DNA extraction, the pellets were resuspended in 3 mL MilliQH$_2$O, and a 400 µL aliquot was transferred to a 96-well plate (Qiagen) with glass beads (2 mm diameter, Merck KGaA, Darmstadt, Germany) using disposable Pasteur pipettes. After overnight preincubation with proteinase K (20 µL, Qiagen), the samples were then processed as described previously [71] following the BS96 DNA Tissue extraction protocol in a BioSprint station (Qiagen). The plates were then stored at −20 °C. For RNA extraction, 400 µL of the supernatant was incubated for 15 min with protease (20 µL, Qiagen) at 70 °C, and the nucleic acids were then extracted as described above (BioSprint 96 DNA in, BioSprint workstation, Qiagen). The total nucleic acids recovered were then subjected to DNA digestion with DNase I (Qiagen) to completely remove any genomic DNA, and the total RNA recovered was used immediately to generate firststrand cDNAs using the QuantiTect Reverse Transcription Kit (Qiagen) according to the manufacturer's instructions. The resultant cDNA was used for subsequent virus analysis with no further dilution [70].

Negative and positive controls were run in parallel for each step: bee maceration, DNA and RNA extraction, and reverse transcription [69].

Published PCR or RT-PCR protocols were considered to screen *Nosema apis* and *N. ceranae* [71], Trypanosomatids and Neogregarines [72,73], *Acarapis woodi* [74], LSV complex [75], AKI complex [76], DWV, and BQCV [77]. Table S2 includes the primers used.

In addition, the proportion of *Nosema* spp. infection was determined by PCR on 25 individual worker honey bees from each colony sampled [71].

Test of Varroa *Mite Resistance to Acaricides*

When possible, mite resistance to acaricides was determined using the respective marketed products, CheckMite® (a.m.: coumaphos), Apistan® (a.m.: tau-fluvalinate), and Apitraz® (a.m.: amitraz), according to the protocol described previously [17] with the following modifications:

1. Inclusion of an additional batch for a 24 h incubation period;
2. Feeding the honey bees with syrup during the incubation periods; and
3. Freezing the honey bees at the end of the incubation period to collect the remaining *Varroa* mites (−80 °C, 15 min).

The honey bees were kept at 35 °C during the test, and after incubation periods, a control test without treatment was used to determine how the basal conditions affected *Varroa* mite mortality.

4.2. Stored Pollen Analysis

Beebread was extracted aseptically from the combs, removing the wax and preparing a composite sample for each colony. Finally, each pollen sample was divided into two 100 g aliquots, for chemical and palynological analyses, and stored at −80 °C.

A multiresidue chemical analysis of 60 substances was carried out following a method described elsewhere [59], assessing acaricides (AC), fungicides (FU), herbicides (HB), and insecticides (IN). In addition, 7 neonicotinoid INs (acetamiprid, clothianidin, dinotefuran, imidacloprid, nitenpyram, thiacloprid, and thiamethoxam) were measured as described previously [78].

Based on the results of the multiresidue analysis and the toxicity data reported previously (Table A2 in [14]), the toxic unit of the mixtures (TUm) was calculated following an approach given elsewhere [14] to assess the risk of the chemical mixture found in each hive sampled. Subsequently, the natural logarithm (Ln(TUm)) was estimated for comparison purposes with [14]. In that work, the possible relationships between TUm and the prevalence of pathogens were studied by using a factor analysis. The natural logarithm of TUm, Ln(TUm), was derived to normalize data. Thus, the higher the value of Ln(TUm), the higher the toxicity risk.

Finally, the type of foraging flora was confirmed by analyzing beebread samples as described elsewhere [11,20,21] and estimating the proportion of pollen from wild (WP) and cultivated (CP) plants. The pollen of the beebread was extracted by diluting 0.5 g in 10 mL of acidulated water (0.5% sulfuric acid) and centrifuging at 2500 r.p.m. for 15 min. The pellet was washed with double-distilled water and centrifuged twice. The sediment was placed onto a glycerine jelly slide and examined microscopically in order to identify the pollen. The frequency of the pollen grains of each taxon is expressed as a percentage of the total pollen grains. Between 300 and 1200 pollen grains were counted in each sample.

The pollen grains were identified and classified on the basis of the identification keys [79,80] and the pollen slide reference collection available at the honey laboratory at the CIAPA.

4.3. Meteorological Data

A Walter–Leith diagram [81] was developed with historical weather data obtained from the meteorological station of Brihuega (lat. 40.765, long. −2.874) from the network of the State Meteorological Agency (AEMET) to compare it with the data for the period of the study. This weather station is located 9 km far away from the studied areas.

Walter and Leith climate diagrams are brief summaries of average climatic variables and their time course. They illustrate precipitation and temperature changes throughout the year in 1 standardized chart. Originally aimed at visualizing those climatic variables and their dynamics, which are particularly important for vegetation, they have proven useful for a wide range of sciences. The diagrams were developed with the diagnostic tool of the Worldwide Bioclimatic Classification System, 1996–2021 [82].

4.4. Statistical Analysis

A 1-tailed Mann–Whitney test ($\alpha = 0.05$) was used to analyze possible differences between the dead and surviving colonies in terms of the different experimental parameters measured (pathogens, % wild pollen in beebread, chemical residues, and TUm). The analysis was carried out with Statgraphics Centurion 18$^\copyright$.

5. Conclusions

The veterinary inspection and analytical evidence presented here indicate that nosemosis C infection was the underlying cause of the colony weakness and collapse of the professional apiary studied, probably accelerated by the presence of high levels of miticides and unusual climatic conditions. In conjunction with the unchecked concentrations of acaricide that accumulated in honey bee hives, *N. ceranae* infection represents a real danger in honey bee colony survival. Therefore, in addition to the correct use of veterinary products to control *V. destructor*, appropriate wax renewal of the combs should be introduced to develop specific preventive strategies aimed at controlling possible infections from prevalent pathogens.

Supplementary Materials: The following are available online at https://www.mdpi.com/article/10.3390/pathogens10080955/s1, Table S1: Details of calculation of TUm values. Table S2: Primers used for each pathogen in PCR reactions. Figure S1: Main taxa identified of the pollen grains found in the beebread samples.

Author Contributions: Conceptualization, M.H., E.A.-P., R.M.-H.; methodology, M.H., E.A.-P., R.M.-H., J.L.B., J.B., A.-V.G.-P.; formal analysis, M.H., E.A.-P., R.M.-H., J.L.B., J.B., A.-V.G.-P.; writing—original draft preparation, M.H., E.A.-P., R.M.-H., J.L.B., J.B., A.-V.G.-P.; writing—review and editing, M.H, E.A.-P., R.M.-H, J.L.B, J.B., A.-V.G.-P.; visualization, M.H., E.A-P., R.M.-H.; supervision, M.H., E.A.-P., R.M.-H.; project administration, M.H., J.L.B.; funding acquisition, M.H., J.L.B. All authors have read and agreed to the published version of the manuscript.

Funding: This research was funded by INIA-FEDER (RTA2017-00004-C02).

Institutional Review Board Statement: In Europe, the EU Directive 2010/63/EU on the protection of animals used for scientific purposes laid the down the ethical framework for the use of animals in scientific experiments. The scope of this directive also includes specific invertebrate species, i.e., cephalopods, but no insects. Thus, according to European legislation no specific permits were required for the described studies.

Informed Consent Statement: Not applicable.

Data Availability Statement: The data presented in this study are available on request from the corresponding author.

Acknowledgments: The authors wish to thank V. Albendea, T. Corrales, M. Gajero and C. Uceta at the honey bee pathology laboratory for their technical support. The authors also want to thank M.D. Moreno for his support in the visit to the apiaries.

Conflicts of Interest: The authors declare no conflict of interest.

References

1. Klein, A.-M.; Vaissière, B.E.; Cane, J.H.; Steffan-Dewenter, I.; Cunningham, S.; Kremen, C.; Tscharntke, T. Importance of pollinators in changing landscapes for world crops. *Proc. R. Soc. B Boil. Sci.* **2006**, *274*, 303–313. [CrossRef] [PubMed]
2. Williams, I.H. The dependence of crop production within the European Union on pollination by honey bees. *Agric. Zool. Rev.* **1994**, *6*, 229–257.
3. Aguilar, R.; Ashworth, L.; Galetto, L.; Aizen, M. Plant reproductive susceptibility to habitat fragmentation: Review and synthesis through a meta-analysis. *Ecol. Lett.* **2006**, *9*, 968–980. [CrossRef]
4. Ashman, T.-L.; Knight, T.M.; Steets, J.A.; Amarasekare, P.; Burd, M.; Campbell, D.; Dudash, M.R.; Johnston, M.O.; Mazer, S.J.; Mitchell, R.J.; et al. Pollen Limitation of Plant Reproduction: Ecological and Evolutionary Causes and Consequences. *Ecology* **2004**, *85*, 2408–2421. [CrossRef]
5. Ernest, S.K.M.; Brown, J.H. Homeostasis and Compensation: The Role of Species and Resources in Ecosystem Stability. *Ecology* **2001**, *82*, 2118–2132. [CrossRef]
6. Chauzat, M.-P.; Cauquil, L.; Roy, L.; Franco, S.; Hendrikx, P.; Ribière-Chabert, M. Demographics of the European Apicultural Industry. *PLoS ONE* **2013**, *8*, e79018. [CrossRef]
7. Vanengelsdorp, D.; Meixner, M.D. A historical review of managed honey bee populations in Europe and the United States and the factors that may affect them. *J. Invertebr. Pathol.* **2010**, *103*, S80–S95. [CrossRef]
8. Hristov, P.; Shumkova, R.; Palova, N.; Neov, B. Factors Associated with Honey Bee Colony Losses: A Mini-Review. *Vet. Sci.* **2020**, *7*, 166. [CrossRef]
9. Ratnieks, F.L.W.; Carreck, N. Clarity on Honey Bee Collapse? *Science* **2010**, *327*, 152–153. [CrossRef]
10. Staveley, J.P.; Law, S.A.; Fairbrother, A.; Menzie, C.A. A Causal Analysis of Observed Declines in Managed Honey Bees (*Apis mellifera*). *Hum. Ecol. Risk Assess. Int. J.* **2013**, *20*, 566–591. [CrossRef]
11. Cepero, A.; Hernández, R.M.; Bartolomé, C.; Gómez-Moracho, T.; Barrios, L.; Bernal, J.; Martín, M.T.; Meana, A.; Higes, M. Passive laboratory surveillance in Spain: Pathogens as risk factors for honey bee colony collapse. *J. Apic. Res.* **2015**, *54*, 525–531. [CrossRef]
12. Cepero, A.; Ravoet, J.; Gómez-Moracho, T.; Bernal, J.L.; Del Nozal, M.J.; Bartolomé, C.; Maside, X.; Meana, A.; González-Porto, A.V.; de Graaf, D.C.; et al. Holistic screening of collapsing honey bee colonies in Spain: A case study. *BMC Res. Notes* **2014**, *7*, 1–10. [CrossRef]
13. Evison, S.E.F.; Roberts, K.E.; Laurenson, L.; Pietravalle, S.; Hui, J.; Biesmeijer, J.C.; Smith, J.E.; Budge, G.; Hughes, W.O.H. Pervasiveness of Parasites in Pollinators. *PLoS ONE* **2012**, *7*, e30641. [CrossRef]
14. Alonso-Prados, E.; Muñoz, I.; De La Rúa, P.; Serrano, J.; Fernández-Alba, A.R.; García-Valcárcel, A.I.; Hernando, M.D.; Alonso, Á.; Alonso-Prados, J.L.; Bartolomé, C.; et al. The toxic unit approach as a risk indicator in honey bees surveillance programmes: A case of study in *Apis mellifera iberiensis*. *Sci. Total. Environ.* **2020**, *698*, 134208. [CrossRef]
15. Martinello, M.; Baratto, C.; Manzinello, C.; Piva, E.; Borin, A.; Toson, M.; Granato, A.; Boniotti, M.B.; Gallina, A.; Mutinelli, F. Spring mortality in honey bees in northeastern Italy: Detection of pesticides and viruses in dead honey bees and other matrices. *J. Apic. Res.* **2017**, *56*, 239–254. [CrossRef]
16. Sánchez-Bayo, F.; Goulson, D.; Pennacchio, F.; Nazzi, F.; Goka, K.; Desneux, N. Are bee diseases linked to pesticides?—A brief review. *Environ. Int.* **2016**, *89–90*, 7–11. [CrossRef]
17. Pettis, J. Test Detecting Varroa Mite Resistance to Apistan, Apivar & Coumaphos. Available online: https://www2.gov.bc.ca/assets/gov/farming-natural-resources-and-industry/agriculture-and-seafood/animal-and-crops/animal-production/bee-assets/api_fs223.pdf (accessed on 25 September 2016).
18. Rosenkranz, P.; Aumeier, P.; Ziegelmann, B. Biology and control of *Varroa destructor*. *J. Invertebr. Pathol.* **2010**, *103*, S96–S119. [CrossRef]
19. Higes, M.; Martín-Hernández, R.; Hernández-Rodríguez, C.S.; González-Cabrera, J. Assessing the resistance to acaricides in *Varroa destructor* from several Spanish locations. *Parasitol. Res.* **2020**, *119*, 3595–3601. [CrossRef] [PubMed]

20. Higes, M.; Hernández, R.M.; Botías, C.; Bailón, E.G.; González-Porto, A.V.; Barrios, L.; Nozal, M.J.; Bernal, J.L.; Jiménez, J.J.; Palencia, P.G.; et al. How natural infection by Nosema ceranae causes honeybee colony collapse. *Environ. Microbiol.* **2008**, *10*, 2659–2669. [CrossRef]
21. Higes, M.; Hernández, R.M.; Garrido-Bailón, E.; González-Porto, A.V.; García-Palencia, P.; Meana, A.; Nozal, M.J.; Mayo, R.; Bernal, J.L. Honeybee colony collapse due to Nosema ceranaein professional apiaries. *Environ. Microbiol. Rep.* **2009**, *1*, 110–113. [CrossRef] [PubMed]
22. Higes, M.; Hernández, R.M.; Meana, A. Nosema ceranaein Europe: An emergent type C nosemosis. *Apidologie* **2010**, *41*, 375–392. [CrossRef]
23. Higes, M.; Meana, A.; Bartolomé, C.; Botías, C.; Hernández, R.M. Nosema ceranae (Microsporidia), a controversial 21st century honey bee pathogen. *Environ. Microbiol. Rep.* **2013**, *5*, 17–29. [CrossRef] [PubMed]
24. Botías, C.; Barrios, L.; Nanetti, A.; Meana, A.; Martín-Hernández, R.; Garrido-Bailón, E.; Higes, M. Nosema spp. parasitization decreases the effectiveness of acaricide strips (Apivar®) in treating varroosis of honey bee (*Apis mellifera iberiensis*) colonies. *Environ. Microbiol. Rep.* **2011**, *4*, 57–65. [CrossRef] [PubMed]
25. Antúnez, K.; Hernández, R.M.; Prieto, L.; Meana, A.; Zunino, P.; Higes, M. Immune suppression in the honey bee (*Apis mellifera*) following infection by Nosema ceranae (Microsporidia). *Environ. Microbiol.* **2009**, *11*, 2284–2290. [CrossRef] [PubMed]
26. Aufauvre, J.; Misme-Aucouturier, B.; Viguès, B.; Texier, C.; Delbac, F.; Blot, N. Transcriptome Analyses of the Honeybee Response to Nosema ceranae and Insecticides. *PLoS ONE* **2014**, *9*, e91686. [CrossRef] [PubMed]
27. Holt, H.L.; Aronstein, K.A.; Grozinger, C.M. Chronic parasitization by Nosema microsporidia causes global expression changes in core nutritional, metabolic and behavioral pathways in honey bee workers (*Apis mellifera*). *BMC Genom.* **2013**, *14*, 1–16. [CrossRef]
28. Kurze, C.; Dosselli, R.; Grassl, J.; Le Conte, Y.; Kryger, P.; Baer, B.; Moritz, R. Differential proteomics reveals novel insights into Nosema–honey bee interactions. *Insect Biochem. Mol. Biol.* **2016**, *79*, 42–49. [CrossRef] [PubMed]
29. Mayack, C.; Naug, D. Energetic stress in the honeybee *Apis mellifera* from Nosema ceranae infection. *J. Invertebr. Pathol.* **2009**, *100*, 185–188. [CrossRef]
30. Dussaubat, C.; Maisonnasse, A.; Crauser, D.; Beslay, D.; Costagliola, G.; Soubeyrand, S.; Kretzschmar, A.; Le Conte, Y. Flight behavior and pheromone changes associated to Nosema ceranae infection of honey bee workers (*Apis mellifera*) in field conditions. *J. Invertebr. Pathol.* **2013**, *113*, 42–51. [CrossRef]
31. Wolf, S.; McMahon, D.; Lim, K.S.; Pull, C.D.; Clark, S.J.; Paxton, R.; Osborne, J.L. So Near and Yet So Far: Harmonic Radar Reveals Reduced Homing Ability of Nosema Infected Honeybees. *PLoS ONE* **2014**, *9*, e103989. [CrossRef]
32. Botías, C.; Martín-Hernández, R.; Barrios, L.; Meana, A.; Higes, M. Nosema spp. infection and its negative effects on honey bees (*Apis mellifera iberiensis*) at the colony level. *Vet. Res.* **2013**, *44*, 25. [CrossRef]
33. Hatjina, F.; Tsoktouridis, G.; Bouga, M.; Charistos, L.; Evangelou, V.; Avtzis, D.; Meeus, I.; Brunain, M.; Smagghe, G.; de Graaf, D.C. Polar tube protein gene diversity among Nosema ceranae strains derived from a Greek honey bee health study. *J. Invertebr. Pathol.* **2011**, *108*, 131–134. [CrossRef]
34. Lodesani, M.; Costa, C.; Besana, A.; Dall'Olio, R.; Franceschetti, S.; Tesoriero, D.; Giacomo, D. Impact of control strategies for *Varroa destructor* on colony survival and health in northern and central regions of Italy. *J. Apic. Res.* **2014**, *53*, 155–164. [CrossRef]
35. Noureddine, A.; Haddad, N. The first data on hygienic behavior of *Apis mellifera intermissa* in Algeria. *J. Biol. Earth Sci.* **2013**, *201*, 1–5.
36. Francis, R.M.; Amiri, E.; Meixner, M.D.; Kryger, P.; Gajda, A.; Andonov, S.; Uzunov, A.; Topolska, G.; Charistos, L.; Costa, C.; et al. Effect of genotype and environment on parasite and pathogen levels in one apiary—a case study. *J. Apic. Res.* **2014**, *53*, 230–232. [CrossRef]
37. Gisder, S.; Hedtke, K.; Möckel, N.; Frielitz, M.-C.; Linde, A.; Genersch, E. Five-Year Cohort Study of Nosema spp. in Germany: Does Climate Shape Virulence and Assertiveness of Nosema ceranae? *Appl. Environ. Microbiol.* **2010**, *76*, 3032–3038. [CrossRef] [PubMed]
38. Stevanovic, J.; Simeunovic, P.; Gajić, B.; Lakic, N.; Radovic, D.; Fries, I.; Stanimirovic, Z. Characteristics of Nosema ceranae infection in Serbian honey bee colonies. *Apidologie* **2013**, *44*, 522–536. [CrossRef]
39. Stevanovic, J.; Stanimirovic, Z.; Genersch, E.; Kovacevic, S.; Ljubenkovic, J.; Radakovic, M.; Aleksic, N. Dominance of Nosema ceranae in honey bees in the Balkan countries in the absence of symptoms of colony collapse disorder. *Apidologie* **2011**, *42*, 49–58. [CrossRef]
40. Dainat, B.; Evans, J.; Chen, Y.P.; Gauthier, L.; Neumann, P. Predictive Markers of Honey Bee Colony Collapse. *PLoS ONE* **2012**, *7*, e32151. [CrossRef]
41. Francis, R.M.; Nielsen, S.L.; Kryger, P. *Varroa*-Virus Interaction in Collapsing Honey Bee Colonies. *PLoS ONE* **2013**, *8*, e57540. [CrossRef]
42. Mordecai, G.J.; Wilfert, L.; Martin, S.; Jones, I.; Schroeder, D. Diversity in a honey bee pathogen: First report of a third master variant of the Deformed Wing Virus quasispecies. *ISME J.* **2015**, *10*, 1264–1273. [CrossRef] [PubMed]
43. Chen, Y.P.; Siede, R. Honey Bee Viruses. *Adv. Virus Res.* **2007**, *70*, 33–80. [CrossRef] [PubMed]
44. Mendoza, Y.; Antúnez, K.; Branchiccela, B.; Anido, M.; Santos, E.; Invernizzi, C. Nosema ceranae and RNA viruses in European and Africanized honeybee colonies (*Apis mellifera*) in Uruguay. *Apidologie* **2013**, *45*, 224–234. [CrossRef]
45. Costa, C.; Tanner, G.; Lodesani, M.; Maistrello, L.; Neumann, P. Negative correlation between Nosema ceranae spore loads and deformed wing virus infection levels in adult honey bee workers. *J. Invertebr. Pathol.* **2011**, *108*, 224–225. [CrossRef] [PubMed]

46. Dussaubat, C.; Brunet, J.-L.; Higes, M.; Colbourne, J.; Lopez, J.; Choi, J.H.; Hernández, R.M.; Botías, C.; Cousin, M.; McDonnell, C.; et al. Gut Pathology and Responses to the Microsporidium Nosema ceranae in the Honey Bee *Apis mellifera*. *PLoS ONE* **2012**, *7*, e37017. [CrossRef] [PubMed]

47. Hedtke, K.; Jensen, P.; Jensen, A.B.; Genersch, E. Evidence for emerging parasites and pathogens influencing outbreaks of stress-related diseases like chalkbrood. *J. Invertebr. Pathol.* **2011**, *108*, 167–173. [CrossRef] [PubMed]

48. Martin, S.J.; Hardy, J.; Villalobos, E.; Hernández, R.M.; Nikaido, S.; Higes, M. Do the honeybee pathogens N osema ceranae and deformed wing virus act synergistically? *Environ. Microbiol. Rep.* **2013**, *5*, 506–510. [CrossRef]

49. Doublet, V.; Natsopoulou, M.; Zschiesche, L.; Paxton, R. Within-host competition among the honey bees pathogens Nosema ceranae and Deformed wing virus is asymmetric and to the disadvantage of the virus. *J. Invertebr. Pathol.* **2015**, *124*, 31–34. [CrossRef]

50. Authority, E.F.S. Conclusion on the peer review of the pesticide risk assessment of the active substance tau-fluvalinate. *EFSA J.* **2010**, *8*, 1645. [CrossRef]

51. Garten, C.; Trabalka, J.R. Evaluation of models for predicting terrestrial food chain behavior of xenobiotics. *Environ. Sci. Technol.* **1983**, *17*, 590–595. [CrossRef]

52. Chauzat, M.-P.; Carpentier, P.; Martel, A.-C.; Bougeard, S.; Cougoule, N.; Porta, P.; Lachaize, J.; Madec, F.; Aubert, M.; Faucon, J.-P. Influence of pesticide residues on honey bee (Hymenoptera: Apidae) colony health in France. *Environ. Èntomol.* **2009**, *38*, 514–523. [CrossRef]

53. Mullin, C.A.; Frazier, M.; Frazier, J.L.; Ashcraft, S.; Simonds, R.; Vanengelsdorp, D.; Pettis, J.S. High Levels of Miticides and Agrochemicals in North American Apiaries: Implications for Honey Bee Health. *PLoS ONE* **2010**, *5*, e9754. [CrossRef]

54. Hörig, K. Development of a toxicokinetic model for the honey bee (Apis mellifera) colony for the use in risk assessment. Ph.D. Thesis, Faculty of Mathematics, Computer Science and Natural Sciences of the RWTH Aachen University, Aachen, Germany, 2015.

55. Bogdanov, S.; Kilchenmann, V.; Imdorf, A. Acaricide residues in some bee products. *J. Apic. Res.* **1998**, *37*, 57–67. [CrossRef]

56. Martel, A.-C.; Zeggane, S.; Aurières, C.; Drajnudel, P.; Faucon, J.-P.; Aubert, M. Acaricide residues in honey and wax after treatment of honey bee colonies with Apivar® or Asuntol® 50. *Apidologie* **2007**, *38*, 534–544. [CrossRef]

57. Sokół, R. The influence of a multimonth persistence of Fluwarol in a hive of a honey bee colony. *Med. Weter.* **1996**, *52*, 718–720.

58. Bogdanov, S. Contaminants of bee products. *Apidologie* **2005**, *37*, 1–18. [CrossRef]

59. Bernal, J.L.; Garrido-Bailon, E.; DEL Nozal, J.B.; González-Porto, A.V.; Hernández, R.M.; Diego, J.C.; Jimenez, J.J.; Higes, M. Overview of Pesticide Residues in Stored Pollen and Their Potential Effect on Bee Colony (*Apis mellifera*) Losses in Spain. *J. Econ. Èntomol.* **2010**, *103*, 1964–1971. [CrossRef]

60. Johnson, R.; Ellis, M.D.; Mullin, C.A.; Frazier, M. Pesticides and honey bee toxicity–USA. *Apidologie* **2010**, *41*, 312–331. [CrossRef]

61. Traynor, K.; Pettis, J.S.; Tarpy, D.; Mullin, C.A.; Frazier, J.L.; Frazier, M.; Vanengelsdorp, D. In-hive Pesticide Exposome: Assessing risks to migratory honey bees from in-hive pesticide contamination in the Eastern United States. *Sci. Rep.* **2016**, *6*, 33207. [CrossRef]

62. Johnson, R.M.; Pollock, H.S.; Berenbaum, M.R. Synergistic Interactions between In-Hive Miticides in *Apis mellifera*. *J. Econ. Èntomol.* **2009**, *102*, 474–479. [CrossRef] [PubMed]

63. Wu, J.Y.; Smart, M.D.; Anelli, C.M.; Sheppard, W.S. Honey bees (*Apis mellifera*) reared in brood combs containing high levels of pesticide residues exhibit increased susceptibility to Nosema (Microsporidia) infection. *J. Invertebr. Pathol.* **2012**, *109*, 326–329. [CrossRef] [PubMed]

64. Di Pasquale, G.; Salignon, M.; Le Conte, Y.; Belzunces, L.P.; Decourtye, A.; Kretzschmar, A.; Suchail, S.; Brunet, J.-L.; Alaux, C. Influence of Pollen Nutrition on Honey Bee Health: Do Pollen Quality and Diversity Matter? *PLoS ONE* **2013**, *8*, e72016. [CrossRef] [PubMed]

65. Alaux, C.; Dantec, C.; Parrinello, H.; Le Conte, Y. Nutrigenomics in honey bees: Digital gene expression analysis of pollen's nutritive effects on healthy and *Varroa*-parasitized bees. *BMC Genom.* **2011**, *12*, 496. [CrossRef] [PubMed]

66. Brodschneider, R.; Crailsheim, K. Nutrition and health in honey bees. *Apidologie* **2010**, *41*, 278–294. [CrossRef]

67. Calovi, M.; Grozinger, C.M.; Miller, D.A.; Goslee, S.C. Summer weather conditions influence winter survival of honey bees (*Apis mellifera*) in the northeastern United States. *Sci. Rep.* **2021**, *11*, 1–12. [CrossRef] [PubMed]

68. Martín-Hernández, R.; Bartolomé, C.; Chejanovsky, N.; Le Conte, Y.; Dalmon, A.; Dussaubat, C.; García-Palencia, P.; Meana, A.; Pinto, M.A.; Soroker, V.; et al. Nosema ceranae in *Apis mellifera*: A 12 years postdetection perspective. *Environ. Microbiol.* **2018**, *20*, 1302–1329. [CrossRef] [PubMed]

69. Buendía-Abad, M.; Martín-Hernández, R.; Ornosa, C.; Barrios, L.; Bartolomé, C.; Higes, M. Epidemiological study of honeybee pathogens in Europe: The results of Castilla-La Mancha (Spain). *Span. J. Agric. Res.* **2018**, *16*, e0502. [CrossRef]

70. Antúnez, K.; Anido, M.; Garrido-Bailón, E.; Botías, C.; Zunino, P.; Martínez-Salvador, A.; Hernández, R.M.; Higes, M. Low prevalence of honeybee viruses in Spain during 2006 and 2007. *Res. Vet. Sci.* **2012**, *93*, 1441–1445. [CrossRef] [PubMed]

71. Martín-Hernández, R.; Botías, C.; Bailón, E.G.; Martínez-Salvador, A.; Prieto, L.; Meana, A.; Higes, M. Microsporidia infecting *Apis mellifera*: Coexistence or competition. Is *Nosema ceranae* replacing *Nosema apis*? *Environ. Microbiol.* **2011**, *14*, 2127–2138. [CrossRef]

72. Meeus, I.; De Graaf, D.; Jans, K.; Smagghe, G. Multiplex PCR detection of slowly-evolving trypanosomatids and neogregarines in bumblebees using broad-range primers. *J. Appl. Microbiol.* **2010**, *109*, 107–115. [CrossRef]

73. Stevanovic, J.; Schwarz, R.S.; Vejnovic, B.; Evans, J.D.; Irwin, R.E.; Glavinic, U.; Stanimirovic, Z. Species-specific diagnostics of *Apis mellifera* trypanosomatids: A nine-year survey (2007–2015) for trypanosomatids and microsporidians in Serbian honey bees. *J. Invertebr. Pathol.* **2016**, *139*, 6–11. [CrossRef] [PubMed]

74. Cepero, A.; Hernández, R.M.; Prieto, L.; Gómez-Moracho, T.; Martínez-Salvador, A.; Bartolomé, C.; Maside, X.; Meana, A.; Higes, M. Is Acarapis woodi a single species? A new PCR protocol to evaluate its prevalence. *Parasitol. Res.* **2014**, *114*, 651–658. [CrossRef] [PubMed]

75. Ravoet, J.; Maharramov, J.; Meeus, I.; De Smet, L.; Wenseleers, T.; Smagghe, G.; De Graaf, D.C. Comprehensive Bee Pathogen Screening in Belgium Reveals Crithidia mellificae as a New Contributory Factor to Winter Mortality. *PLoS ONE* **2013**, *8*, e72443. [CrossRef] [PubMed]

76. Francis, R.; Kryger, P. Single Assay Detection of Acute Bee Paralysis Virus, Kashmir Bee Virus and Israeli Acute Paralysis Virus. *J. Apic. Sci.* **2012**, *56*, 137–146. [CrossRef]

77. Chantawannakul, P.; Ward, L.; Boonham, N.; Brown, M. A scientific note on the detection of honeybee viruses using real-time PCR (TaqMan) in *Varroa* mites collected from a Thai honeybee (*Apis mellifera*) apiary. *J. Invertebr. Pathol.* **2006**, *91*, 69–73. [CrossRef] [PubMed]

78. Valverde, S.; Bernal, J.L.; Martín, M.T.; Nozal, M.J. Fast determination of neonicotinoid insecticides in bee pollen using QuECHERS and ultra high performance liquid chromatography coupled to quadrupole time-of-flight mass spectrometry. *Electrophoresis* **2016**, *37*, 2470–2477. [CrossRef] [PubMed]

79. Faegri, K.; Iversen, J. *Textbook of Pollen Analysis*, 4th ed.; John Wiley & Sons: Chichester, West Sussex, UK, 1989.

80. Valdés, B.; Díez, M.J.; Fernández, I. *Atlas polínico de Andalucía Occidental*; Instituto de Desarrollo Regional Nº 43, Universidad de Sevilla y Excma: Diputación de Cádiz: Cádiz, Spain, 1987.

81. Walter, H.; Lieth, H. *Klimadiagramm–Weltatlas*; VEB G. Fisher Verlag: Jena, Germany, 1967.

82. Rivas-Martinez, S.S.R.-S.S. Worldwide Bioclimatic Classification System, 1996–2021. Available online: http://www.globalbioclimatics.org/default.htm (accessed on 5 July 2021).

Systematic Review

Pathogens Spillover from Honey Bees to Other Arthropods

Antonio Nanetti, Laura Bortolotti * and Giovanni Cilia

Council for Agricultural Research and Agricultural Economics Analysis, Centre for Agriculture and Environment Research (CREA-AA), Via di Saliceto 80, 40128 Bologna, Italy; antonio.nanetti@crea.gov.it (A.N.); giovanni.cilia@crea.gov.it (G.C.)
* Correspondence: laura.bortolotti@crea.gov.it; Tel.: +39-051-353103

Abstract: Honey bees, and pollinators in general, play a major role in the health of ecosystems. There is a consensus about the steady decrease in pollinator populations, which raises global ecological concern. Several drivers are implicated in this threat. Among them, honey bee pathogens are transmitted to other arthropods populations, including wild and managed pollinators. The western honey bee, *Apis mellifera*, is quasi-globally spread. This successful species acted as and, in some cases, became a maintenance host for pathogens. This systematic review collects and summarizes spillover cases having in common *Apis mellifera* as the mainteinance host and some of its pathogens. The reports are grouped by final host species and condition, year, and geographic area of detection and the co-occurrence in the same host. A total of eighty-one articles in the time frame 1960–2021 were included. The reported spillover cases cover a wide range of hymenopteran host species, generally living in close contact with or sharing the same environmental resources as the honey bees. They also involve non-hymenopteran arthropods, like spiders and roaches, which are either likely or unlikely to live in close proximity to honey bees. Specific studies should consider host-dependent pathogen modifications and effects on involved host species. Both the plasticity of bee pathogens and the ecological consequences of spillover suggest a holistic approach to bee health and the implementation of a One Health approach.

Keywords: spillover; inter-species transmission; honey bee diseases; pathogens; virus; bacteria; microsporidia; *Nosema*; trypanosomatids; wild bees; arthropods; Hymenoptera

Citation: Nanetti, A.; Bortolotti, L.; Cilia, G. Pathogens Spillover from Honey Bees to Other Arthropods. *Pathogens* **2021**, *10*, 1044. https://doi.org/10.3390/pathogens10081044

Academic Editor: Nemat O. Keyhani

Received: 9 July 2021
Accepted: 12 August 2021
Published: 17 August 2021

Publisher's Note: MDPI stays neutral with regard to jurisdictional claims in published maps and institutional affiliations.

1. Introduction

Interspecific transmission may occur from a definite maintenance host (aka "reservoir") to an incidental or non-maintenance species (aka "spillover host"). Spillover cases are crucial to pathogen dynamics [1,2].

In a single-host scenario, reservoirs are sufficient and pathogen replication does not need other host species [3]. The basic reproduction number (R0) defines the frequency of new cases originating from each primary event, where R0 = 1 is the threshold between declining infections (R0 < 1) and pathogen persistence within the population by intra-specific transmission (R0 > 1) [4]. When multiple host species are involved, the presence of new maintenance or incidental hosts may result in an increased pathogen transmission [1]. In this case, R0 >> 0 denotes multi-host pathogen scenarios that may be respectively true or apparent, depending on the high or low interspecies transmission. When the R0 is between 0 and 1, the event is called "apparent multi-host pathogen", while "true multi-host pathogen" indicates an event in which there are two different maintenance hosts and the occurrence of interspecies transmission is higher than 1 [5].

Strictly speaking, spillover only occurs when the recipient species is characterized by R0 ≈ 0 [5]. However, in this review, we follow the use of the term *sensu lato*, commonly indicating a multifaceted range of host shift events [2].

Pollinators are crucial to the generation of crops contributing to the human diet [6]. These agroecosystem service is provided by a range of different species, including honey bees, wild bees, wasps, hoverflies, and butterflies [7–10]. However, different factors contribute to a decline of pollinating entomofauna, in terms of population size, biodiversity, abundance and distribution [11–19].

Pathogens and parasites are deemed drivers of this decline, together with other factors including pesticides and global warming. Nonetheless, the global picture is certainly far from complete, since data may misrepresent the actual distribution and gaps remain in our understanding of both epidemiological features and invasion dynamics of many pathogens [12,20–25]. *Apis mellifera* is known to share pathogens with bumblebee species, including viruses, bacteria, fungi and protozoa [26–34]. After acting as incidental hosts, western honey bees may become the primary maintenance host, as occurred in the cases of *Nosema ceranae*, *Crithidia bombi*, and *Apicystis bombi* [35]. Pathogens may also genetically adapt to a range of new species [12,13,23,30,36], acting as incidental or maintenance hosts [37–39].

Interspecific transmission to arthropods sharing the same environment as honey bees may occur orofecally, via direct contact and by pollen contamination [38]. Besides, infected foragers may contaminate pollen, nectar and floral organs with pathogens [40–44]. Spillover could also involve species not expected to come into direct contact with the bees. Wasps predating infected bees [45–48] and cannibalizing their carcasses [49–51] are likely to become contaminated with pathogens.

Honey bees and other insects have natural immune defense systems against bacteria, protozoa, mites, and viruses. They include antimicrobial peptides like apidaecin, defensin, abaecin, hymenoptaecin, and lysozyme, which are regulated by the immune pathways Toll, IMD, JAK/STAT and JNK [52]. Those defenses are challenged by insecticides and other pesticides used in modern agriculture [53].

Spillover events are difficult to prove. Indeed, viral infection and replication in new hosts, which may not develop under artificial conditions, can occur in nature [36,54,55]. The increasing number of reports about honey bee pathogens found in new hosts contributes to depict a scenario including one reservoir species and multiple spillover events. Indeed, population studies might elucidate those aspects [56] which, in the specific case of wild bees, are complicated by the peculiar characteristics of those species [17,57,58]. This makes spillover routes generally unknown and undetermined [59], albeit each report deserves further research to illustrate thoroughly their respective epidemiological scenarios.

This systematic review is intended to collect, group, and summarize the spillover cases *sensu lato* reported by the literature and involving honey bee pathogens. Other arthropods were also considered as alternative hosts. The spillover cases are grouped by: (i) host species, condition and stage, (ii) geographical region and year of the report, and (iii) co-occurrence in the same host.

2. Results

In total, from 1960 to 2021, 81 studies investigated spillover cases of honey bee pathogens to wild and/or managed arthropods (Figure 1). Some of the studies considered more than one species. In detail, they considered the spillover to other bee species (Supplementary Table S1), other Hymenoptera (Supplementary Table S1) and other arthropods (Supplementary Table S2).

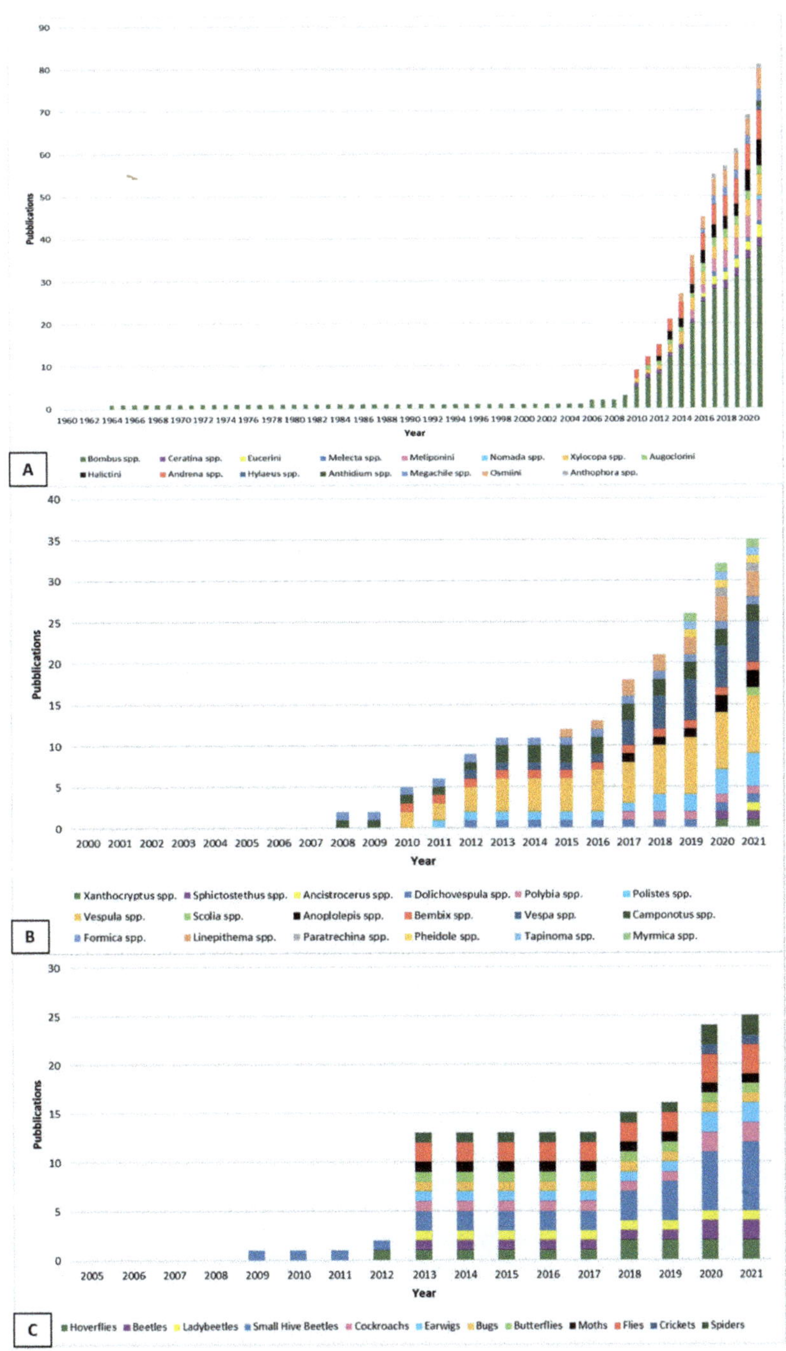

Figure 1. Cumulative number of spillover studies of honey bee pathogens available in the literature between 1960 and 2021 involving other bees (**A**), non-bee Hymenoptera (**B**), and other arthropods (**C**).

As shown in Figure 1A, the first article about spillover of honey bee pathogens to other bees was published in 1964, but the number of articles on this topic steadily increased from the year 2020, likely due to the quick development of molecular genetic tools for pathogen detection. Considering other hymenopteran species, the first detection of spillover cases dates back to 2008, with a rapid increase of cases in the following years (Figure 1B). The first spillover case to other arthropods was assessed in 2009, but later the frequency increased, covering a wide range of species (Figure 1C).

The geographical distribution of spillover studies present in the literature (Figure 2) shows a high number of studies in both North and South America, Europe and New Zealand, whereas the reports from other countries were less frequent.

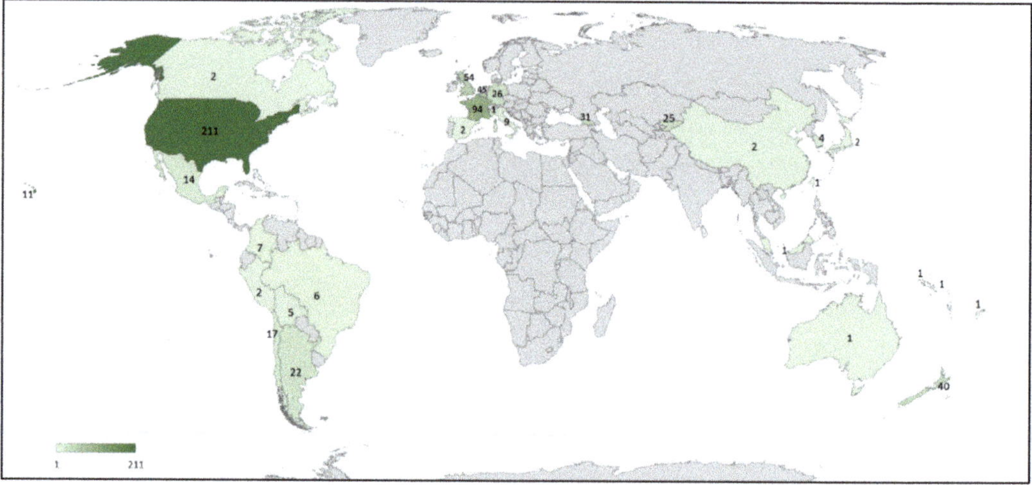

Figure 2. Geographical distribution of the honey bee pathogen spillover studies reported in the literature. The number of published cases is indicated for each country and highlighted by different shades of green.

Figures 3 and 4 summarize the spillover cases for each honey bee pathogen in relation to arthropods groups. In events encompassing at least 20 spillover cases, DWV was the most frequently detected (158 cases). BQCV, SBV, IAPV, ABPV, KBV, *N. ceranae*, SBPV and LSV resulted implicated with progressively decreasing frequency.

The chord graph (Figure 3) shows all spillover cases described in this review, evaluating the relationship to the investigated arthropod genus. Additionally, Figure 4 highlights the reported frequency of honey bee pathogens in the investigated arthropod communities, to emphazise their plasticity to the host.

Some individuals were found infected with multiple honey bee pathogens (Figure 5). The highest incidence of coinfections was found in bumblebees, followed by mason bees, mining bees and the honey bee pest *Aethina tumida*. A high number of co-infections was reported for *Eucera nigrescens*, *Osmia bicornis* and *Osmia cornuta*, for which 6 pathogens were found in the same individuals. Besides, the most abundant coinfecting pathogens able to co-infect the arthropods hosts were DWV, BQCV, SBV, ABPV and *N. ceranae*.

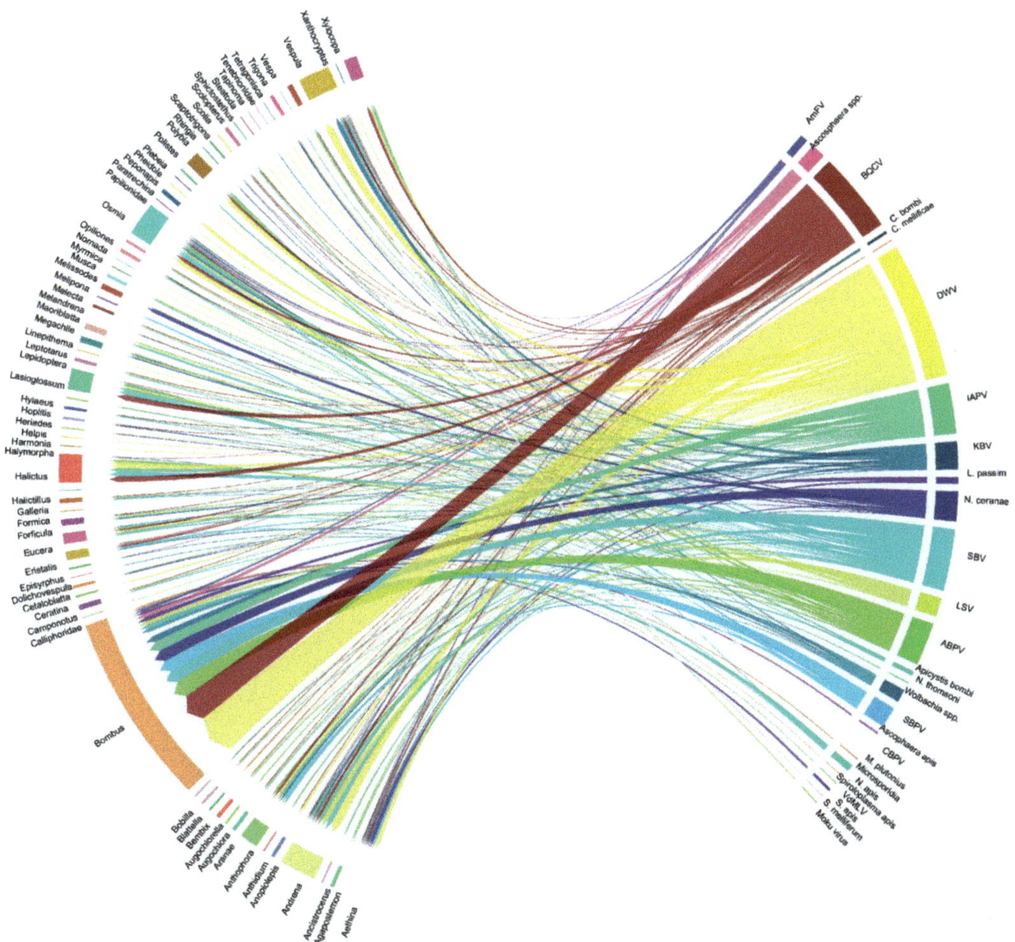

Figure 3. Visual schematization of honey bee pathogen spillover to alternative host arthropods reported in the literature. Different colors denote distinct pathogens or host genera. Legend: ABPV: Acute Bee Paralysis Virus; IAPV: Israeli Acute Paralysis Virus; BQCV: Black Queen Cell Virus; SBV: Sacbrood Virus; DWV: Deforming Wing Virus; LSV; Lake Sinai Virus; *Am*FV: *Apis mellifera* Filamentous Virus; KBV: Kashmir Bee Virus; SBPV: Slow Bee Paralysis Virus; CBPV: Chronic Bee Paralysis Virus; VdMLV: *Varroa destructor* Macula-like Virus.

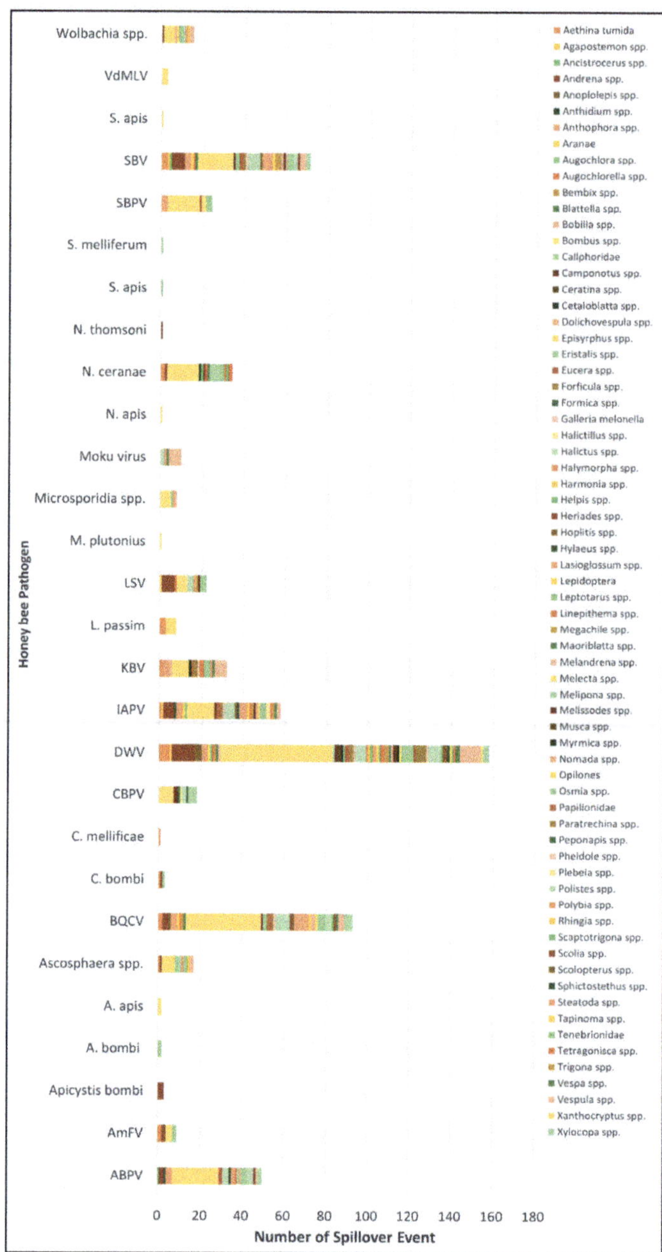

Figure 4. Frequency of spillover events involving single honey bee pathogens and the range of arthropods found infected with them. Different colors denote distinct host groups. Legend: ABPV: Acute Bee Paralysis Virus; IAPV: Israeli Acute Paralysis Virus; BQCV: Black Queen Cell Virus; SBV: Sacbrood Virus; DWV: Deforming Wing Virus; LSV; Lake Sinai Virus; *Am*FV: *Apis mellifera* Filamentous Virus; KBV: Kashmir Bee Virus; SBPV: Slow Bee Paralysis Virus; CBPV: Chronic Bee Paralysis Virus; VdMLV: *Varroa destructor* Macula-like Virus.

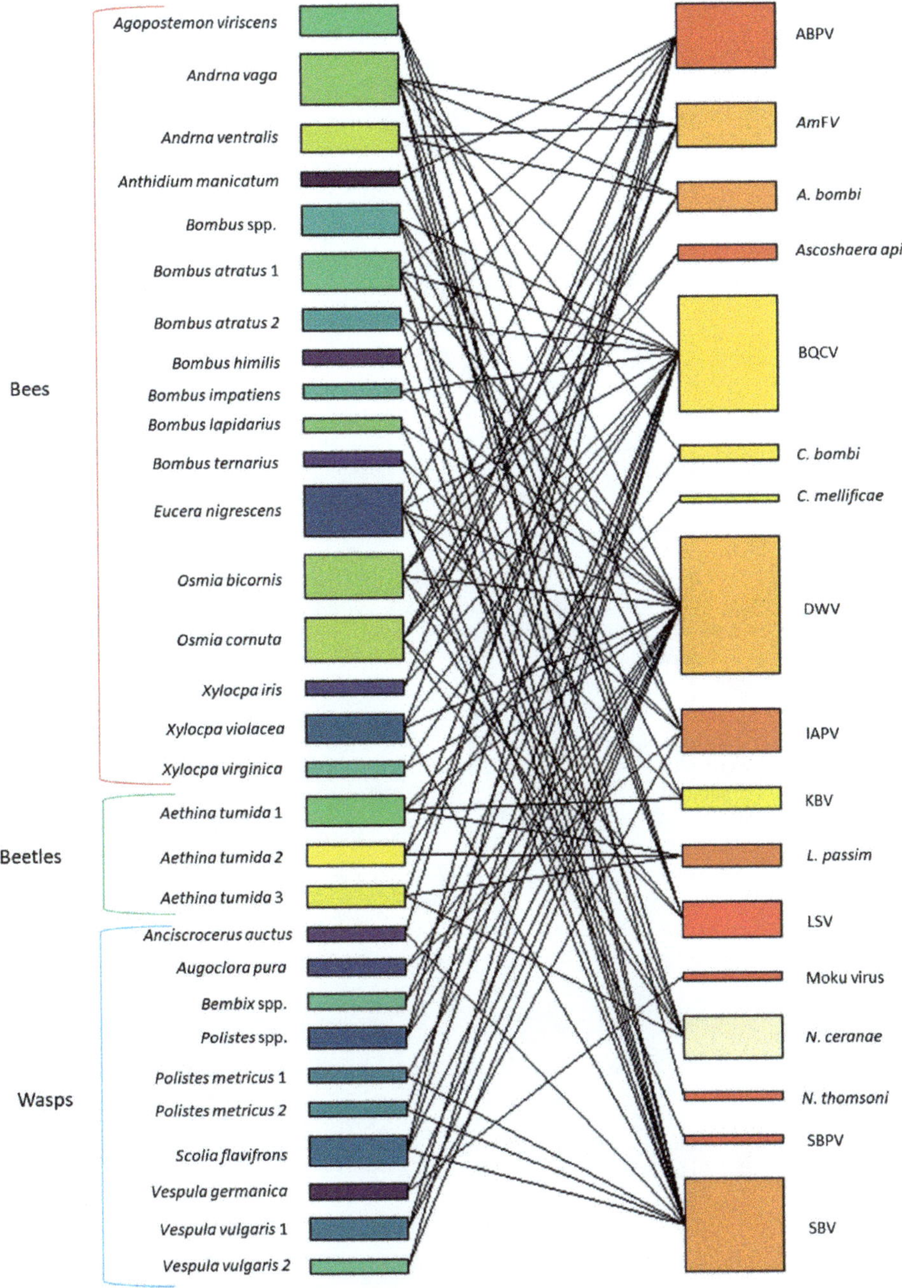

Figure 5. Co-occurrence of honey bee pathogens in individual hosts. These are grouped as bees, beetles, and wasps. Box size is indicative of the frequency. Legend: ABPV: Acute Bee Paralysis Virus; IAPV: Israeli Acute Paralysis Virus; BQCV: Black Queen Cell Virus; SBV: Sacbrood Virus; DWV: Deforming Wing Virus; LSV; Lake Sinai Virus; *Am*FV: *Apis mellifera* Filamentous Virus; KBV: Kashmir Bee Virus; SBPV: Slow Bee Paralysis Virus; CBPV: Chronic Bee Paralysis Virus; VdMLV: *Varroa destructor* Macula-like Virus.

3. Discussion

The results of this systematic review highlights that the case history of spillover events involving honey bee pathogens increased over the past six decades. This is consistent with the growing interest of the scientific community in understanding the underlying factors [12,20,22,54]. The higher incidence of spillover cases recorded in Europe, New Zealand, and the Americas may reflect their advances in research and apiculture compared to other regions [60–67].

Viruses vectored by *V. destructor* (DWV, KBV, and IAPV) and quasi-ubiquitous pathogens (BQCV, SBV and *N. ceranae*) were among the most frequently reported cases.

Bumblebees, mason bees and leafcutter bees were the species in which the spillover was studied more intensely, possibly because of their use in crops and fruit pollination. The fact that some of the surveys were carried out on arthropods ranging freely in the same environment as the managed honey bees is indicative of a pathogen circulation in their common environment. Despite honey bee pathogens were detected in other arthropods, symptoms and other effects on the alternative host populations remain unknown—except for some publications reporting individual bumblebees with crippled wings and scoring positive to DWV [68,69].

The importance of investigating the spillover of honey bee pathogens is also indicated by the discovery of active coinfections in wild hymenopteran individuals. As for the honey bees [30,70,71], multiple infections were found in wild bees, wasps and *Aethina tumida* individuals, which shows the importance of other arthropods as incidental hosts. The multiple infections that were identified (Figure 5 and Supplementary Table S1) have both the effect to increase the circulation of pathogens within the arthropod communities, and to recirculate them to the managed honey bee colonies, so generating damage at individual and colony levels.

All of these aspects, including their modifications and effects encompass the implementation of a One Health approach to bee health [72,73]. The health of managed honey bees is dependent on the health of wild bees and other arthropods, and vice versa. This approach is essential to provide suitable ecosystems to pollinators and other arthropods contributing to human livelihoods and environmental health, and for understanding the eco-immunology to prevent the transmission of pathogens and pests, thereby limiting damages in managed and wild insect populations [73–75]. Therefore, the circulation/recirculation and the possible impact of honey bee pathogens to the arthropod communities are crucial to build the basis for the One Health approach to the bee health. Here we provide a brief discussion of each of the honey bee pathogens reported in Supplementary Tables S1 and S2, in relation to their spillover hosts.

3.1. Viruses

3.1.1. Deformed Wing Virus (DWV)

DWV is a non-enveloped ssRNA (+) virus belonging to *Iflavirus* genus within the *Picornaviridae* family [76]. The DWV is a pathogen including three distinct genomotypes: A, B and C [77,78].

The DWV is probably the most known, spread, prevalent, and studied honey bee pathogen, often associated to *V. destructor* [79]. The DWV can be asymptomatically replicated in *V. destructor* mites [80].

The impact of DWV on honey bees leads to increased interspecific transmission, reaching several species of hymenopterans and other arthropods (Supplementary Tables S1 and S2).

The virus was identified not only in species living in close contact with the honey bees, like *A. tumida*, *G. mellonella*, *Vespa* spp. [39,45,47,81,82], but also in *Apis* and non-*Apis* species that may act as incidental hosts [38,39]. DWV was found in naturally and artificially infected asymptomatic arthropods [38,39,54,59,79], although some commercial and wild *B. terrestris* and *B. pascuorum* individuals were found with crippled wings [68,69]. Besides, artificial infection experiments highlighted that DWV reduced the individual lifespan in some *Bombus* species [40,83–85] or generate reinfection in the honey bees [38,39,81,82,86].

3.1.2. Kashmir Bee Virus (KBV)

KBV is a non-enveloped ssRNA (+) virus belonging to the *Cripavirus* genus within the *Dicistroviridae* family [28,81,87]. The genome of KBV is strictly related to ABPV (Acute Bee Paralysis Virus) and IAPV (Israeli Acute Paralysis Virus) [88–90].

Although the virus is considered endemic in America and New Zealand, it has been rarely reported in other regions, both in honey bees and other arthropods (Supplementary Tables S1 and S2).

KBV was found in various Hymenoptera species, like *Bombus* spp. [38,91–93], *Eucera* spp., *Anthophora* spp., *Osmia* spp. [38], wasps, hornets [46,91,94,95], and ants [49,91]. It was also detected in *A. tumida* [39,81], *Galleria melonella*, earwigs, roaches and crickets [39,91].

3.1.3. Acute Bee Paralysis Virus (ABPV)

ABPV is a non-enveloped virus and widespread ssRNA (+) virus belonging to *Apavirus* genus within the *Dicistroviridae* family [28,96]. As reported above, ABPV is genetically linked to KBV and IAPV [88]. ABPV was detected in *V. destructor*, where is is reported incapable to replicate [97,98]. ABPV spillover is not recent (Figure 1 and Supplementary Table S1) as in 1964 various *Bombus* species were found infected in the United Kingdom [99]. The list of bees in which ABPV was found increases constantly, including many *Bombus* species as well as a wide range of other bee species [54,100,101]. In non-bee Hymenoptera, ABPV was detected in *Ancistrocerus auctus*, *Polistes* spp., *V. germanica*, *Scolia flavifrons* and *Linepithema humile* [49,54,94].

3.1.4. Israeli Acute Paralysis Virus (IAPV)

Israeli acute paralysis virus (IAPV) is a non-enveloped ssRNA (+) virus, belonging to *Apavirus* genus within the *Dicistroviridae* family, whose genome shows high homology to ABPV and KBV [88,102,103]. The virus has been isolated in Israel, but there are several known strains [102]. In honey bees, it induces disorientation, shivering wings, crawling, progressive paralysis and death within or ourside the nest [104].

The IAPV is widespread [102] and spillover cases were studied in a wide range of non-*Apis* bee species (Supplementary Table S1). Furthermore, the virus was found in the wasps, *V. germanica* and *V. vulgaris* [38,94] and in the ants, *Camponatus* spp. and *L. humile* [39,49]. Outside Hymenoptera, earwigs, spiders, moths, small hive beetles [39], and *V. velutina* [48] showed to act as IAPV incidental hosts.

3.1.5. Slow Bee Paralysis Virus (SBPV)

Slow bee paralysis virus (SBPV) is an icosahedral non-enveloped ssRNA(+) virus from the *Iflavirus* genus within the *Iflaviridae* family [105,106]. The infection is responsible for paralysis of the first and second pairs of legs in roughly 12-day old honey bees and their sudden death [107,108].

Recently, it was found in wild *Bombus* spp., *E. nigriscens* and *O. bicornis* in Kyrgyzstan, Germany and Georgia [109], in the United Kingdom [110,111] and Belgium [112]. Furthermore, *E. nigriscens* and *O. bicornis* species in Kyrgyzstan, Germany and Georgia scored positive for SBPV infection [109] (Supplementary Table S1). No further spillover events have been reported so far in other arthropods.

3.1.6. Chronic Bee Paralysis Virus (CBPV)

Chronic bee paralysis virus (CBPV) is an unclassified enveloped ssRNA (+) virus carachterized by articulate genome and association to a satellite virus (CBPSV) [113,114]. The infection causes a multifaceted disease encompassing different combinations of symptoms evidencing neurotropism like ataxia, incapability to fly, and trembling, as well as hairlessness and dark colour in the infected bees [114,115].

CBPV is capable to infect other insects. Spillover was reported in the wild in individuals of *B. dalhbomii* [92], *B. impatiens* and *B. ruderatus* [116], *B. pauloensis* [117], *B. terrestris* [92,116],

X. augusti and *X. nigrocinta* [117], *X. dissimilis* [109], *H. amplilobus* [117] and *H. parallelum* [109]. Replicative CBPV was found in two ant species, *C. vagus* and *F. rufa* also [118].

3.1.7. Sacbrood Virus (SBV)

SBV is a non-enveloped ssRNA (+) virus belonging to *Iflavirus* genus and *Dicistroviridae* family [27,28,119]. It is very common in honey bees, that exhibit symptoms in pre-imaginal stages coming into contact with the virus during the brood tending [27]. The virus is spread worldwide and genetic variants were identified in Korea (K-SBV), China (C-SBV), Thailand (T-SBV), Europe (E-SBV) and New Guinea (G-SBV) [120–124]. SBV was detected in a wide range of non-*Apis* bees (Supplementary Table S1) and other hymenopteran species (Supplementary Table S2) [38,39,54,110,116,125–128]. CBPV was detected in hoverflies, small hive beetles, spiders and lepidopterans also [39,82,129,130].

3.1.8. Black Queen Cell Virus (BQCV)

BQCV belongs to the *Cripavirus* genus within *Dicistroviridae* family. As the other *Dicistroviridae*, BQCV is non-enveloped and ssRNA (+) virus [22,131,132].

Despite a high prevalence in adult honey bees [28,97,133–135], symptomatic infections occur in queen pupae and/or pre-pupae, that decompose in irregular, black cells [133,136].

BQCV is spread worldwideand affects several honey bee species and subspecies, like *A. mellifera*, *A. cerana indica*, *A. cerana japonica*, *A. dorsata* and *A. florea* [137]. The range of possible hosts is very wide and includes several wild hymenopteran species [37–39,46,54,109,110,125,126,138,139]. Small hive beetles, hoverflies, roaches, spiders and wax moths scored positive to BQCV also [39,129,130].

3.1.9. Lake Sinai Virus (LSV)

LSV is an ssRNA(+) belonging to the *Sinhaliviridae* family and *Sinaivirus* genus, of which two strains have been identified so far: LSV-1 and LSV-2 [140]. The virus was discovered in honey bees sampled during a colony transhumance near the Lake Sinai, South Dakota, USA. LSV was reported as involved in the colony collapse disorder, despite both pathogenicity and epidemiology have not been clarified yet [70,141].

Cases of LSV spillover have been reported in *Andrena* spp. [37,127], *Bombus* spp., [85,112,127], and species belonging to the families of *Halictidae* and *Megachilidae* [127]. LSV has never been detected outside the *Apoidea* superfamily so far.

3.1.10. *Apis mellifera* Filamentous Virus (*Am*FV)

*Am*FV is an unclassified dsDNA isolated from honey bees, whose relationship with the host and epidemiology are poorly studied. Originally, the pathogen was described as a rickettsia disease, but recently it has been recognized as a virus [142,143]. Severe infections of adult honey bees are associated to milk white hemolymph as a consequence of the high virion concentration. The infected bees show signs of weakness and tend to gather at the hive entrance. Nevertheless, the virus is weakly pathogenic and has low impact on bee lifespan [143–146].

Few spillover cases have been reported so far. They involved as alternative hosts *Andrena* spp. [37], *Bombus* spp. [147], *Osmia* spp. [37] and in *A. tumida* [148] (Supplementary Table S1).

3.1.11. *Varroa destructor* Macula-like Virus (*Vd*MLV)

*Vd*MLV is an unclassified ssRNA(+) virus of the *Tymoviridae* family. The mite *V. destructor* is its primary host and the virus was found in the honey bees as a likely result of the trophic activity of the parasite [149]. Little knowledge is available for this virus. Few spillover cases have been reported so far about *Vd*MLV (Supplementary Table S1), all of them in the wild. Those involved *B. lapidarius*, *B. pascuorum* and *B. pratorum* as host species [112].

3.1.12. Moku Virus

Moku virus is an unclassified ssRNA (+) *Iflavirus*. The virus was first discovered in *Vespula pensylvanica* in Hawaii, but it spread in honey bees too, often associated to *V. destructor* [150]. Since its discovery, Moku virus findings increased rapidly until the detection in a wide range of Hymenoptera species (Supplementary Table S1), that includes *Polistes* spp. [91], *Vespula* spp. [91,95,130,150], *V. velutina* [151] and *L. humile* [91]. Besides, Moku virus was found capable to infect the spiders *H. minitabunda* and *S. capensis* [91] (Supplementary Table S2).

3.2. Fungi

3.2.1. Nosema ceranae

Nosema ceranae is a microsporidium that causes nosemosis type C in western honey bees [152,153]. It is an intracellular obligate parasite, infecting the ventricular epithelial cells [154,155]. The effects of *N. ceranae* infections can be recognized both at individual and colony levels, impacting the bee lifespan, inducing lethargic behaviour, reducing the pollen and honey harvest, and causing colony dwindling [156–159].

The main known spillover event occurred when the pathogen jumped from the Asian honey bee *A. ceranae*, which is deemed as the original host, to the western honey bee *A. mellifera* [152,153].

In addition to *A. cerana* and *A. mellifera*, the microsporidium was reported in several other Hymenoptera (Supplementary Table S1), including *A. ventralis*, *H. truncorum* and *Osmia* spp. [37], commercial and wild *Bombus* species [36,37,83,125,147,160–162], stingless bees, and *Polybya* spp. [163]. Besides, it was detected in the small hive beetle as well as in *A. tumida* [148,164]. Finally, the microsporidium was found in the regurgitated pellets of the European bee-eater *Merops apiaster* [165].

3.2.2. Nosema apis

Nosema apis is the classic microsporidium infecting *A. mellifera*, which is responsible for the nosemosis Type A [166]. Like the other microsporidians, it is an intracellular obligate parasite. It causes, in contrast to *N. ceranae*, severe dysentery that impacts mainly the colony foragers [166–168]. Presently, its spread is limited to specific ecological niches as a possible consequence of the competition with the predominant *N. ceranae* [71,157]. *N. apis* was detected in commercial *B. terrestris* colonies [20], but the transmission route remained unclarified.

3.2.3. Ascosphaera apis

The fungus *Ascosphaera apis* is a honey bee pathogen responsible for the mycosis called chalkbrood disease [169,170]. The infection occurs by spore ingestion in bee larvae, especially in those of the fifth instar, that reduces food consumption and prevents eating [169,170]. The proliferating mycelium invades the larval body, which is transformed into a chalk-like "mummy", so the disease name [169,171,172].

Despite the disease is typical to the honey bees, artificial infections showed the pathogen capability to colonize the intestine of *B. terrestris* adults and larvae [20].

3.3. Bacteria

3.3.1. Melissococcus plutonius

The bacterium *M. plutonius* is the Gram-negative coccus representing the etiological agent of the European foulbrood disease [173,174].

The pathogen is spread worldwide and infects the brood, which dies by undernutrition [175,176]. The infected larvae become flaccid and yellowish by 5 days after infection [173,175,176].

In the United Kingdom, *M. plutonius* was found to impair the development of *B. terrestris* colonies [20].

3.3.2. *Spiroplasma apis*

Spiroplasma apis is a small, helical and motile Gram-positive *Eubacterium* deprived of a cell wall [177,178]. The bacterium was isolated in France from colonies showing symptoms of "May disease" [179]. *S. apis* is lethal to the honey bees when ingested, and the infection may spread by faecal contamination [179].

Strains of *S. apis* were isolated and detected in wild specimens belonging to *B. atratus* [125] and *O. bicornis* [37], with unknown effects.

3.3.3. *Spiroplasma melliferum*

Spiroplasma melliferum is another Eubacterium isolated from the honey bees [180]. The *S. melliferum* infection has similar symptoms and transmission route as *S. apis*, although less virulent [179,180]. As for *S. apis*, *S. melliferum* spillover was observed occasionally (Supplementary Table S1). This is the case of *O. bicornis* individuals, that were found infected in Belgium [37].

3.3.4. *Wolbachia* spp.

Wolbachia spp. are Gram-negative intracellular bacterial symbionts, which can infect the cells of both female honey bees and drones [181,182]. *Wolbachia* spp. impacts the host reproduction. The vertical transmission via the eggs represents the main transmission route to persist in honey bee populations [183,184].

During a national survey in the U.S.A. (Supplementary Tables S1 and S2), several arthropods scored positive to *Wolbachia* spp.: *Andrena* spp., several *Bombus* species, *Lasioglossum* spp., *Halictus* spp., *D. sylvestris*, *V. germanica* and *V. vulgaris*, and two hoverflies [128].

3.4. *Trypanosomatidae*

3.4.1. *Lotmaria passim*

Lotmaria passim is a trypanosomatid with a single flagellum, capable to colonize the digestive tract of *A. mellifera* [185,186]. The parasite spreads within the colony by fecal contact, and the transmission occur via the oro-faecal route [187,188]. The infection impacts the colony by altering behaviour and lifespan of the infected bees [141,189]. *L. passim* is spread worldwide. The colonization implied the replacing of the other honey bee trypanosomatids *Crithidia mellificae* [190,191].

Besides, *L. passim* is present in bumblebee species (Supplementary Table S1) namely th South American *B. funebris*, *B. dalhbomii*, *B. opifex*, *B. ruderatus* and *B. terrestris* [92,147].

L. passim was found also in the small hive beetle, *A. tumida*, as a possible result of the feeding behaviour of this scavenger [81,148].

3.4.2. *Crithidia mellificae*

Crithidia mellificae is another trypanosomatid which can replicate in the honey bee intestine to survive [185,186]. Transmission route and impact on bees are very similar to the other parasite *L. passim* [187,188,192]. *C. mellificae* was almost completely replaced by *L. passim* and its infection has been rarely observed [185,193,194].

Despite that, one spillover case was observed in *A. tumida*, that live in contact with bee colony debris [81].

3.4.3. *Crithidia bombi*

Crithidia bombi is a trypanosomatid infecting *B. terrestris* colonies [195,196]. The infection occurs during the external activity of the forager bumblebees [197,198] and, back to the nest, it spreads by fecal contamination to the other workers [199,200]. *C. bombi* may harm the bumblebee populations as hibernating queens may reduce the success in founding the colonies and remarkably lower their fitness [201]. On queen emergence from the diapause, *C. bombi* infections grow together with the colony that is being established [200].

C. bombi is transmitted during the foraging activity. The pathogen was detected in the wild on *A. vaga* and *O. bicornis* individuals [37] and in small hive beetles collected from the nest of honey bee colonies [148]. Artificial infections showed that *C. bombi* can replicate in *O. lignaria*, *M. rotundata* and *H. ligatus* [202,203].

3.5. Neogregarine

Apicystis bombi

Apicystis bombi is a parasite found primarily in bumblebees. It was found to occur also in honey bees from Europe and North America [204–206]. Upon the ingestion of the oocytes by the bee, the sporozoites develop and migrate to the fat body, where they develop, multiply and disrupt the adipose tissue. The infection increases the worker mortality rate and, due to the fat body disruption, both queen survival to hibernation and colony foundation success are impaired [84,207,208].

Likely, the infection occurs via contact on contaminated flowers [208]. Indeed, *A. bombi* was found in wild species also, namely *A. vaga*, *A. ventralis*, *H. truncorum*, *O. bicornis* and *O. cornuta* [37].

4. Materials and Methods

Protocol and Literature Search

This systematic review was carried out according to the Preferred Reporting Items for Systematic Review and Meta-Analysis (PRISMA) protocols [209]. The research question to be reviewed was: "Which honey bee pathogens may generate spillover to managed and wild Hymenoptera species and, more in general, to the arthropofauna?"

The search intentionally excluded arthropods living in close contact with the honey bees that, like *V. destructor*, are obligate parasites.

The article search was carried out in PubMed, Web of Science, Science Direct, Google Scholar, and Scopus scientific databases for studies aimed to assess the detection of spillover cases of honey bee pathogens. Filters were used to select articles published from January 1960 to April 2021. The last search date was 31 May 2021.

The following search strategy was designed and utilized: "Honey Bee Pathogens" OR "Bumblebee Pathogens" OR "Spillover" OR "Spill-over " OR "Inter species transmission" OR "Inter taxa transmission" OR "Host species transmission" OR "*Apis mellifera*" OR "Honey Bee Diseases" OR "Honey Bee Virus" OR "Honey Bee Bacteria" OR "Honey Bee Microsporidia" OR "Honey Bee Protozoa" OR "Managed Bees" OR "Wild Bees" OR "Commercial Bees" OR "Artificial Infection" OR "Replicative Virus" OR "Bumblebees" OR "Colony Collapse Disorder" OR "Deformed Wing Virus" OR "Acute Bee Paralysis Virus" OR "Israeli Acute Paralysis Virus" OR "Black Queen Cell Virus" OR "Sacbrood Virus" OR "*Apis mellifera* Filamentous Virus" OR "Kashmir Bee Virus" OR "Slow Bee Paralysis Virus" OR "Lake Sinai Virus" OR "*Varroa destructor*" OR "Macula-like Virus" OR "*Nosema apis*" OR "*Nosema ceranae*" OR "*Nosema bombi*" OR "*Spiroplasma*" OR "*Ascosphaera*" OR "*Apicystis*" OR "Arthopods" OR "Entomofauna" OR "Hive Hosts" OR "Hive" OR "Free-Ranging Insect" OR "Bee Interaction" OR "*Varroa destructor*". The logical operator "OR" was used to combine the descriptors.

Studies carried out both in field and laboratory conditions were selected. Besides, studies that did not assess whether the presence of the honey bee pathogens could be related to external contamination were not included. The detected active replication of honey bee viruses was also reported in the Supplementary Materials with an asterisk.

Duplicate studies were excluded. The search and screening for titles, abstracts and results were carried out independently by the authors, including all articles, letters, notes, scientific notes and communications aimed to assess a spillover case of honey bee pathogens and excluding reviews, books, book chapters and theses.

The potentially eligible research articles were read and reviewed independently by the authors and the data were compared to ensure integrity and reliability.

For each article included in this review, relevant information related to the authors, publication year, host species, host conditions, host stage, pathogens and prevalence were extracted. The data from the eligible studies are expressed in the Supplementary Materials and Figures. The authors provided a narrative synthesis of the results for each pathogen capable to generate a spillover case, according to the main characteristics and results related to the topic addressed.

5. Conclusions

This review shows that, in recent years, the frequency of recorded spillover cases of honey bee pathogens to other arthropods, including wild bees, has dramatically increased. Certainly, human movements and globalization have fostered the inflow of novel pathogenic microorganisms, often with detrimental consequences. However, it should also be considered that the analytical methods currently available give impulse to the research on bee pathology, increasing the chance to identify interspecific transmission events.

The host plasticity shown by some honey bee pathogens raises ecological concern for the potential negative consequences on the pollinating entomofauna and ecosystems in general. Despite the fact that research on these pathogens has significantly improved, we have limited knowledge of their potential impact on other bees, insects, and arthropods in general and the cascade of environmental effects. Laboratory studies are not sufficient to cover this gap, for the intricate interaction of the involved biotic and abiotic factors. For the same reasons, the exploitation of these pathogens in the control of arthropods considered as pests (e.g., *A. tumida*, *G. melonella*, *V. velutina*, *L. humile*) should be considered with extreme carefulness.

The tight interaction between honey bees and the other environmental components suggests a holistic approach to the study of bee diseases, including their control. Indeed, pathogens may survive in alternate hosts, generating spillback events and possibly jeopardizing the efficacy of the treatments. This emphasizes the beekeeper's responsibility to maintain healthy colonies to benefit both their production and the environment.

Spillover of honey bee pathogens may have undetected yet important repercussions on the health and functioning of an ecosystem. Health management of honey bee colonies is of high importance in this context. Honey bees and the beekeeping industry should, therefore, undertake an essential role in the One Health concept. This requires the adoption of dedicated research actions.

Supplementary Materials: The following are available online at https://www.mdpi.com/article/10.3390/pathogens10081044/s1, Table S1: Bee pathogen spillover and prevalence identified in hymenopteran hosts, of which are reported condition, stage, geographical area and year, Table S2: Bee pathogen spillover and prevalence identified in arthropod hosts, of which are reported condition, stage, geographical are and year [210–232].

Author Contributions: Conceptualization, A.N., L.B. and G.C.; methodology, G.C.; writing—original draft preparation, G.C.; writing—review and editing, A.N., L.B. and G.C.; project administration, L.B. All authors have read and agreed to the published version of the manuscript.

Funding: This research was funded by the project BeeNet (Italian National Fund under FEASR 2014-2020 from MIPAAF "Ministero delle politiche agricole alimentari e forestali" to CREA "Consiglio per la Ricerca in agricoltura e l'analisi dell'economia agraria").

Institutional Review Board Statement: Not applicable.

Informed Consent Statement: Not applicable.

Data Availability Statement: MDPI Research Data Policies.

Acknowledgments: The Authors acknowledge Laura Zavatta (CREA-AA) for her kind support in generating some of the charts published in this paper.

Conflicts of Interest: The authors declare no conflict of interest.

References

1. Power, A.G.; Mitchell, C.E. Pathogen spillover in disease epidemics. *Am. Nat.* **2004**, *164*, S79–S89. [CrossRef] [PubMed]
2. Nugent, G. Maintenance, spillover and spillback transmission of bovine tuberculosis in multi-host wildlife complexes: A New Zealand case study. *Vet. Microbiol.* **2011**, *151*, 34–42. [CrossRef]
3. Thrusfield, M. *Veterinary Epidemiology*, 4th ed.; Wiley-Blackwell: Hoboken, NJ, USA, 2018.
4. Roberts, M.G. The pluses and minuses of $\mathcal{R}0$. *J. R. Soc. Interface* **2007**, *4*, 949–961. [CrossRef]
5. Fenton, A.; Pedersen, A.B. Community epidemiology framework for classifying disease threats. *Emerg. Infect. Dis.* **2005**, *11*, 1815–1821. [CrossRef]
6. Klein, A.M.; Vaissière, B.E.; Cane, J.H.; Steffan-Dewenter, I.; Cunningham, S.A.; Kremen, C.; Tscharntke, T. Importance of pollinators in changing landscapes for world crops. *Proc. R. Soc. B Biol. Sci.* **2007**, *274*, 303–313. [CrossRef]
7. Gallai, N.; Salles, J.M.; Settele, J.; Vaissière, B.E. Economic valuation of the vulnerability of world agriculture confronted with pollinator decline. *Ecol. Econ.* **2009**, *68*, 810–821. [CrossRef]
8. Rucker, R.R.; Thurman, W.N.; Burgett, M. Honey bee pollination markets and the internalization of reciprocal benefits. *Am. J. Agric. Econ.* **2012**, *94*, 956–977. [CrossRef]
9. Rader, R.; Bartomeus, I.; Garibaldi, L.A.; Garratt, M.P.D.; Howlett, B.G.; Winfree, R.; Cunningham, S.A.; Mayfield, M.M.; Arthur, A.D.; Andersson, G.K.S.; et al. Non-bee insects are important contributors to global crop pollination. *Proc. Natl. Acad. Sci. USA* **2016**, *113*, 146–151. [CrossRef]
10. Velthuis, H.H.W.; van Doorn, A. A century of advances in bumblebee domestication and the economic and environmental aspects of its commercialization for pollination. *Apidologie* **2006**, *37*, 421–451. [CrossRef]
11. Potts, S.G.; Roberts, S.P.M.; Dean, R.; Marris, G.; Brown, M.A.; Jones, R.; Neumann, P.; Settele, J. Declines of managed honey bees and beekeepers in Europe. *J. Apic. Res.* **2010**, *49*, 15–22. [CrossRef]
12. Tehel, A.; Brown, M.J.; Paxton, R.J. Impact of managed honey bee viruses on wild bees. *Curr. Opin. Virol.* **2016**, *19*, 16–22. [CrossRef] [PubMed]
13. Meeus, I.; Brown, M.J.F.; De Graaf, D.C.; Smagghe, G. Effects of Invasive Parasites on Bumble Bee Declines. *Conserv. Biol.* **2011**, *25*, 662–671. [CrossRef]
14. Baldock, K.C. Opportunities and threats for pollinator conservation in global towns and cities. *Curr. Opin. Insect Sci.* **2020**, *38*, 63–71. [CrossRef] [PubMed]
15. Herrera, C.M. Gradual replacement of wild bees by honeybees in flowers of the Mediterranean Basin over the last 50 years. *Proc. R. Soc. B Biol. Sci.* **2020**, *287*. [CrossRef] [PubMed]
16. Kerr, J.T.; Pindar, A.; Galpern, P.; Packer, L.; Potts, S.G.; Roberts, S.M.; Rasmont, P.; Schweiger, O.; Colla, S.R.; Richardson, L.L.; et al. Climate change impacts on bumblebees converge across continents. *Science* **2015**, *349*, 177–180. [CrossRef]
17. Koh, I.; Lonsdorf, E.V.; Williams, N.M.; Brittain, C.; Isaacs, R.; Gibbs, J.; Ricketts, T.H. Modeling the status, trends, and impacts of wild bee abundance in the United States. *Proc. Natl. Acad. Sci. USA* **2016**, *113*, 140–145. [CrossRef]
18. Biesmeijer, J.C.; Roberts, S.P.M.; Reemer, M.; Ohlemüller, R.; Edwards, M.; Peeters, T.; Schaffers, A.P.; Potts, S.G.; Kleukers, R.; Thomas, C.D.; et al. Parallel declines in pollinators and insect-pollinated plants in Britain and the Netherlands. *Science* **2006**, *313*, 351–354. [CrossRef] [PubMed]
19. Soroye, P.; Newbold, T.; Kerr, J. Climate change contributes to widespread declines among bumble bees across continents. *Science* **2020**, *367*, 685–688. [CrossRef]
20. Graystock, P.; Yates, K.; Evison, S.E.F.; Darvill, B.; Goulson, D.; Hughes, W.O.H. The Trojan hives: Pollinator pathogens, imported and distributed in bumblebee colonies. *J. Appl. Ecol.* **2013**, *50*, 1207–1215. [CrossRef]
21. Brutscher, L.M.; McMenamin, A.J.; Flenniken, M.L. The Buzz about Honey Bee Viruses. *PLoS Pathog.* **2016**, *12*, e1005757. [CrossRef]
22. Gisder, S.; Genersch, E. Viruses of commercialized insect pollinators. *J. Invertebr. Pathol.* **2017**, *147*, 51–59. [CrossRef]
23. Graystock, P.; Goulson, D.; Hughes, W.O.H. The relationship between managed bees and the prevalence of parasites in bumblebees. *PeerJ* **2014**, *2*, e522. [CrossRef]
24. Grupe, A.C.; Quandt, C.A. A growing pandemic: A review of Nosema parasites in globally distributed domesticated and native bees. *PLoS Pathog.* **2020**, *16*, e1008580. [CrossRef]
25. McMenamin, A.J.; Genersch, E. Honey bee colony losses and associated viruses. *Curr. Opin. Insect Sci.* **2015**, *8*, 121–129. [CrossRef]
26. Martin, S.J.; Highfield, A.C.; Brettell, L.; Villalobos, E.M.; Budge, G.E.; Powell, M.; Nikaido, S.; Schroeder, D.C. Global honey bee viral landscape altered by a parasitic mite. *Science* **2012**, *336*, 1304–1306. [CrossRef]
27. Chen, Y.P.; Siede, R. Honey Bee Viruses. *Adv. Virus Res.* **2007**, *70*, 33–80.
28. Chen, Y.P.; Pettis, J.S.; Collins, A.; Feldlaufer, M.F. Prevalence and Transmission of Honeybee Viruses. *Appl. Environ. Microbiol.* **2006**, *72*, 606–611. [CrossRef]
29. Martín-Hernández, R.; Bartolomé, C.; Chejanovsky, N.; Le Conte, Y.; Dalmon, A.; Dussaubat, C.; García-Palencia, P.; Meana, A.; Pinto, M.A.; Soroker, V.; et al. *Nosema ceranae* in *Apis mellifera*: A 12 years postdetection perspective. *Environ. Microbiol.* **2018**, *20*, 1302–1329. [CrossRef]
30. Bromenshenk, J.J.; Henderson, C.B.; Wick, C.H.; Stanford, M.F.; Zulich, A.W.; Jabbour, R.E.; Deshpande, S.V.; McCubbin, P.E.; Seccomb, R.A.; Welch, P.M.; et al. Iridovirus and Microsporidian Linked to Honey Bee Colony Decline. *PLoS ONE* **2010**, *5*, e13181. [CrossRef] [PubMed]

31. Cilia, G.; Sagona, S.; Giusti, M.; Jarmela dos Santos, P.E.; Nanetti, A.; Felicioli, A. *Nosema ceranae* infection in honeybee samples from Tuscanian Archipelago (Central Italy) investigated by two qPCR methods. *Saudi J. Biol. Sci.* **2019**, *26*, 1553–1556. [CrossRef]

32. Beaurepaire, A.; Piot, N.; Doublet, V.; Antunez, K.; Campbell, E.; Chantawannakul, P.; Chejanovsky, N.; Gajda, A.; Heerman, M.; Panziera, D.; et al. Diversity and Global Distribution of Viruses of the Western Honey Bee, *Apis mellifera*. *Insects* **2020**, *11*, 239. [CrossRef]

33. Laomettachit, T.; Liangruksa, M.; Termsaithong, T.; Tangthanawatsakul, A.; Duangphakdee, O. A model of infection in honeybee colonies with social immunity. *PLoS ONE* **2021**, *16*, e0247294. [CrossRef]

34. Hinshaw, C.; Evans, K.C.; Rosa, C.; López-Uribe, M.M. The Role of Pathogen Dynamics and Immune Gene Expression in the Survival of Feral Honey Bees. *Front. Ecol. Evol.* **2021**, *8*, 505. [CrossRef]

35. Evans, J.D.; Schwarz, R.S. Bees brought to their knees: Microbes affecting honey bee health. *Trends Microbiol.* **2011**, *19*, 614–620. [CrossRef]

36. Manley, R.; Temperton, B.; Doyle, T.; Gates, D.; Hedges, S.; Boots, M.; Wilfert, L. Knock-on community impacts of a novel vector: Spillover of emerging DWV-B from *Varroa* -infested honeybees to wild bumblebees. *Ecol. Lett.* **2019**, *22*, ele.13323. [CrossRef]

37. Ravoet, J.; De Smet, L.; Meeus, I.; Smagghe, G.; Wenseleers, T.; de Graaf, D.C. Widespread occurrence of honey bee pathogens in solitary bees. *J. Invertebr. Pathol.* **2014**, *122*, 55–58. [CrossRef]

38. Singh, R.; Levitt, A.L.; Rajotte, E.G.; Holmes, E.C.; Ostiguy, N.; vanEngelsdorp, D.; Lipkin, W.I.; dePamphilis, C.W.; Toth, A.L.; Cox-Foster, D.L. RNA Viruses in Hymenopteran Pollinators: Evidence of Inter-Taxa Virus Transmission via Pollen and Potential Impact on Non-*Apis* Hymenopteran Species. *PLoS ONE* **2010**, *5*, e14357. [CrossRef]

39. Levitt, A.L.; Singh, R.; Cox-Foster, D.L.; Rajotte, E.; Hoover, K.; Ostiguy, N.; Holmes, E.C. Cross-species transmission of honey bee viruses in associated arthropods. *Virus Res.* **2013**, *176*, 232–240. [CrossRef]

40. Mazzei, M.; Carrozza, M.L.; Luisi, E.; Forzan, M.; Giusti, M.; Sagona, S.; Tolari, F.; Felicioli, A. Infectivity of DWV Associated to Flower Pollen: Experimental Evidence of a Horizontal Transmission Route. *PLoS ONE* **2014**, *9*, e113448. [CrossRef]

41. Chen, Y.; Evans, J.; Feldlaufer, M. Horizontal and vertical transmission of viruses in the honey bee, *Apis mellifera*. *J. Invertebr. Pathol.* **2006**, *92*, 152–159. [CrossRef]

42. Graystock, P.; Goulson, D.; Hughes, W.O.H. Parasites in bloom: Flowers aid dispersal and transmission of pollinator parasites within and between bee species. *Proc. R. Soc. B Biol. Sci.* **2015**, *282*. [CrossRef] [PubMed]

43. Schittny, D.; Yañez, O.; Neumann, P. Honey Bee Virus Transmission via Hive Products. *Vet. Sci.* **2020**, *7*, 96. [CrossRef]

44. Alger, S.A.; Burnham, P.A.; Brody, A.K. Flowers as viral hot spots: Honey bees (*Apis mellifera*) unevenly deposit viruses across plant species. *PLoS ONE* **2019**, *14*, e0221800. [CrossRef]

45. Forzan, M.; Sagona, S.; Mazzei, M.; Felicioli, A. Detection of deformed wing virus in *Vespa crabro*. *Bull. Insectology* **2017**, *70*, 261–265.

46. Mazzei, M.; Cilia, G.; Forzan, M.; Lavazza, A.; Mutinelli, F.; Felicioli, A. Detection of replicative Kashmir Bee Virus and Black Queen Cell Virus in Asian hornet *Vespa velutina* (Lepelieter 1836) in Italy. *Sci. Rep.* **2019**, *9*, 1–9. [CrossRef]

47. Mazzei, M.; Forzan, M.; Cilia, G.; Sagona, S.; Bortolotti, L.; Felicioli, A. First detection of replicative deformed wing virus (DWV) in *Vespa velutina* nigrithorax. *Bull. Insectology* **2018**, *71*, 211–216.

48. Yañez, O.; Zheng, H.-Q.; Hu, F.-L.; Neumann, P.; Dietemann, V. A scientific note on Israeli acute paralysis virus infection of Eastern honeybee *Apis cerana* and vespine predator *Vespa velutina*. *Apidologie* **2012**, *43*, 587–589. [CrossRef]

49. Gruber, M.A.M.; Cooling, M.; Baty, J.W.; Buckley, K.; Friedlander, A.; Quinn, O.; Russell, J.F.E.J.; Sébastien, A.; Lester, P.J. Single-stranded RNA viruses infecting the invasive *Argentine ant, Linepithema humile*. *Sci. Rep.* **2017**, *7*, 1–10. [CrossRef] [PubMed]

50. Sébastien, A.; Lester, P.J.; Hall, R.J.; Wang, J.; Moore, N.E.; Gruber, M.A.M. Invasive ants carry novel viruses in their new range and form reservoirs for a honeybee pathogen. *Biol. Lett.* **2015**, *11*, 20150610. [CrossRef]

51. Cooling, M.; Gruber, M.A.M.; Hoffmann, B.D.; Sébastien, A.; Lester, P.J. A metatranscriptomic survey of the invasive yellow crazy ant, *Anoplolepis gracilipes*, identifies several potential viral and bacterial pathogens and mutualists. *Insectes Soc.* **2017**, *64*, 197–207. [CrossRef]

52. Ilyasov, R.A.; Gaifullina, L.R.; Saltykova, E.S.; Poskryakov, A.V.; Nikolenko, A.G. Review of the expression of antimicrobial peptide defensin in honey bees *Apis mellifera* L. *J. Apic. Sci.* **2012**, *56*, 115–124. [CrossRef]

53. Collison, E.J.; Hird, H.; Tyler, C.R.; Cresswell, J.E. Effects of neonicotinoid exposure on molecular and physiological indicators of honey bee immunocompetence. *Apidologie* **2018**, *49*, 196–208. [CrossRef]

54. Dalmon, A.; Diévart, V.; Thomasson, M.; Fouque, R.; Vaissière, B.E.; Guilbaud, L.; Le Conte, Y.; Henry, M. Possible Spillover of Pathogens between Bee Communities Foraging on the Same Floral Resource. *Insects* **2021**, *12*, 122. [CrossRef]

55. Gusachenko, O.N.; Woodford, L.; Balbirnie-Cumming, K.; Ryabov, E.V.; Evans, D.J. Evidence for and against deformed wing virus spillover from honey bees to bumble bees: A reverse genetic analysis. *Sci. Rep.* **2020**, *10*, 16847. [CrossRef]

56. Benjamin-Chung, J.; Arnold, B.F.; Berger, D.; Luby, S.P.; Miguel, E.; Colford, J.M.; Hubbard, A.E. Spillover effects in epidemiology: Parameters, study designs and methodological considerations. *Int. J. Epidemiol.* **2018**, *47*, 332–347. [CrossRef]

57. Drossart, M.; Gérard, M. Beyond the decline of wild bees: Optimizing conservation measures and bringing together the actors. *Insects* **2020**, *11*, 649. [CrossRef]

58. Prendergast, K.S.; Menz, M.H.M.; Dixon, K.W.; Bateman, P.W. The relative performance of sampling methods for native bees: An empirical test and review of the literature. *Ecosphere* **2020**, *11*, e03076. [CrossRef]

59. Yañez, O.; Piot, N.; Dalmon, A.; de Miranda, J.R.; Chantawannakul, P.; Panziera, D.; Amiri, E.; Smagghe, G.; Schroeder, D.; Chejanovsky, N. Bee Viruses: Routes of Infection in Hymenoptera. *Front. Microbiol.* **2020**, *11*, 943. [CrossRef]

60. Crane, E. (Ed.) *The World History of Beekeeping and Honey Hunting*, 1st ed.; Routledge: Oxford, UK, 1999; ISBN 9780415924672.

61. Winston, M.L. (Ed.) *From Where I Sit: Essays on Bees, Beekeeping, and Science*, 1st ed.; Comstock Publishing Associates: Rio de Janeiro, Brazil, 1998; ISBN 0801484782.

62. Kritsky, G. Beekeeping from Antiquity through the Middle Ages. *Annu. Rev. Entomol.* **2017**, *62*, 249–264. [CrossRef]

63. Lodesani, M.; Costa, C. Bee breeding and genetics in Europe. *Bee World* **2003**, *84*, 69–85. [CrossRef]

64. Crane, E. Beekeeping. In *Encyclopedia of Insects*; Elsevier Inc.: Amsterdam, The Netherlands, 2009; pp. 66–71, ISBN 9780123741448.

65. Requier, F.; Antúnez, K.; Morales, C.L.; Aldea Sánchez, P.; Castilhos, D.; Garrido, P.M.; Giacobino, A.; Reynaldi, F.J.; Rosso Londoño, J.M.; Santos, E.; et al. Trends in beekeeping and honey bee colony losses in Latin America. *J. Apic. Res.* **2018**, *57*, 657–662. [CrossRef]

66. López-Uribe, M.M.; Simone-Finstrom, M. Special Issue: Honey Bee Research in the US: Current State and Solutions to Beekeeping Problems. *Insects* **2019**, *10*, 22. [CrossRef]

67. Ingrid, H.W.; Corbet, S.A.; Osborne, J.L. Beekeeping, wild bees and pollination in the european community. *Bee World* **1991**, *72*, 170–180. [CrossRef]

68. Cilia, G.; Zavatta, L.; Ranalli, R.; Nanetti, A.; Bortolotti, L. Replicative Deformed Wing Virus found in the head of adults from symptomatic commercial bumblebee (*Bombus terrestris*) colonies. *Vet. Sci.* **2021**, *8*, 117. [CrossRef] [PubMed]

69. Genersch, E.; Yue, C.; Fries, I.; de Miranda, J.R. Detection of Deformed wing virus, a honey bee viral pathogen, in bumble bees (*Bombus terrestris* and *Bombus pascuorum*) with wing deformities. *J. Invertebr. Pathol.* **2006**, *91*, 61–63. [CrossRef]

70. Cornman, R.S.; Tarpy, D.R.; Chen, Y.; Jeffreys, L.; Lopez, D.; Pettis, J.S.; vanEngelsdorp, D.; Evans, J.D. Pathogen Webs in Collapsing Honey Bee Colonies. *PLoS ONE* **2012**, *7*, e43562. [CrossRef]

71. Martín-Hernández, R.; Botías, C.; Bailón, E.G.; Martínez-Salvador, A.; Prieto, L.; Meana, A.; Higes, M. Microsporidia infecting *Apis mellifera*: Coexistence or competition. Is *Nosema ceranae* replacing *Nosema apis*? *Environ. Microbiol.* **2012**, *14*, 2127–2138. [CrossRef]

72. Mahefarisoa, K.L.; Simon Delso, N.; Zaninotto, V.; Colin, M.E.; Bonmatin, J.M. The threat of veterinary medicinal products and biocides on pollinators: A One Health perspective. *One Health* **2021**, *12*, 100237. [CrossRef]

73. Wilfert, L.; Brown, M.J.F.; Doublet, V. OneHealth implications of infectious diseases of wild and managed bees. *J. Invertebr. Pathol.* **2020**, 107506. [CrossRef] [PubMed]

74. Sadd, B.M.; Schmid-Hempel, P. Principles of ecological immunology. *Evol. Appl.* **2009**, *2*, 113–121. [CrossRef]

75. Manley, R.; Boots, M.; Wilfert, L. Emerging viral disease risk to pollinating insects: Ecological, evolutionary and anthropogenic factors. *J. Appl. Ecol.* **2015**, *52*, 331–340. [CrossRef]

76. De Miranda, J.R.; Genersch, E. Deformed wing virus. *J. Invertebr. Pathol.* **2010**, *103*, S48–S61. [CrossRef]

77. Mordecai, G.J.; Wilfert, L.; Martin, S.J.; Jones, I.M.; Schroeder, D.C. Diversity in a honey bee pathogen: First report of a third master variant of the Deformed Wing Virus quasispecies. *ISME J.* **2016**, *10*, 1264–1273. [CrossRef]

78. McMahon, D.P.; Natsopoulou, M.E.; Doublet, V.; Fürst, M.; Weging, S.; Brown, M.J.F.; Gogol-Döring, A.; Paxton, R.J. Elevated virulence of an emerging viral genotype as a driver of honeybee loss. *Proc. R. Soc. B Biol. Sci.* **2016**, *283*, 20160811. [CrossRef] [PubMed]

79. Martin, S.J.; Brettell, L.E. Deformed Wing Virus in Honeybees and Other Insects. *Annu. Rev. Virol.* **2019**, *6*, 49–69. [CrossRef] [PubMed]

80. Ryabov, E.V.; Wood, G.R.; Fannon, J.M.; Moore, J.D.; Bull, J.C.; Chandler, D.; Mead, A.; Burroughs, N.; Evans, D.J. A Virulent Strain of Deformed Wing Virus (DWV) of Honeybees (*Apis mellifera*) Prevails after Varroa destructor-Mediated, or In Vitro, Transmission. *PLoS Pathog.* **2014**, *10*, e1004230. [CrossRef]

81. Nanetti, A.; Ellis, J.D.; Cardaio, I.; Cilia, G. Detection of *Lotmaria passim*, *Crithidia mellificae* and Replicative Forms of Deformed Wing Virus and Kashmir Bee Virus in the Small Hive Beetle (*Aethina tumida*). *Pathogens* **2021**, *10*, 372. [CrossRef] [PubMed]

82. Eyer, M.; Chen, Y.P.; Schäfer, M.O.; Pettis, J.; Neumann, P. Small hive beetle, *Aethina tumida*, as a potential biological vector of honeybee viruses. *Apidologie* **2009**, *40*, 419–428. [CrossRef]

83. Fürst, M.A.; McMahon, D.P.; Osborne, J.L.; Paxton, R.J.; Brown, M.J.F.F. Disease associations between honeybees and bumblebees as a threat to wild pollinators. *Nature* **2014**, *506*, 364–366. [CrossRef]

84. Graystock, P.; Meeus, I.; Smagghe, G.; Goulson, D.; Hughes, W.O.H. The effects of single and mixed infections of *Apicystis bombi* and deformed wing virus in *Bombus terrestris*. *Parasitology* **2016**, *143*, 358–365. [CrossRef]

85. Lucia, M.; Reynaldi, F.J.; Sguazza, G.H.; Abrahamovich, A.H. First detection of deformed wing virus in *Xylocopa augusti* larvae (Hymenoptera: Apidae) in Argentina. *J. Apic. Res.* **2014**, *53*, 466–468. [CrossRef]

86. Huwiler, M.; Papach, A.; Cristina, E.; Yañez, O.; Williams, G.R.; Neumann, P. Deformed wings of small hive beetle independent of virus infections and mites. *J. Invertebr. Pathol.* **2020**, *172*, 107365. [CrossRef] [PubMed]

87. De Miranda, J.R.; Drebot, M.; Tyler, S.; Shen, M.; Cameron, C.E.; Stoltz, D.B.; Camazine, S.M. Complete nucleotide sequence of Kashmir bee virus and comparison with acute bee paralysis virus. *J. Gen. Virol.* **2004**, *85*, 2263–2270. [CrossRef]

88. De Miranda, J.R.; Cordoni, G.; Budge, G. The Acute bee paralysis virus–Kashmir bee virus–Israeli acute paralysis virus complex. *J. Invertebr. Pathol.* **2010**, *103*, S30–S47. [CrossRef]

89. Anderson, D.L. Kashmir bee virus: A relatively harmless virus of honey bee colonies. *Am. Bee J.* **1991**, *131*, 767–771.

90. Evans, J.D. Genetic Evidence for Coinfection of Honey Bees by Acute Bee Paralysis and Kashmir Bee Viruses. *J. Invertebr. Pathol.* **2001**, *78*, 189–193. [CrossRef]

91. Dobelmann, J.; Felden, A.; Lester, P.J. Genetic strain diversity of multi-host RNA viruses that infect a wide range of pollinators and associates is shaped by geographic origins. *Viruses* **2020**, *12*, 358. [CrossRef]

92. Arismendi, N.; Riveros, G.; Zapata, N.; Smagghe, G.; González, C.; Vargas, M. Occurrence of bee viruses and pathogens associated with emerging infectious diseases in native and non-native bumble bees in southern Chile. *Biol. Invasions* **2021**, 1–15. [CrossRef]

93. Sachman-Ruiz, B.; Narváez-Padilla, V.; Reynaud, E. Commercial *Bombus* impatiens as reservoirs of emerging infectious diseases in central México. *Biol. Invasions* **2015**, *17*, 2043–2053. [CrossRef]

94. Brenton-Rule, E.C.; Dobelmann, J.; Baty, J.W.; Brown, R.L.; Dvorak, L.; Grangier, J.; Masciocchi, M.; McGrannachan, C.; Shortall, C.R.; Schmack, J.; et al. The origins of global invasions of the German wasp (*Vespula germanica*) and its infection with four honey bee viruses. *Biol. Invasions* **2018**, *20*, 3445–3460. [CrossRef]

95. Quinn, O.; Gruber, M.A.M.; Brown, R.L.; Baty, J.W.; Bulgarella, M.; Lester, P.J. A metatranscriptomic analysis of diseased social wasps (*Vespula vulgaris*) for pathogens, with an experimental infection of larvae and nests. *PLoS ONE* **2018**, *13*, e0209589. [CrossRef] [PubMed]

96. Benjeddou, M.; Leat, N.; Allsopp, M.; Davison, S. Detection of Acute Bee Paralysis Virus and Black Queen Cell Virus from Honeybees by Reverse Transcriptase PCR. *Appl. Environ. Microbiol.* **2001**, *67*, 2384–2387. [CrossRef] [PubMed]

97. Berenyi, O.; Bakonyi, T.; Derakhshifar, I.; Koglberger, H.; Nowotny, N. Occurrence of Six Honeybee Viruses in Diseased Austrian Apiaries. *Appl. Environ. Microbiol.* **2006**, *72*, 2414–2420. [CrossRef]

98. Genersch, E.; Evans, J.D.; Fries, I. Honey bee disease overview. *J. Invertebr. Pathol.* **2010**, *103*, S2–S4. [CrossRef]

99. Bailey, L.; Gibbs, A.J. Acute infection of bees with paralysis virus. *J. Insect Pathol.* **1964**, *6*, 395–407.

100. Melathopoulos, A.; Ovinge, L.; Veiga, P.W.; Castillo, C.; Ostermann, D.; Hoover, S. Viruses of managed alfalfa leafcutting bees (*Megachille rotundata* Fabricus) and honey bees (*Apis mellifera* L.) in Western Canada: Incidence, impacts, and prospects of cross-species viral transmission. *J. Invertebr. Pathol.* **2017**, *146*, 24–30. [CrossRef] [PubMed]

101. Alvarez, L.J.; Reynaldi, F.J.; Ramello, P.J.; Garcia, M.L.G.; Sguazza, G.H.; Abrahamovich, A.H.; Lucia, M. Detection of honey bee viruses in Argentinian stingless bees (Hymenoptera: Apidae). *Insectes Soc.* **2018**, *65*, 191–197. [CrossRef]

102. Chen, Y.P.; Pettis, J.S.; Corona, M.; Chen, W.P.; Li, C.J.; Spivak, M.; Visscher, P.K.; DeGrandi-Hoffman, G.; Boncristiani, H.; Zhao, Y.; et al. Israeli Acute Paralysis Virus: Epidemiology, Pathogenesis and Implications for Honey Bee Health. *PLoS Pathog.* **2014**, *10*, e1004261. [CrossRef]

103. Palacios, G.; Hui, J.; Quan, P.L.; Kalkstein, A.; Honkavuori, K.S.; Bussetti, A.V.; Conlan, S.; Evans, J.; Chen, Y.P.; van Engelsdorp, D.; et al. Genetic Analysis of Israel Acute Paralysis Virus: Distinct Clusters Are Circulating in the United States. *J. Virol.* **2008**, *82*, 6209–6217. [CrossRef] [PubMed]

104. Hou, C.; Rivkin, H.; Slabezki, Y.; Chejanovsky, N. Dynamics of the Presence of Israeli Acute Paralysis Virus in Honey Bee Colonies with Colony Collapse Disorder. *Viruses* **2014**, *6*, 2012–2027. [CrossRef]

105. Kalynych, S.; Přidal, A.; Pálková, L.; Levdansky, Y.; de Miranda, J.R.; Plevka, P. Virion Structure of Iflavirus Slow Bee Paralysis Virus at 2.6-Angstrom Resolution. *J. Virol.* **2016**, *90*, 7444–7455. [CrossRef] [PubMed]

106. Santillán-Galicia, M.T.; Ball, B.V.; Clark, S.J.; Alderson, P.G. Slow bee paralysis virus and its transmission in honey bee pupae by varroa destructor. *J. Apic. Res.* **2014**, *53*, 146–154. [CrossRef]

107. De Miranda, J.R.; Dainat, B.; Locke, B.; Cordoni, G.; Berthoud, H.; Gauthier, L.; Neumann, P.; Budge, G.E.; Ball, B.V.; Stoltz, D.B. Genetic characterization of slow bee paralysis virus of the honeybee (*Apis mellifera* L.). *J. Gen. Virol.* **2010**, *91*, 2524–2530. [CrossRef]

108. Bailey, L.; Woods, R.D. Three previously undescribed viruses from the honey bee. *J. Gen. Virol.* **1974**, *25*, 175–186. [CrossRef]

109. Radzevičiūtė, R.; Theodorou, P.; Husemann, M.; Japoshvili, G.; Kirkitadze, G.; Zhusupbaeva, A.; Paxton, R.J. Replication of honey bee-associated RNA viruses across multiple bee species in apple orchards of Georgia, Germany and Kyrgyzstan. *J. Invertebr. Pathol.* **2017**, *146*, 14–23. [CrossRef] [PubMed]

110. McMahon, D.P.; Fürst, M.A.; Caspar, J.; Theodorou, P.; Brown, M.J.F.; Paxton, R.J. A sting in the spit: Widespread cross-infection of multiple RNA viruses across wild and managed bees. *J. Anim. Ecol.* **2015**, *84*, 615–624. [CrossRef]

111. Manley, R.; Boots, M.; Wilfert, L. Condition-dependent virulence of slow bee paralysis virus in *Bombus terrestris*: Are the impacts of honeybee viruses in wild pollinators underestimated? *Oecologia* **2017**, *184*, 305–315. [CrossRef]

112. Parmentier, L.; Smagghe, G.; de Graaf, D.C.; Meeus, I. Varroa destructor Macula-like virus, Lake Sinai virus and other new RNA viruses in wild bumblebee hosts (*Bombus pascuorum*, *Bombus lapidarius* and *Bombus pratorum*). *J. Invertebr. Pathol.* **2016**, *134*, 6–11. [CrossRef]

113. Ribiere, M.; Triboulot, C.; Mathieu, L.; Aurieres, C.; Faucon, J.-P.; Pepin, M. Molecular diagnosis of chronic bee paralysis virus infection. *Apidologie* **2002**, *33*, 339–351. [CrossRef]

114. Ribière, M.; Olivier, V.; Blanchard, P. Chronic bee paralysis: A disease and a virus like no other? *J. Invertebr. Pathol.* **2010**, *103*, S120–S131. [CrossRef]

115. Olivier, V.; Massou, I.; Celle, O.; Blanchard, P.; Schurr, F.; Ribière, M.; Gauthier, M. In situ hybridization assays for localization of the chronic bee paralysis virus in the honey bee (*Apis mellifera*) brain. *J. Virol. Methods* **2008**, *153*, 232–237. [CrossRef]

116. Choi, Y.S.; Lee, M.Y.; Hong, I.P.; Kim, N.S.; Byeon, K.H.; Yoon, H. Detection of Honeybee Virus from Bumblebee (*Bombus terrestris* L. and *Bombus ignitus*). *Korean J. Apic.* **2010**, *25*, 259–266.

117. Fernandez de Landa, G.; Revainera, P.; Brasesco, C.; di Gerónimo, V.; Plischuk, S.; Meroi, F.; Maggi, M.; Eguaras, M.; Quintana, S. Chronic bee paralysis virus (CBPV) in South American non-Apis bees. *Arch. Virol.* **2020**, *165*, 2053–2056. [CrossRef]

118. Celle, O.; Blanchard, P.; Olivier, V.; Schurr, F.; Cougoule, N.; Faucon, J.P.; Ribière, M. Detection of Chronic bee paralysis virus (CBPV) genome and its replicative RNA form in various hosts and possible ways of spread. *Virus Res.* **2008**, *133*, 280–284. [CrossRef] [PubMed]

119. Bailey, L.; Gibbs, A.J.; Woods, R.D. Sacbrood virus of the larval honey bee (*Apis mellifera* linnaeus). *Virology* **1964**, *23*, 425–429. [CrossRef]

120. Chang, J.-C.; Chang, Z.-T.; Ko, C.-Y.; Chen, Y.-W.; Nai, Y.-S. Genomic Sequencing and Comparison of Sacbrood Viruses from *Apis cerana* and *Apis mellifera* in Taiwan. *Pathogens* **2020**, *10*, 14. [CrossRef]

121. Reddy, K.E.; Yoo, M.S.; Kim, Y.H.; Kim, N.H.; Ramya, M.; Jung, H.N.; Thao, L.T.B.; Lee, H.S.; Kang, S.W. Homology differences between complete Sacbrood virus genomes from infected *Apis mellifera* and *Apis cerana* honeybees in Korea. *Virus Genes* **2016**, *52*, 281–289. [CrossRef] [PubMed]

122. Rana, R.; Bana, B.S.; Kaushal, N.; Kaundal, P.; Rana, K.; Khan, M.A.; Gwande, S.J.; Sharma, H.K. Identification of sacbrood virus disease in honeybee, *Apis mellifera* L. by using ELISA and RT-PCR techniques. *Indian J. Biotechnol.* **2011**, *10*, 274–284.

123. Roberts, J.M.K.; Anderson, D.L. A novel strain of sacbrood virus of interest to world apiculture. *J. Invertebr. Pathol.* **2014**, *118*, 71–74. [CrossRef]

124. Zhang, Y.; Huang, X.; Xu, Z.; Han, R.; Chen, J. Differential Gene Transcription in Honeybee (*Apis cerana*) Larvae Challenged by Chinese Sacbrood Virus (CSBV). *Sociobiology* **2013**, *60*, 413–420. [CrossRef]

125. Gamboa, V.; Ravoet, J.; Brunain, M.; Smagghe, G.; Meeus, I.; Figueroa, J.; Riaño, D.; de Graaf, D.C. Bee pathogens found in *Bombus atratus* from Colombia: A case study. *J. Invertebr. Pathol.* **2015**, *129*, 36–39. [CrossRef] [PubMed]

126. Reynaldi, F.; Sguazza, G.; Albicoro, F.; Pecoraro, M.; Galosi, C. First molecular detection of co-infection of honey bee viruses in asymptomatic *Bombus atratus* in South America. *Braz. J. Biol.* **2013**, *73*, 797–800. [CrossRef]

127. Dolezal, A.G.; Hendrix, S.D.; Scavo, N.A.; Carrillo-Tripp, J.; Harris, M.A.; Wheelock, M.J.; O'Neal, M.E.; Toth, A.L. Honey Bee Viruses in Wild Bees: Viral Prevalence, Loads, and Experimental Inoculation. *PLoS ONE* **2016**, *11*, e0166190. [CrossRef] [PubMed]

128. Evison, S.E.F.; Roberts, K.E.; Laurenson, L.; Pietravalle, S.; Hui, J.; Biesmeijer, J.C.; Smith, J.E.; Budge, G.; Hughes, W.O.H. Pervasiveness of Parasites in Pollinators. *PLoS ONE* **2012**, *7*, e30641. [CrossRef] [PubMed]

129. Bailes, E.J.; Deutsch, K.R.; Bagi, J.; Rondissone, L.; Brown, M.J.F.; Lewis, O.T. First detection of bee viruses in hoverfly (syrphid) pollinators. *Biol. Lett.* **2018**, *14*, 20180001. [CrossRef]

130. Brettell, L.E.; Schroeder, D.C.; Martin, S.J. RNAseq analysis reveals virus diversity within hawaiian apiary insect communities. *Viruses* **2019**, *11*, 397. [CrossRef]

131. Mayo, M.A. A summary of taxonomic changes recently approved by ICTV. *Arch. Virol.* **2002**, *147*, 1655–1656. [CrossRef] [PubMed]

132. Mayo, M.A. Virology Division News. *Arch. Virol.* **2002**, *147*, 1071–1076. [CrossRef]

133. Leat, N.; Ball, B.; Govan, V.; Davison, S. Analysis of the complete genome sequence of black queen-cell virus, a picorna-like virus of honey bees. *J. Gen. Virol.* **2000**, *81*, 2111–2119. [CrossRef] [PubMed]

134. Tentcheva, D.; Gauthier, L.; Zappulla, N.; Dainat, B.; Cousserans, F.; Colin, M.E.; Bergoin, M. Prevalence and seasonal variations of six bee viruses in *Apis mellifera* L. and *Varroa destructor* mite populations in France. *Appl. Environ. Microbiol.* **2004**, *70*, 7185–7191. [CrossRef] [PubMed]

135. Kajobe, R.; Marris, G.; Budge, G.; Laurenson, L.; Cordoni, G.; Jones, B.; Wilkins, S.; Cuthbertson, A.G.S.; Brown, M.A. First molecular detection of a viral pathogen in Ugandan honey bees. *J. Invertebr. Pathol.* **2010**, *104*, 153–156. [CrossRef] [PubMed]

136. Siede, R.; Büchler, R. Symptomatic Black Queen Cell Virus infection of drone brood in Hessian apiaries. *Berl. Munch. Tierarztl. Wochenschr.* **2003**, *116*, 130–133. [PubMed]

137. Mookhploy, W.; Kimura, K.; Disayathanoowat, T.; Yoshiyama, M.; Hondo, K.; Chantawannakul, P. Capsid Gene Divergence of Black Queen Cell Virus Isolates in Thailand and Japan Honey Bee Species. *J. Econ. Entomol.* **2015**, *108*, 1460–1464. [CrossRef] [PubMed]

138. Choi, N.R.; Jung, C.; Lee, D.-W. Optimization of detection of black queen cell virus from *Bombus terrestris* via real-time PCR. *J. Asia. Pac. Entomol.* **2015**, *18*, 9–12. [CrossRef]

139. Alger, S.A.; Burnham, P.A.; Boncristiani, H.F.; Brody, A.K. RNA virus spillover from managed honeybees (*Apis mellifera*) to wild bumblebees (*Bombus* spp.). *PLoS ONE* **2019**, *14*, e0217822. [CrossRef]

140. Daughenbaugh, K.F.; Martin, M.; Brutscher, L.M.; Cavigli, I.; Garcia, E.; Lavin, M.; Flenniken, M.L. Honey bee infecting Lake Sinai viruses. *Viruses* **2015**, *7*, 3285–3309. [CrossRef] [PubMed]

141. Runckel, C.; Flenniken, M.L.; Engel, J.C.; Ruby, J.G.; Ganem, D.; Andino, R.; DeRisi, J.L. Temporal analysis of the honey bee microbiome reveals four novel viruses and seasonal prevalence of known viruses, Nosema, and Crithidia. *PLoS ONE* **2011**, *6*. [CrossRef] [PubMed]

142. Gauthier, L.; Cornman, S.; Hartmann, U.; Cousserans, F.; Evans, J.; de Miranda, J.; Neumann, P. The *Apis mellifera* Filamentous Virus Genome. *Viruses* **2015**, *7*, 3798–3815. [CrossRef] [PubMed]

143. Hartmann, U.; Forsgren, E.; Charrière, J.D.; Neumann, P.; Gauthier, L. Dynamics of *Apis mellifera* filamentous virus (AmFV) infections in honey bees and relationships with other parasites. *Viruses* **2015**, *7*, 2654–2667. [CrossRef]

144. Varis, A.L.; Ball, B.V.; Allen, M. The incidence of pathogens in honey bee (*Apis mellifera* L) colonies in Finland and Great Britain. *Apidologie* **1992**, *23*, 133–137. [CrossRef]

145. Quintana, S.; Brasesco, C.; Porrini, L.P.; Di Gerónimo, V.; Eguaras, M.J.; Maggi, M. First molecular detection of *Apis mellifera* filamentous virus (AmFV) in honey bees (*Apis mellifera*) in Argentina. *J. Apic. Res.* **2021**, *60*, 111–114. [CrossRef]

146. Hou, C.; Li, B.; Luo, Y.; Deng, S.; Diao, Q. First detection of *Apis mellifera* filamentous virus in *Apis cerana* cerana in China. *J. Invertebr. Pathol.* **2016**, *138*, 112–115. [CrossRef] [PubMed]

147. Plischuk, S.; Fernández de Landa, G.; Revainera, P.; Quintana, S.; Pocco, M.E.; Cigliano, M.M.; Lange, C.E. Parasites and pathogens associated with native bumble bees (Hymenoptera: Apidae: *Bombus* spp.) from highlands in Bolivia and Peru. *Stud. Neotrop. Fauna Environ.* **2020**. [CrossRef]

148. de Landa, G.F.; Porrini, M.P.; Revainera, P.; Porrini, D.P.; Farina, J.; Correa-Benítez, A.; Maggi, M.D.; Eguaras, M.J.; Quintana, S. Pathogens Detection in the Small Hive Beetle (*Aethina tumida* (Coleoptera: Nitidulidae)). *Neotrop. Entomol.* **2020**, 1–5. [CrossRef] [PubMed]

149. De Miranda, J.R.; Scott Cornman, R.; Evans, J.D.; Semberg, E.; Haddad, N.; Neumann, P.; Gauthier, L. Genome characterization, prevalence and distribution of a macula-like virus from *Apis mellifera* and *Varroa destructor*. *Viruses* **2015**, *7*, 3586–3602. [CrossRef]

150. Mordecai, G.J.; Brettell, L.E.; Pachori, P.; Villalobos, E.M.; Martin, S.J.; Jones, I.M.; Schroeder, D.C. Moku virus; a new Iflavirus found in wasps, honey bees and Varroa. *Sci. Rep.* **2016**, *6*, 34983. [CrossRef] [PubMed]

151. Garigliany, M.; Taminiau, B.; El Agrebi, N.; Cadar, D.; Gilliaux, G.; Hue, M.; Desmecht, D.; Daube, G.; Linden, A.; Farnir, F.; et al. Moku Virus in Invasive Asian Hornets, Belgium, 2016. *Emerg. Infect. Dis.* **2017**, *23*, 2109–2112. [CrossRef]

152. Fries, I.; Feng, F.; da Silva, A.; Slemenda, S.B.; Pieniazek, N.J. *Nosema ceranae* n. sp. (Microspora, Nosematidae), morphological and molecular characterization of a microsporidian parasite of the Asian honey bee *Apis cerana* (Hymenoptera, Apidae). *Eur. J. Protistol.* **1996**, *32*, 356–365. [CrossRef]

153. Paxton, R.J.; Klee, J.; Korpela, S.; Fries, I. *Nosema ceranae* has infected *Apis mellifera* in Europe since at least 1998 and may be more virulent than *Nosema apis*. *Apidologie* **2007**, *38*, 558–565. [CrossRef]

154. Higes, M.; García-Palencia, P.; Urbieta, A.; Nanetti, A.; Martín-Hernández, R. *Nosema apis* and *Nosema ceranae* Tissue Tropism in Worker Honey Bees (*Apis mellifera*). *Vet. Pathol.* **2020**, *57*, 132–138. [CrossRef] [PubMed]

155. Fries, I.; Martín, R.; Meana, A.; García-Palencia, P.; Higes, M. Natural infections of *Nosema ceranae* in European honey bees. *J. Apic. Res.* **2006**, *45*, 230–233. [CrossRef]

156. Cilia, G.; Cabbri, R.; Maiorana, G.; Cardaio, I.; Dall'Olio, R.; Nanetti, A. A novel TaqMan® assay for *Nosema ceranae* quantification in honey bee, based on the protein coding gene Hsp70. *Eur. J. Protistol.* **2018**, *63*, 44–50. [CrossRef]

157. Michalczyk, M.; Sokół, R. Estimation of the influence of selected products on co-infection with *N. apis/N. ceranae* in *Apis mellifera* using real-time PCR. *Invertebr. Reprod. Dev.* **2018**, *62*, 92–97. [CrossRef]

158. Botías, C.; Martín-Hernández, R.; Meana, A.; Higes, M. Screening alternative therapies to control Nosemosis type C in honey bee (*Apis mellifera* iberiensis) colonies. *Res. Vet. Sci.* **2013**, *95*, 1041–1045. [CrossRef] [PubMed]

159. Cilia, G.; Garrido, C.; Bonetto, M.; Tesoriero, D.; Nanetti, A. Effect of Api-Bioxal® and ApiHerb® Treatments against *Nosema ceranae* Infection in *Apis mellifera* Investigated by Two qPCR Methods. *Vet. Sci.* **2020**, *7*, 125. [CrossRef] [PubMed]

160. Plischuk, S.; Martín-Hernández, R.; Prieto, L.; Lucía, M.; Botías, C.; Meana, A.; Abrahamovich, A.H.; Lange, C.; Higes, M. South American native bumblebees (Hymenoptera: Apidae) infected by *Nosema ceranae* (*Microsporidia*), an emerging pathogen of honeybees (*Apis mellifera*). *Environ. Microbiol. Rep.* **2009**, *1*, 131–135. [CrossRef] [PubMed]

161. Plischuk, S.; Lange, C.E. *Bombus brasiliensis* Lepeletier (Hymenoptera, Apidae) infected with *Nosema ceranae* (Microsporidia). *Rev. Bras. Entomol.* **2016**, *60*, 347–351. [CrossRef]

162. Graystock, P.; Yates, K.; Darvill, B.; Goulson, D.; Hughes, W.O.H. Emerging dangers: Deadly effects of an emergent parasite in a new pollinator host. *J. Invertebr. Pathol.* **2013**, *114*, 114–119. [CrossRef] [PubMed]

163. Porrini, M.P.; Porrini, L.P.; Garrido, P.M.; de Melo e Silva Neto, C.; Porrini, D.P.; Muller, F.; Nuñez, L.A.; Alvarez, L.; Iriarte, P.F.; Eguaras, M.J. *Nosema ceranae* in South American Native Stingless Bees and Social Wasp. *Microb. Ecol.* **2017**, *74*, 761–764. [CrossRef] [PubMed]

164. Cilia, G.; Cardaio, I.; dos Santos, P.E.J.; Ellis, J.D.; Nanetti, A. The first detection of *Nosema ceranae* (Microsporidia) in the small hive beetle, *Aethina tumida* Murray (Coleoptera: Nitidulidae). *Apidologie* **2018**, *49*, 619–624. [CrossRef]

165. Higes, M.; Martín-Hernández, R.; Garrido-Bailón, E.; Botías, C.; García-Palencia, P.; Meana, A. Regurgitated pellets of *Merops apiaster* as fomites of infective *Nosema ceranae* (Microsporidia) spores. *Environ. Microbiol.* **2008**, *10*, 1374–1379. [CrossRef]

166. Fries, I. *Nosema apis*—A Parasite in the Honey Bee Colony. *Bee World* **1993**, *74*, 5–19. [CrossRef]

167. Fries, I. Infectivity ans multiplication of *Nosema apis* Z. in the ventriculus of the honey bee. *Apidologie* **1988**, *19*, 319–328. [CrossRef]

168. Forsgren, E.; Fries, I. Comparative virulence of *Nosema ceranae* and *Nosema apis* in individual European honey bees. *Vet. Parasitol.* **2010**, *170*, 212–217. [CrossRef]

169. Heath, L.A.F. Chalk Brood Pathogens: A Review. *Bee World* **1982**, *63*, 130–135. [CrossRef]

170. Spiltoir, C.F. Life Cycle of *Ascosphaera apis* (*Pericystis apis*). *Am. J. Bot.* **1955**, *42*, 501. [CrossRef]

171. Heath, L.A.F.; Gaze, B.M. Carbon dioxide activation of spores of the chalkbrood fungus *Ascosphaera apis*. *J. Apic. Res.* **1987**, *26*, 243–246. [CrossRef]

172. Flores, J.M.; Spivak, M.; Gutiérrez, I. Spores of *Ascosphaera apis* contained in wax foundation can infect honeybee brood. *Vet. Microbiol.* **2005**, *108*, 141–144. [CrossRef]

173. Grossar, D.; Kilchenmann, V.; Forsgren, E.; Charrière, J.D.; Gauthier, L.; Chapuisat, M.; Dietemann, V. Putative determinants of virulence in *Melissococcus plutonius*, the bacterial agent causing European foulbrood in honey bees. *Virulence* **2020**, *11*, 554–567. [CrossRef]
174. Forsgren, E. European foulbrood in honey bees. *J. Invertebr. Pathol.* **2010**, *103*, S5–S9. [CrossRef]
175. Budge, G.E.; Shirley, M.D.F.; Jones, B.; Quill, E.; Tomkies, V.; Feil, E.J.; Brown, M.A.; Haynes, E.G. Molecular epidemiology and population structure of the honey bee brood pathogen *Melissococcus plutonius*. *ISME J.* **2014**, *8*, 1588–1597. [CrossRef]
176. Arai, R.; Tominaga, K.; Wu, M.; Okura, M.; Ito, K.; Okamura, N.; Onishi, H.; Osaki, M.; Sugimura, Y.; Yoshiyama, M.; et al. Diversity of *Melissococcus plutonius* from honeybee larvae in Japan and experimental reproduction of European foulbrood with cultured atypical isolates. *PLoS ONE* **2012**, *7*. [CrossRef]
177. Regassa, L.B.; Gasparich, G.E. Spiroplasmas: Evolutionary relationships and biodiversity. *Front. Biosci.* **2006**, *11*, 2983–3002. [CrossRef] [PubMed]
178. Mouches, C.; Bové, J.M.; Tully, J.G.; Rose, D.L.; McCoy, R.E.; Carle-Junca, P.; Garnier, M.; Saillard, C. *Spiroplasma apis*, a new species from the honey-bee *Apis mellifera*. *Ann. l'Institut Pasteur Microbiol.* **1983**, *134*, 383–397. [CrossRef]
179. Mouches, C.; Bové, J.M.; Albisetti, J.; Clark, T.B.; Tully, J.G. A spiroplasma of serogroup IV causes a May-disease-like disorder of honeybees in Southwestern France. *Microb. Ecol.* **1982**, *8*, 387–399. [CrossRef] [PubMed]
180. Clark, T.B.; Whitcomb, R.F.; Tully, J.G. *Spiroplasma melliferum*, a new species from the honeybee (*Apis mellifera*). *Int. J. Syst. Bacteriol.* **1985**, *35*, 296–408. [CrossRef]
181. Stahlhut, J.K.; Gibbs, J.; Sheffield, C.S.; Smith, M.A.; Packer, L. *Wolbachia* (Rickettsiales) infections and bee (Apoidea) barcoding: A response to Gerth et al. *Syst. Biodivers.* **2012**, *10*, 395–401. [CrossRef]
182. Gerth, M.; Geibler, A.; Bleidorn, C. *Wolbachia* infections in bees (Anthophila) and possible implications for DNA barcoding. *Syst. Biodivers.* **2011**, *9*, 319–327. [CrossRef]
183. Werren, J.H.; Baldo, L.; Clark, M.E. *Wolbachia*: Master manipulators of invertebrate biology. *Nat. Rev. Microbiol.* **2008**, *6*, 741–751. [CrossRef]
184. Saridaki, A.; Bourtzis, K. *Wolbachia*: More than just a bug in insects genitals. *Curr. Opin. Microbiol.* **2010**, *13*, 67–72. [CrossRef] [PubMed]
185. Arismendi, N.; Bruna, A.; Zapata, N.; Vargas, M. PCR-specific detection of recently described *Lotmaria passim* (Trypanosomatidae) in Chilean apiaries. *J. Invertebr. Pathol.* **2016**, *134*, 1–5. [CrossRef]
186. Schwarz, R.S.; Bauchan, G.R.; Murphy, C.A.; Ravoet, J.; de Graaf, D.C.; Evans, J.D. Characterization of Two Species of Trypanosomatidae from the Honey Bee *Apis mellifera*: *Crithidia mellificae* Langridge and McGhee, and *Lotmaria passim* n. gen., n. sp. *J. Eukaryot. Microbiol.* **2015**, *62*, 567–583. [CrossRef] [PubMed]
187. Ruiz- Gonzalez, M.X.; Brown, M.J.F. Honey bee and bumblebee trypanosomatids: Specificity and potential for transmission. *Ecol. Entomol.* **2006**, *31*, 616–622. [CrossRef]
188. Runckel, C.; DeRisi, J.; Flenniken, M.L. A Draft Genome of the Honey Bee Trypanosomatid Parasite *Crithidia mellificae*. *PLoS ONE* **2014**, *9*, e95057. [CrossRef] [PubMed]
189. Schwarz, R.S.; Evans, J.D. Single and mixed-species trypanosome and microsporidia infections elicit distinct, ephemeral cellular and humoral immune responses in honey bees. *Dev. Comp. Immunol.* **2013**, *40*, 300–310. [CrossRef] [PubMed]
190. Buendía, M.; Martín-Hernández, R.; Ornosa, C.; Barrios, L.; Bartolomé, C.; Higes, M. Epidemiological study of honeybee pathogens in Europe: The results of Castilla-La Mancha (Spain). *Span. J. Agric. Res.* **2018**, *16*, e0502. [CrossRef]
191. Boncristiani, H.; Ellis, J.D.; Bustamante, T.; Graham, J.; Jack, C.; Kimmel, C.B.; Mortensen, A.; Schmehl, D.R. World Honey Bee Health: The Global Distribution of Western Honey Bee (*Apis mellifera* L.) Pests and Pathogens. *Bee World* **2021**, *98*, 2–6. [CrossRef]
192. Ravoet, J.; Maharramov, J.; Meeus, I.; De Smet, L.; Wenseleers, T.; Smagghe, G.; de Graaf, D.C. Comprehensive Bee Pathogen Screening in Belgium Reveals *Crithidia mellificae* as a New Contributory Factor to Winter Mortality. *PLoS ONE* **2013**, *8*, e72443. [CrossRef]
193. Arismendi, N.; Caro, S.; Castro, M.P.; Vargas, M.; Riveros, G.; Venegas, T. Impact of Mixed Infections of Gut Parasites *Lotmaria passim* and *Nosema ceranae* on the Lifespan and Immune-related Biomarkers in *Apis mellifera*. *Insects* **2020**, *11*, 420. [CrossRef]
194. Vargas, M.; Arismendi, N.; Riveros, G.; Zapata, N.; Bruna, A.; Vidal, M.; Rodríguez, M.; Gerding, M. Viral and intestinal diseases detected in *Apis mellifera* in central and southern Chile. *Chil. J. Agric. Res.* **2017**, *77*, 243–249. [CrossRef]
195. Schmid-Hempel, P.; Aebi, M.; Barribeau, S.; Kitajima, T.; Du Plessis, L.; Schmid-Hempel, R.; Zoller, S. The genomes of *Crithidia bombi* and *C. expoeki*, common parasites of bumblebees. *PLoS ONE* **2018**, *13*, e0189738. [CrossRef]
196. Yourth, C.P.; Schmid-Hempel, P. Serial passage of the parasite *Crithidia bombi* within a colony of its host, *Bombus terrestris*, reduces success in unrelated hosts. *Proc. R. Soc. B Biol. Sci.* **2006**, *273*, 655–659. [CrossRef]
197. Jones, C.M.; Brown, M.J.F. Parasites and genetic diversity in an invasive bumblebee. *J. Anim. Ecol.* **2014**, *83*, 1428–1440. [CrossRef] [PubMed]
198. Ruiz-González, M.X.; Bryden, J.; Moret, Y.; Reber-Funk, C.; Schmid-Hempel, P.; Brown, M.J.F. Dynamic transmission, host quality, and population structure in a multihost parasite of bumblebees. *Evolution* **2012**, *66*, 3053–3066. [CrossRef]
199. Folly, A.J.; Koch, H.; Stevenson, P.C.; Brown, M.J.F. Larvae act as a transient transmission hub for the prevalent bumblebee parasite *Crithidia bombi*. *J. Invertebr. Pathol.* **2017**, *148*, 81–85. [CrossRef]

200. Popp, M.; Erler, S.; Lattorff, H.M.G. Seasonal variability of prevalence and occurrence of multiple infections shape the population structure of *Crithidia bombi*, an intestinal parasite of bumblebees (*Bombus* spp.). *Microbiologyopen* **2012**, *1*, 362–372. [CrossRef]

201. Brown, M.J.F.; Schmid-Hempel, R.; Schmid-Hempel, P. Strong context-dependent virulence in a host-parasite system: Reconciling genetic evidence with theory. *J. Anim. Ecol.* **2003**, *72*, 994–1002. [CrossRef]

202. Figueroa, L.L.; Grincavitch, C.; McArt, S.H. *Crithidia bombi* can infect two solitary bee species while host survivorship depends on diet. *Parasitology* **2021**, *148*, 435–442. [CrossRef]

203. Ngor, L.; Palmer-Young, E.C.; Burciaga Nevarez, R.; Russell, K.A.; Leger, L.; Giacomini, S.J.; Pinilla-Gallego, M.S.; Irwin, R.E.; McFrederick, Q.S. Cross-infectivity of honey and bumble bee-associated parasites across three bee families. *Parasitology* **2020**, *147*, 1290–1304. [CrossRef]

204. Lipa, J.J.; Triggiani, O. *Apicystis* gen nov and *Apicystis bombi* (Liu, Macfarlane & Pengelly) comb nov (Protozoa: Neogregarinida), a cosmopolitan parasite of *Bombus* and *Apis* (Hymenoptera: Apidae). *Apidologie* **1996**, *27*, 29–34. [CrossRef]

205. Colla, S.R.; Otterstatter, M.C.; Gegear, R.J.; Thomson, J.D. Plight of the bumble bee: Pathogen spillover from commercial to wild populations. *Biol. Conserv.* **2006**, *129*, 461–467. [CrossRef]

206. Plischuk, S.; Lange, C.E. Invasive *Bombus terrestris* (Hymenoptera: Apidae) parasitized by a flagellate (Euglenozoa: Kinetoplastea) and a neogregarine (Apicomplexa: Neogregarinorida). *J. Invertebr. Pathol.* **2009**, *102*, 263–265. [CrossRef]

207. Plischuk, S.; Meeus, I.; Smagghe, G.; Lange, C.E. *Apicystis bombi* (Apicomplexa: Neogregarinorida) parasitizing *Apis mellifera* and *Bombus terrestris* (Hymenoptera: Apidae) in Argentina. *Environ. Microbiol. Rep.* **2011**, *3*, 565–568. [CrossRef]

208. Plischuk, S.; Antúnez, K.; Haramboure, M.; Minardi, G.M.; Lange, C.E. Long-term prevalence of the protists *C rithidia bombi* and *A picystis bombi* and detection of the microsporidium *Nosema bombi* in invasive bumble bees. *Environ. Microbiol. Rep.* **2017**, *9*, 169–173. [CrossRef]

209. Moher, D.; Liberati, A.; Tetzlaff, J.; Altman, D.G. Preferred reporting items for systematic reviews and meta-analyses: The PRISMA statement. *J. Clin. Epidemiol.* **2009**, *62*, 1006–1012. [CrossRef]

210. Murray, E.A.; Burand, J.; Trikoz, N.; Schnabel, J.; Grab, H.; Danforth, B.N. Viral transmission in honey bees and native bees, supported by a global black queen cell virus phylogeny. *Environ. Microbiol.* **2018**, *21*, 972–983. [CrossRef]

211. Peng, W.; Li, J.; Boncristiani, H.; Strange, J.; Hamilton, M.; Chen, Y. Host range expansion of honey bee Black Queen Cell Virus in the bumble bee, *Bombus huntii*. *Apidologie* **2011**, *42*, 650–658. [CrossRef]

212. Li, J.; Peng, W.; Wu, J.; Strange, J.; Boncristiani, H.; Chen, Y. Cross-Species Infection of Deformed Wing Virus Poses a New Threat to Pollinator Conservation. *J. Econ. Èntomol.* **2011**, *104*, 732–739. [CrossRef]

213. Toplak, I.; Šimenc, L.; Ocepek, M.P.; Bevk, D. Determination of Genetically Identical Strains of Four Honeybee Viruses in Bumblebee Positive Samples. *Viruses* **2020**, *12*, 1310. [CrossRef]

214. Jabal-Uriel, C.; Martín-Hernández, R.; Ornosa, C.; Higes, M.; Berriatua, E.; De la Rua, P. Short communication: First data on the prevalence and distribution of pathogens in bumblebees (*Bombus terrestris* and *Bombus pascuorum*) from Spain. *Span. J. Agric. Res.* **2017**, *15*, 5. [CrossRef]

215. Salvarrey, S.; Antúnez, K.; Arredondo, D.; Plischuk, S.; Revainera, P.; Maggi, M.; Invernizzi, C. Parasites and RNA viruses in wild and laboratory reared bumble bees *Bombus pauloensis* (Hymenoptera: Apidae) from Uruguay. *PLoS ONE* **2021**, *16*, e0249842. [CrossRef] [PubMed]

216. Meeus, I.; de Miranda, J.R.; de Graaf, D.C.; Wäckers, F.; Smagghe, G. Effect of oral infection with Kashmir bee virus and Israeli acute paralysis virus on bumblebee (*Bombus terrestris*) reproductive success. *J. Invertebr. Pathol.* **2014**, *121*, 64–69. [CrossRef]

217. Piot, N.; Snoeck, S.; Vanlede, M.; Smagghe, G.; Meeus, I. The Effect of Oral Administration of dsRNA on Viral Replication and Mortality in *Bombus terrestris*. *Viruses* **2015**, *7*, 3172–3185. [CrossRef] [PubMed]

218. Niu, J.; Smagghe, G.; De Coninck, D.I.; Van Nieuwerburgh, F.; Deforce, D.; Meeus, I. In vivo study of Dicer-2-mediated immune response of the small interfering RNA pathway upon systemic infections of virulent and avirulent viruses in *Bombus terrestris*. *Insect Biochem. Mol. Biol.* **2015**, *70*, 127–137. [CrossRef] [PubMed]

219. Wang, H.; Meeus, I.; Piot, N.; Smagghe, G. Systemic Israeli acute paralysis virus (IAPV) infection in bumblebees (*Bombus terrestris*) through feeding and injection. *J. Invertebr. Pathol.* **2017**, *151*, 158–164. [CrossRef]

220. Pereira, K.D.S.; Meeus, I.; Smagghe, G. Honey bee-collected pollen is a potential source of *Ascosphaera apis* infection in managed bumble bees. *Sci. Rep.* **2019**, *9*, 1–9. [CrossRef]

221. Tapia-González, J.M.; Morfin, N.; Macías-Macías, J.O.; De La Mora, A.; Tapia-Rivera, J.C.; Ayala, R.; Contreras-Escareño, F.; Gashout, H.A.; Guzman-Novoa, E. Evidence of presence and replication of honey bee viruses among wild bee pollinators in subtropical environments. *J. Invertebr. Pathol.* **2019**, *168*, 107256. [CrossRef]

222. Santamaria, J.; Villalobos, E.M.; Brettell, L.; Nikaido, S.; Graham, J.R.; Martin, S. Evidence of Varroa-mediated deformed wing virus spillover in Hawaii. *J. Invertebr. Pathol.* **2018**, *151*, 126–130. [CrossRef]

223. Ueira-Vieira, C.; Almeida, L.O.; De Almeida, F.C.; Amaral, I.M.R.; Brandeburgo, M.A.M.; Bonetti, A.M. Scientific note on the first molecular detection of the acute bee paralysis virus in Brazilian stingless bees. *Apidologie* **2015**, *46*, 628–630. [CrossRef]

224. Morfin, N.; Gashout, H.A.; Macías-Macías, J.O.; De la Mora, A.; Tapia-Rivera, J.C.; Tapia-González, J.M.; Contreras-Escareño, F.; Guzman-Novoa, E. Detection, replication and quantification of deformed wing virus-A, deformed wing virus-B, and black queen cell virus in the endemic stingless bee, *Melipona colimana*, from Jalisco, Mexico. *Int. J. Trop. Insect Sci.* **2020**, *41*, 1285–1292. [CrossRef]

225. Guzman-Novoa, E.; Hamiduzzaman, M.M.; Anguiano-Baez, R.; Correa-Benítez, A.; Castañeda-Cervantes, E.; Arnold, N.I. First detection of honey bee viruses in stingless bees in North America. *J. Apic. Res.* **2015**, *54*, 93–95. [CrossRef]

226. Purkiss, T.; Lach, L. Pathogen spillover from *Apis mellifera* to a stingless bee. *Proc. R. Soc. B Boil. Sci.* **2019**, *286*, 20191071. [CrossRef] [PubMed]

227. Ravoet, J.; De Smet, L.; Wenseleers, T.; de Graaf, D.C. Genome sequence heterogeneity of Lake Sinai Virus found in honey bees and Orf1/RdRP-based polymorphisms in a single host. *Virus Res.* **2015**, *201*, 67–72. [CrossRef]

228. Gabín-García, L.B.; Bartolomé, C.; Guerra-Tort, C.; Rojas-Nossa, S.V.; Llovo, J.; Maside, X. Identification of pathogens in the invasive hornet *Vespa velutina* and in native Hymenoptera (Apidae, Vespidae) from SW-Europe. *Sci. Rep.* **2021**, *11*, 1–12. [CrossRef] [PubMed]

229. Marzoli, F.; Forzan, M.; Bortolotti, L.; Pacini, M.I.; Rodríguez-Flores, M.S.; Felicioli, A.; Mazzei, M. Next generation sequencing study on RNA viruses of *Vespa velutina* and *Apis mellifera* sharing the same foraging area. *Transbound. Emerg. Dis.* **2020**. [CrossRef] [PubMed]

230. Lin, C.-Y.; Lee, C.-C.; Nai, Y.-S.; Hsu, H.-W.; Lee, C.-Y.; Tsuji, K.; Yang, C.-C.S. Deformed Wing Virus in Two Widespread Invasive Ants: Geographical Distribution, Prevalence, and Phylogeny. *Viruses* **2020**, *12*, 1309. [CrossRef]

231. Schläppi, D.; Lattrell, P.; Yanez, O.; Chejanovsky, N.; Neumann, P. Foodborne Transmission of Deformed Wing Virus to Ants (*Myrmica rubra*). *Insects* **2019**, *10*, 394. [CrossRef] [PubMed]

232. Eyer, M.; Chen, Y.P.; Schäfer, M.; Pettis, J.; Neuman, P. Honey bee sacbrood virus infects adult small hive beetles, *Aethina tumida* (Coleoptera: Nitidulidae). *J. Apic. Res.* **2009**, *48*, 296–297. [CrossRef]

Article

The Pathogens Spillover and Incidence Correlation in Bumblebees and Honeybees in Slovenia

Metka Pislak Ocepek [1,*], **Ivan Toplak** [2], **Urška Zajc** [3] **and Danilo Bevk** [4]

1 Institute of Pathology, Wild Animals, Fish and Bees, Veterinary Faculty, University of Ljubljana, Gerbičeva 60, 1000 Ljubljana, Slovenia
2 Virology Unit, Institute of Microbiology and Parasitology, Veterinary Faculty, University of Ljubljana, Gerbičeva 60, 1000 Ljubljana, Slovenia; ivan.toplak@vf.uni-lj.si
3 Bacteriology Unit, Institute of Microbiology and Parasitology, Veterinary Faculty, University of Ljubljana, Gerbičeva 60, 1000 Ljubljana, Slovenia; urska.zajc@vf.uni-lj.si
4 Department of Organisms and Ecosystems Research, National Institute of Biology, Večna pot 111, 1000 Ljubljana, Slovenia; danilo.bevk@nib.si
* Correspondence: metka.pislakocepek@vf.uni-lj.si

Abstract: Slovenia has a long tradition of beekeeping and a high density of honeybee colonies, but less is known about bumblebees and their pathogens. Therefore, a study was conducted to define the incidence and prevalence of pathogens in bumblebees and to determine whether there are links between infections in bumblebees and honeybees. In 2017 and 2018, clinically healthy workers of bumblebees (*Bombus* spp.) and honeybees (*Apis mellifera*) were collected on flowers at four different locations in Slovenia. In addition, bumblebee queens were also collected in 2018. Several pathogens were detected in the bumblebee workers using PCR and RT-PCR methods: 8.8% on acute bee paralysis virus (ABPV), 58.5% on black queen cell virus (BQCV), 6.8% on deformed wing virus (DWV), 24.5% on sacbrood bee virus (SBV), 15.6% on Lake Sinai virus (LSV), 16.3% on *Nosema bombi*, 8.2% on *Nosema ceranae*, 15.0% on *Apicystis bombi* and 17.0% on *Crithidia bombi*. In bumblebee queens, only the presence of BQCV, *A. bombi* and *C. bombi* was detected with 73.3, 26.3 and 33.3% positive samples, respectively. This study confirmed that several pathogens are regularly detected in both bumblebees and honeybees. Further studies on the pathogen transmission routes are required.

Keywords: bumblebees; honeybees; viruses; *Nosema* spp.; *Crithidia bombi*; *Apicystis bombi*; *Lotmaria passim*; pathogens transmission

Citation: Pislak Ocepek, M.; Toplak, I.; Zajc, U.; Bevk, D. The Pathogens Spillover and Incidence Correlation in Bumblebees and Honeybees in Slovenia. *Pathogens* **2021**, *10*, 884. https://doi.org/10.3390/pathogens10070884

Academic Editor: Giovanni Cilia

Received: 30 May 2021
Accepted: 6 July 2021
Published: 12 July 2021

1. Introduction

Honeybees and wild pollinators play an essential role in plant pollination, which is important for both agricultural production and biodiversity conservation [1,2]. In addition to the honeybees, the role of wild pollinators is also very important, as they are in many cases even more effective than honeybees and it is now known that honeybees can complement but not replace wild pollinators [3]. Evidence of pollinator decline and disappearance is alarming in many countries around the world [4]. However, the importance of bumblebees has only been increasingly researched in recent years, when the proportion of publications on bumblebee conservation began to grow exponentially [5]. In Europe, more than 20% of bumblebees are threatened with extinction and populations are declining in nearly 50% of species [4]. Important reasons for the decline of pollinator populations and diversity are not only habitat degradation and loss mainly due to urbanisation [6], intensive agriculture, which also involves the use of pesticides [7,8] and climate changes [9,10], but also various pathogens that affect wild pollinators [5,11–14].

Many diseases occur in both honeybees and wild bees, but less is known about pathogen transmission routes between them [12,15–17]. The collection of nectar and pollen by pollinators on flowers allows transmission of pathogens between different pollinator species [18–21]. Typically, pathogen transmission occurs from farmed species, such as honeybees and commercial bumblebee farms, to wild species [22–25]. Two main mechanisms of parasite spread between managed and wild populations are spillover and spillback. Facilitation also leads to a decline in wild bees, while a high density of managed bees leads to wild bees being stressed and more susceptible to infection [24], which is also influenced by adequate food in the environment [26].

The commonly found pathogens in bumblebees are *Apicystis bombi* (*A. bombi*), *Crithidia bombi* (*C. bombi*), *Nosema bombi* (*N. bombi*), *Nosema ceranae* (*N. ceranae*) and several viruses such a Deformed wing virus (DWV), Acute bee paralysis virus (ABPV), Black queen cell virus (BQCV), Sacbrood bee virus (SBV) and Lake Sinai virus (LSV) [11,12,27]. Many of these pathogens primarily or commonly infect honeybees. Infection with *C. bombi*, an intestinal trypanosome, can cause behavioural changes and disturbances in flower colour perception in bumblebees, reducing their ability to forage [28]. *C. bombi* infection can also reduce the survival of bumblebee queens in hibernation [29] and their ability to successfully establish colonies [30]. *Lotmaria passim* (*L. passim*) is a highly prevalent trypanosomatid in honeybees but its pathogenicity/impact on the bumblebee health is not yet clear [31,32]. *Nosema* spp. is a microsporidian that is widely distributed among honeybees and bumblebees. *N. bombi* has been shown to have negative effects on the vitality of bumblebees by impairing the ability of queens to form new colonies, affecting colony size and the vitality of young queens and drones and shortening the life span of workers and drones [33–35]. *N. ceranae* is primarily a pathogen of the Asian bee *Apis ceranae* and is also found in bumblebee colonies worldwide. As previously confirmed, *N. ceranae* can infect bumblebees, affecting their longevity [12,36], but this has not always been demonstrated [37]. *A. bombi* is a neogregarine pathogen that is recognised as one of the causes of bumblebee declines [38]. Experimentally, high virulence on bumblebee queens has been found in individuals with *A. bombi* infection [39]. Spillover from farmed bumblebees to wild bumblebees is suspected due to higher pathogens prevalence around greenhouses with commercial bumblebees [40].

Several viruses infect honeybees and can cause pathological changes at different stages. The use of molecular methods in recent studies has confirmed the occurrence of these honeybee viruses in commercial and wild bumblebees as well [12,14,27]. In bumblebees, virus replication with some clinical changes has also been recognised [41,42]. Clinical signs of DWV infection have been found in *B. terrestris* and *B. pascuorum* [43] and an association between wing deformities and virus localisation in the bumblebee head has been confirmed in commercially bred colonies of *B. terrestris* [44]. By sequencing and phylogenetic analysis of individual viruses, the same strains of ABPV, BQCV, SBV and LSV were identified in bumblebees and honeybees, confirming the assumption that viruses are successfully transmitted between different species [27]. Very little is known about the impact of bee viruses on bumblebee decline and the occurrence of other, unexplored viruses in bumblebees [45].

Over 500 different species of wild bees have been found in Slovenia, of which 35 are bumblebees (*Bombus* spp.). The trend that wild bee populations are declining has also been observed in Slovenia [46]. It is extremely important to identify all threatening factors in order to determine the appropriate protection measures. On the other hand, Slovenia is a country with a long beekeeping tradition and is home to the Carniolan honeybee (*Apis mellifera carnica*), which is protected by law. Slovenia is only 20,271 square kilometres in size and has a population of about 2 million, but according to the national register of apiaries there are about 11,000 beekeepers with more than 200,000 honeybee colonies and the number of honeybee colonies in Slovenia has grown rapidly in recent years (data from Ministry of agriculture, forestry and food for year 2020). There is a lack of knowledge about the presence of some pathogens in wild pollinators in Slovenia and about the possible effects of high honeybee density on wild pollinators.

The aim of this research is to monitor the health status of the bumblebee population and to compare the prevalence of pathogens in bumblebees with the status of honeybees at four different locations in Slovenia. We were also interested in how bumblebees become infected, whether with transmission from queen to nest, or whether bumblebee workers become infected when collecting food on flowers. Therefore, the study of prevalence of some pathogens on bumblebees compared to honeybees was conducted in Slovenia in 2017 and 2018.

2. Results

A total of 147 bumblebees and 8 pooled samples of honeybees from four different locations in Slovenia and 15 bumblebee queens from two locations were analysed. Overall, 8.8% of bumblebee workers were detected positive on ABPV, 58.5% were positive on BQCV, 6.8% were positive on DWV, 24.5% were positive on SBV, 15.6% were positive on LSV, 16.3% were positive on *N. bombi*, 8.2% were positive on *N. ceranae*, 15.0% were positive on *A. bombi* and 17.0% were positive on *C. bombi* (Table 1). For honeybee samples, 62.5% were positive on ABPV, 100% were positive on BQCV, 12.5% were positive on chronic bee paralysis virus (CBPV), 25% were positive on DWV, 50% were positive on SBV, 87.5% were positive on LSV, 87.5% were positive on *N. ceranae*, 12.5% were positive on *A. bombi*, 75.0% were positive on *C. bombi* and 100% were positive on *L. passim*. There were no positive honeybee samples on *N. bombi* and *N. apis* (Table 2). In bumblebee queens, the presence of BQCV, *A. bombi* and *C. bombi* was detected with 73.3, 26.7 and 33.3% positive bumblebee queens, respectively. There were no bumblebee queens positive on CBPV, *N. apis* and *L. passim* (Table 3).

Table 1. Results of laboratory tests, obtained by RT-PCR and PCR methods of bumblebee worker samples at four locations (Sevno, Lukovica, Naklo and Ljubljana), in two years (2017, 2018) and for different bumblebee species (BT = *Bombus terrestris/lucorum*, BL = *Bombus lapidarious*, BS = *Bombus sylvarum*, BP = *Bombus pascuorum*, BHO = *Bombus hortorum*, BHU = *Bombus humilis*). Results are presented as number of positive samples/number of tested samples and % of positive samples (in bracket) for each pathogen, year, bumblebee species and location of sampling.

		ABPV	BQCV	CBPV	DWV	SBV	LSV	*Nosema ceranae*	*Nosema bombi*	*Nosema apis*	*Crithidia bombi*	*Apicystis bombi*	*Lotmaria passim*
Sevno 2017	BT	1/9 (11.1%)	0/9 (0%)	0/9 (0%)	4/9 (44.4%)	0/9 (0%)	0/9 (0%)	0/9 (0%)	1/9 (11.1%)	0/9 (0%)	1/9 (11.1%)	0/9 (0%)	0/9 (0%)
	BL	3/10 (30%)	6/10 (60%)	0/10 (0%)	4/10 (40%)	0/10 (0%)	0/10 (0%)	0/10 (0%)	0/10 (0%)	0/10 (0%)	0/10 (0%)	0/10 (0%)	0/10 (0%)
	BP	2/5 (40%)	0/5 (0%)	0/5 (0%)	0/5 (0%)	0/5 (0%)	0/5 (0%)	0/5 (0%)	0/5 (0%)	0/5 (0%)	0/5 (0%)	0/5 (0%)	0/5 (0%)
Sevno 2018	BT	0/10 (0%)	7/10 (70%)	0/10 (0%)	0/10 (0%)	0/10 (0%)	2/10 (20%)	4/10 (40%)	4/10 (40%)	0/10 (0%)	7/10 (70%)	0/10 (0%)	0/10 (0%)
	BL	0/10 (0%)	9/10 (90%)	0/10 (0%)	0/10 (0%)	0/10 (0%)	1/10 (10%)	0/10 (0%)	1/10 (10%)	0/10 (0%)	1/10 (10%)	0/10 (0%)	0/10 (0%)
	BS	0/10 (0%)	9/10 (90%)	0/10 (0%)	0/10 (0%)	0/10 (0%)	1/10 (10%)	0/10 (0%)	0/10 (0%)	0/10 (0%)	0/10 (0%)	0/10 (0%)	0/10 (0%)
Lukovica 2017	BT	1/5 (20%)	0/5 (0%)	0/5 (0%)	0/5 (0%)	1/5 (20%)	0/5 (0%)	0/5 (0%)	1/5 (20%)	0/5 (0%)	1/5 (20%)	1/5 (20%)	0/5 (0%)
	BP	0/5 (0%)	0/5 (0%)	0/5 (0%)	0/5 (0%)	0/5 (0%)	2/5 (40%)	0/5 (0%)	0/5 (0%)	0/5 (0%)	0/5 (0%)	0/5 (0%)	0/5 (0%)
Lukovica 2018	BT	0/5 (0%)	5/5 (100%)	0/5 (0%)	0/5 (0%)	0/5 (0%)	2/5 (40%)	1/5 (20%)	0/5 (0%)	0/5 (0%)	0/5 (0%)	0/5 (0%)	0/5 (0%)
	BP	0/10 (0%)	6/10 (60%)	0/10 (0%)	0/10 (0%)	0/10 (0%)	5/10 (50%)	1/10 (10%)	8/10 (80%)	0/10 (0%)	0/10 (0%)	0/10 (0%)	0/10 (0%)
Naklo 2017	BT	2/10 (20%)	6/10 (60%)	0/10 (0%)	0/10 (0%)	8/10 (80%)	2/10 (20%)	1/10 (10%)	2/10 (20%)	0/10 (0%)	0/10 (0%)	2/10 (20%)	0/10 (0%)
	BP	0/10 (0%)	2/10 (20%)	0/10 (0%)	1/10 (10%)	8/10 (80%)	0/10 (0%)	0/10 (0%)	0/10 (0%)	0/10 (0%)	1/10 (10%)	1/10 (10%)	0/10 (0%)

Table 1. *Cont.*

		ABPV	BQCV	CBPV	DWV	SBV	LSV	*Nosema ceranae*	*Nosema bombi*	*Nosema apis*	*Crithidia bombi*	*Apicystis bombi*	*Lotmaria passim*
Naklo 2018	BS	0/2 (0%)	2/2 (100%)	0/2 (0%)	0/2 (0%)	2/2 (100%)	0/2 (0%)	0/2 (0%)	0/2 (0%)	0/2 (0%)	1/2 (50%)	1/2 (50%)	0/2 (0%)
	BP	0/10 (0%)	7/10 (70%)	0/10 (0%)	0/10 (0%)	7/10 (70%)	1/10 (10%)	0/10 (0%)	1/10 (10%)	0/10 (0%)	0/10 (0%)	2/10 (20%)	0/10 (0%)
	BHO	2/3 (66.7%)	3/3 (100%)	0/3 (0%)	0/3 (0%)	1/3 (33.3%)	1/3 (33.3%)	0/3 (0%)	1/3 (33.3%)	0/3 (0%)	0/3 (0%)	1/3 (33.3%)	0/3 (0%)
	BHU	0/2 (0%)	2/2 (100%)	0/2 (0%)	1/2 (50%)	0/2 (0%)	0/2 (0%)	0/2 (0%)	0/2 (0%)	0/2 (0%)	0/2 (0%)	1/2 (50%)	0/2 (0%)
Ljubljana 2017	BT	0/5 (0%)	2/5 (40%)	0/5 (0%)	0/5 (0%)	1/5 (20%)	1/5 (20%)	0/5 (0%)	0/5 (0%)	0/5 (0%)	1/5 (20%)	0/5 (0%)	0/5 (0%)
	BP	0/6 (0%)	4/6 (66.7%)	0/6 (0%)	0/6 (0%)	0/6 (0%)	1/6 (16.7%)	1/6 (16.7%)	0/6 (0%)	0/6 (0%)	1/6 (16.7%)	1/6 (16.7%)	0/6 (0%)
Ljubljana 2018	BT	1/10 (10%)	9/10 (90%)	0/10 (0%)	0/10 (0%)	1/10 (10%)	2/10 (20%)	4/10 (40%)	5/10 (50%)	0/10 (0%)	9/10 (90%)	8/10 (80%)	0/10 (0%)
	BP	1/10 (10%)	7/10 (70%)	0/10 (0%)	0/10 (0%)	7/10 (70%)	2/10 (20%)	0/10 (0%)	0/10 (0%)	0/10 (0%)	2/10 (20%)	4/10 (40%)	0/10 (0%)

Table 2. Results of laboratory tests of honeybee worker samples collected in 2017 and 2018 at four locations (Sevno, Lukovica, Naklo and Ljubljana). Results obtained by RT-PCR and PCR methods are presented as positive (+) or negative (−) sample for each pathogen tested.

Location and Year of Sampling	ABPV	BQCV	CBPV	DWV	SBV	LSV	*Nosema ceranae*	*Nosema bombi*	*Nosema apis*	*Crithidia bombi*	*Apicystis bombi*	*Lotmaria passim*
Sevno 2017	+	+	−	−	−	−	+	−	−	−	−	+
Sevno 2018	−	+	−	−	−	+	+	−	−	+	−	+
Lukovica 2017	−	+	−	−	−	+	+	−	−	+	−	+
Lukovica 2018	−	+	−	−	−	+	+	−	−	−	−	+
Naklo 2017	+	+	−	−	+	+	−	−	−	+	+	+
Naklo 2018	+	+	+	+	+	+	+	−	−	+	−	+
Ljubljana 2017	+	+	−	+	+	+	+	−	−	+	−	+
Ljubljana 2018	+	+	−	−	+	+	+	−	−	+	−	+

Table 3. Results of laboratory testing obtained by RT-PCR and PCR methods of 15 bumblebee queens of three species (BT = *Bombus terrestris/lucorum*, BL = *Bombus lapidarious*, BP = *Bombus pascuorum*), collected in 2018 from two locations (Sevno n = 10 samples and Ljubljana n = 5 samples). Results are presented as number of positive queen samples/ number of tested samples and % of positive samples (in bracket) for each pathogen, bumblebee species and location of sampling.

		ABPV	BQCV	CBPV	DWV	*Nosema ceranae*	*Nosema bombi*	*Nosema apis*	*Crithidia bombi*	*Apicystis bombi*	*Lotmaria passim*
Sevno	BT	0/5 (0%)	5/5 (100%)	0/5 (0%)	0/5 (0%)	0/5 (0%)	0/5 (0%)	0/5 (0%)	2/5 (40%)	2/5 (40%)	0/5 (0%)
	BL	0/5 (0%)	3/5 (60%)	0/5 (0%)	0/5 (0%)	0/5 (0%)	0/5 (0%)	0/5 (0%)	1/5 (20%)	2/5 (40%)	0/5 (0%)
Ljubljana	BP	0/5 (0%)	3/5 (60%)	0/5 (0%)	0/5 (0%)	0/5 (0%)	0/5 (0%)	0/5 (0%)	2/5 (40%)	0/5 (0%)	0/5 (0%)

Results for bumblebee workers were presented together for samplings in 2017 and 2018 and calculated as a percentage of total positive samples for each location (Figure 1). To compare the results for different species, results for honeybees were also summed for 2017 and 2018 and calculated as a percentage of positive samples. Results for bumblebee queens sampled in 2018 were calculated as a percentage of positive samples. Calculated honeybee and bumblebee queen results were compared to bumblebee worker results (Figure 2).

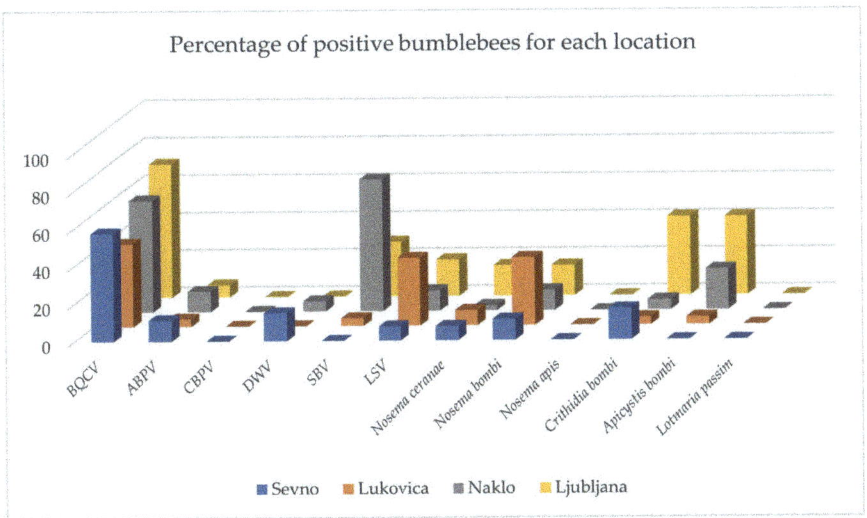

Figure 1. Comparison of positive bumblebee samples at four locations (Sevno, Lukovica, Naklo and Ljubljana). Results are presented together as a percentage of positive samples for samplings in 2017 and 2018 for all tested pathogens (BQCV = black queen cell virus, ABPV = acute bee paralysis virus, CBPV = chronic bee paralysis virus, DWV = deformed wing virus, SBV = sacbrood bee virus, LSV = Lake Sinai virus).

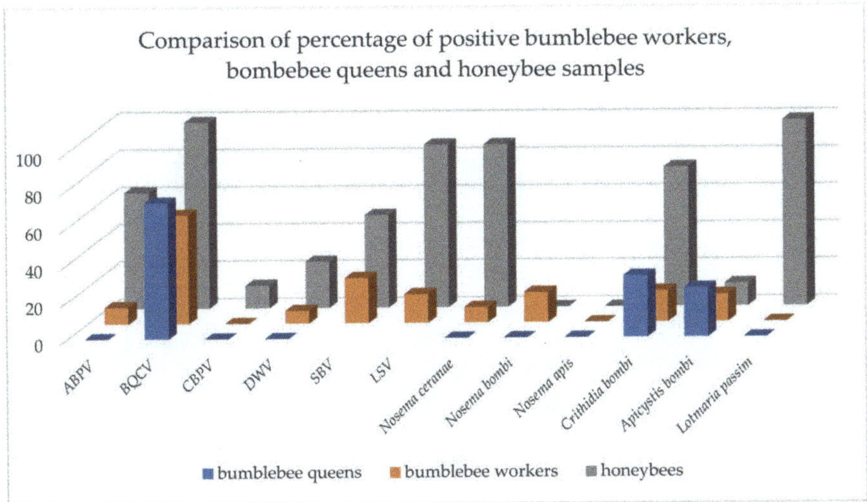

Figure 2. Comparison of percentage of positive bumblebee worker, bumblebee queen and honeybee samples. Results for both years of sampling and for all four locations are presented together for all tested pathogens (BQCV = black queen cell virus, ABPV = acute bee paralysis virus, CBPV = chronic bee paralysis virus, DWV = deformed wing virus, SBV = sacbrood bee virus, LSV = Lake Sinai virus).

The results of determining the presence of pathogens in worker bumblebees were also analysed according to the bumblebee species. Since only three specimens of *B. hortorum* and two of *B. humilis* were collected in this study; these two species were excluded from the analysis. The results of analysis are shown in Figure 3 as percentage of positive samples for each pathogen. As no positive samples were determined for CBPV, *N. apis* and *L. passim*, these pathogens are not included.

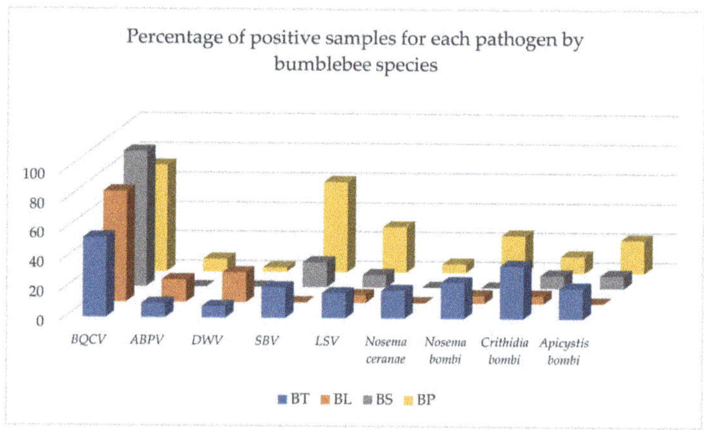

Figure 3. Comparison of percentage of positive bumblebee worker samples for each pathogen (BQCV= black queen cell virus, ABPV= acute bee paralysis virus, DWV= deformed wing virus, SBV= sacbrood bee virus, LSV= Lake Sinai virus), analysed by bumblebee species (BT = *Bombus terrestris/lucorum*, BL = *Bombus lapidarious*, BS = *Bombus sylvarum*, BP= *Bombus pascuorum*).

For all sampled bumblebee workers, the analysis of the number of individual pathogens confirmed in each bumblebee was done. At the Sevno location, no pathogen was detected in 10 (18.5%) samples, while one, two, three and four pathogens were detected in 27 (50%), 12 (22.2%), 3 (5.6%) and 2 (3.7%) samples, respectively. At Lukovica location, 4 (16%) of the bumblebee workers were without pathogens, while one, two, three and four pathogens were detected in 10 (40%), 9 (36%), 1 (4%) and 1 (4%) sample, respectively. At Naklo location no pathogen was detected in 2 (5.4%) samples, one, two, three and four pathogens were detected in 12 (32.4%), 11 (29.7%), 9 (24.3%) and 3 (8.1%) samples, respectively. At Ljubljana location, no pathogen was detected in 3 (9.7%) samples, one, two, three and four pathogens were detected in 6 (19.4%), 6 (19.4%), 10 (32.3%) and 4 (12.9%) samples, respectively, while at this location 1 (3.2%) bumblebee worker was infected with five pathogens and 1 (3.2%) with six pathogens (Figure 4). All four bumblebee sampling sites are in the same category in terms of the honeybee colonies density of 11.6–13.5 honeybee colonies per km^2 (data from national register of apiaries for year 2020, Ministry of agriculture, forestry and food).

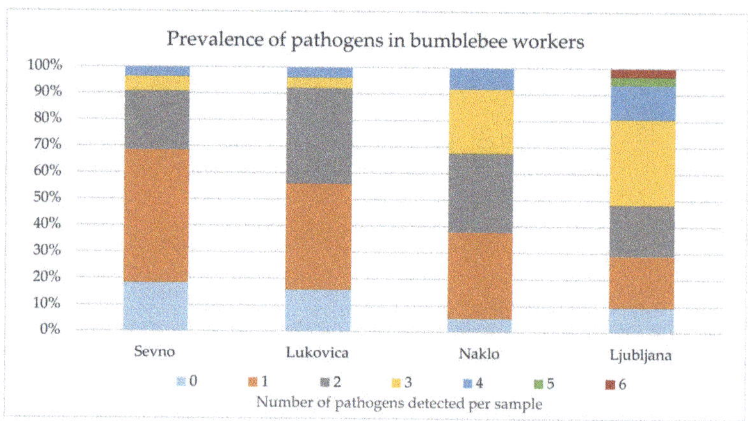

Figure 4. Results of pathogen prevalence analysis for each bumblebee worker. Numbers 0–6 represent the number of pathogens detected per sample; results are presented as percentage of samples with number of pathogens detected.

3. Discussion

This study is the first comprehensive investigation of the occurrence and prevalence of various pathogens in bumblebees in Slovenia. Our results also show that bumblebees can be simultaneously infected with several pathogens and that many of them are shared with honeybees. In bumblebees, ABPV, BQCV, DWV, SBV and LSV were confirmed, while no CBPV was detected in healthy bumblebees, as expected, since CBPV is one of the most pathogenic viruses of adult honeybees [47]. Among the detected viruses, most of the bumblebees were positive for BQCV (58.5%), followed by SBV (24.5%), LSV (15.6%), ABPV (8.8%) and DWV (6.8%). The collected 10 honeybees at each of the four locations were pooled into one pool sample per sampling day, thus eight honeybee samples were included in this study to prove the presence or absence of the individual pathogen at the time of collection also among the honeybees. Although precise data on the proportion of positive individual honeybees were not obtained, the results of these samples serve as good evidence that the detected pathogen was present locally at the time of samplings and a comparison with prevalence in bumblebees was possible. Among the honeybee samples, BQCV (100%) was the most frequently detected virus, followed by LSV (87.5%), ABPV (62.5%), SBV (50.0%), DWV (25.0%) and CBPV. (12.5%). When comparing these results with previously published data [48], the most frequently detected virus in honeybees in Slovenia was also BQCV (83.3%), followed by DWV (70%), ABPV (40%), CBPV (18.3%) and SBV (8.3%). In data interpretation, it should be noted that this time we collected clinically healthy specimens of honeybees on flowers, whereas in a previous study we collected samples from honeybee colonies with some notable pathology, and this is the main reason for the observed differences in prevalence. The results of this study showed that CBPV was not detected in any of the bumblebee samples, although this virus is present and regularly/yearly detected in Slovenia mainly in clinically diseased honeybees. The same observation for CBPV was also reported by some other authors in their previous studies [11,13,14].

In addition to viruses, *N. ceranae*, *N. bombi*, *C. bombi* and *A. bombi* have also been detected in bumblebee workers, confirming the observation of previous studies [11,29,38]. The highest prevalence was found for *C. bombi* (17%), followed by *N. bombi* (16.3%), *A. bombi* (15%) and *N. ceranae* (8.2%), while we could not confirm *N. apis* and *L. passim*. In our experience, *L. passim* is very common in honeybees in Slovenia, as well as in some other countries [31,32]. Therefore, we included it in our study to see if it is also transmitted to bumblebees, but we could not detect it in any bumblebee sample, although all collected honeybee samples at the same locations were positive. It could be concluded that bumblebees are not susceptible to infection with *L. passim* and that various honeybee pathogens are not present in bumblebees only as a result of contamination on flowers. In Slovenia, *N. apis* was not detected in honeybee samples for years, so it was not surprising that all bumblebee samples were negative. *N. ceranae*, on the other hand, is frequently diagnosed in honeybees in Slovenia, which is also evident in this study (7 out of 8 pool samples were detected positive). Our results show that *N. ceranae* also infects bumblebees. Despite the fact, that *N. bombi*, *C. bombi* and *A. bombi* are pathogens, known to infect bumblebees [11,28,29,38], we found *C. bombi* and *A. bombi* also in honeybees.

Comparing the data presented for bumblebee workers, bumblebee queens and honeybee samples, a correlation of the occurrence and prevalence of pathogens between honeybees and bumblebees can already be evident for individual pathogens (Figure 2). According to these results and studies by other authors [11–14,24,25,27] there is spillover of pathogens between managed honeybees and wild bumblebees. We found the bumblebee pathogens *C. bombi* (6 positive pool samples out of 8) and *A. bombi* (1 positive pool sample out of 8) also in honeybees. This suggests that the spillback effect is probably also present, mainly due to the high density of honeybee colonies/apiaries in Slovenia, which was more than 10 colonies per km^2 in 2020, according to the national register of apiaries.

The results are also analysed by bumblebee species. In the Figure 3 same differences between species are evident, but we cannot say that there is one species of bumblebee that is more or less healthy, since there are several pathogens present in each species. The different number of samples for each species must be taken into account. Since we had only three and two samples of *B. hortorum* and *B. humilis*, respectively, these two species were excluded from the analysis. It is obvious, that in species with a higher number of examined samples (*B. terrestis/lucorum* and *B. pascuorum*) more pathogens were detected, as the probability of collecting infected specimen is higher. Despite a few differences between species, there are no significant results, except for a slightly higher percentage of positive *B. pascuorum* samples in the SBV, for which we do not know the reason. However, it may be useful to examine *B. pascuorum* nests for the presence of clinical signs of SBV in the future.

To monitor the health status of the bumblebee population in Slovenia, the results were analysed and interpreted according to the number of pathogens detected in each bumblebee. Between zero and four pathogens were detected in most of bumblebee samples, while at the location of Ljubljana one bumblebee was identified with five pathogens and another with six pathogens. The presented data on prevalence analysis for each bumblebee worker sample (Figure 5) showed important differences between four locations, the least pathogens in individual bumblebees were found at the location Sevno, followed by Lukovica, Naklo and the most at the location Ljubljana. We do not know the real reason for this result, as the locations do not differ significantly in terms of honeybee colony density, perhaps the proximity of urban area (Ljubljana is the capital of Slovenia) is more stressful for bumblebees.

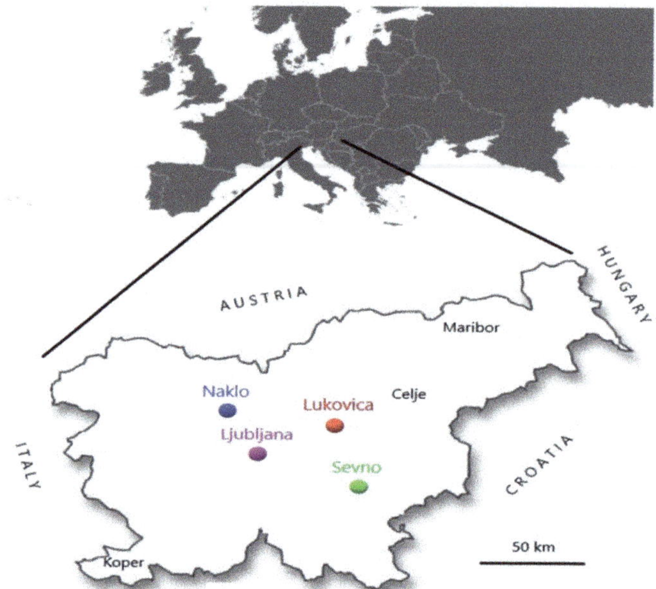

Figure 5. Four locations (Sevno, Lukovica, Naklo and Ljubljana) on the map of Slovenia where samples were collected in 2017 and 2018.

In our previously published research study based on a molecular epidemiological approach and phylogenetic comparison of detected ABPV, BQCV, SBV and LSV in different species, we confirmed that several viruses are undoubtedly transmitted between bees and bumblebees [27]. To identify the possible ways of transmitting pathogens between different species, 15 bumblebee queens were sampled in April of the second year of the study and included in the comparison. Since bumblebee queens are the only ones that overwinter and form a new colony during the season, these samples were tested for ABPV, BQCV, CBPV, DWV, *N. bombi*, *N. ceranae*, *N. apis*, *A. bombi*, *C. bombi* and *L. passim* for the first time. Only the presence of BQCV, *A. bombi* and *C. bombi* was detected in queens.

Regarding this observation, it seems that *Nosema* spp. and some viruses are not transmitted by queens, indicating the possibility of infection of bumblebee workers by indirect contacts on flowers. If pathogens are transmitted between different species in this way, they may also be transmitted among honeybees, especially if there is a high density of honeybee colonies in areas with a rich honey flow. When diseases spread among managed bees, pathogens multiply more easily and are transmitted even more to wild pollinators [25]. This also adds a new dimension to the health of honeybees and other pollinators that should be considered by beekeepers and policy-makers. Even more, effects in the nature are so closely related that we cannot separate the care of honeybees from the conservation of wild pollinators. However, further studies are needed to confirm the possibility of disease transmission between honeybees and various pollinators during their pollination activities.

4. Materials and Methods

In August 2017 and August 2018, a total of 147 clinically healthy bumblebee workers of different species: *Bombus terrestis/ lucorum, B. lapidarious, B. sylvarum, B. pascuorum, B. hortorum* and *B. humilis* (Table 4) were individually collected on flowers in nature. Sampling was carried out at four different locations in Slovenia (Figure 5): 24 and 30 bumblebees were collected in Sevno in 2017 and 2018, 10 and 15 in Lukovica, 20 and 17 in Naklo and 11 and 20 in Ljubljana, respectively. At the same locations (Sevno, Lukovica, Naklo and Ljubljana) on the day of bumblebee sampling, also 10 clinically healthy honeybee workers (*Apis mellifera carnica*) were collected on flowers, for a total of 80 clinically healthy honeybee workers. In April 2018, also 15 clinically healthy bumblebee queens were collected on the same way as bumblebee workers on flowers, 10 samples at the Sevno site (5 samples of *B. terrestris/lucorum* and 5 samples of *B. lapidarious*) and 5 samples of *B. pascuorum* at the Ljubljana site. All samples were frozen and stored at minus 60 °C until use.

Table 4. Number of tested bumblebee workers of six different species collected in Slovenia in 2017 and 2018.

Species/Year	2017	2018	Total
Bombus terrestris/lucorum	29	25	54
Bombus lapidarious	10	10	20
Bombus sylvarum	0	12	12
Bombus pascuorum	26	30	56
Bombus hortorum	0	3	3
Bombus humilis	0	2	2
All collected samples of *Bombus* spp.	65	82	147

In the laboratory, each bumblebee was placed in an Ultra-Turrax DT-20 tube (IKA, Germany) and 3 mL of RPMI 1640 medium was added. Clinically healthy honeybees collected on the same day at each location were pooled (10 bees from the same location and at the same time of sampling in one pool) and in laboratory 5 mL of RPMI 1640 medium (Gibco, UK) was added to each sample. The samples were homogenised, and 1 mL of the suspension was taken for isolation of DNA before centrifugation. The remainder was centrifuged at $2500\times g$ for 5 min. Total RNA was isolated from each sample using the QIAamp viral RNA mini kit (Qiagen, Germany) according to the manufacturer's instructions.

DNA was isolated using a commercial isolation kit (Institute of Metagenomics and Microbial Technologies-IMMT, Slovenia). Briefly, 1 mL of the mixture was added to a 2-mL tube containing $\leq$106-µm-diameter glass beads (Sigma-Aldrich, St. Louis, MI, USA) and centrifuged at $10,000\times g$ for 5 min. The pellet was resuspended in 392 µL of lysis buffer and 8 µL of proteinase K (Sigma-Aldrich, St. Louis, MI, USA). This was followed by bead beating on a MagNALyser device (Roche, Basel, Switzerland), at 6400 rpm for 60 s and incubation at 56 °C for 15 min. Bead beating and incubation were repeated three times and twice, respectively. The rest of the isolation was performed according to the manufacturer's protocol.

The RNA of six honeybee viruses in bumblebee workers: ABPV, BQCV, CBPV, DWV, SBV and LSV was detected by specific reverse transcription and polymerase chain reaction method (RT-PCR) as previously described [27,48]. Results were considered positive based on the size of the RT-PCR products in the agarose gel when the expected product size was present (ABPV 452 nt, BQCV 770 nt, CBPV 570 nt, DWV 504 nt, SBV 814 nt and LSV 603 nt). Isolated RNA from bumblebee queens was tested for ABPV, BQCV, CBPV and DWV as described above.

Isolated DNA from each bumblebee worker and queen was used to detect *N. bombi*, *N. ceranae*, *N. apis*, *A.s bombi*, *C. bombi* and *L. passim*. Polymerase chain reactions (PCR) and real-time polymerase chain reaction (qPCR) were performed according to previously published protocols [49–51].

5. Conclusions

In 147 bumblebees tested, the prevalence of ABPV, BQCV, DWV, SBV, LSV and *N. ceranae*, *N. bombi*, *C. bombi* and *A. bombi* was detected for the first time in Slovenia, while honeybees sampled at the same time and locations were positive for ABPV, BQCV, CBPV, DWV, SBV, LSV, *N. ceranae*, *C. bombi*, *A. bombi* and *L. passim*. In bumblebee queens, only BQCV, *C. bombi* and *A. bombi* were diagnosed.

The study raised some new questions regarding the transmission of pathogens between honeybees and bumblebees. However, it must be kept in mind that many factors can have an impact on surviving pollinators, including the transmission of pathogens from managed bees to wild pollinators. The evident spillover is why we need to put more attention also in good care of managed bees in order to preserve wild bees.

Author Contributions: Conceptualisation: M.P.O. and D.B.; methodology: M.P.O., I.T., U.Z. and D.B.; investigation: M.P.O., I.T., U.Z. and D.B., in-field activity: D.B.; laboratory activity: I.T. and U.Z.; formal analysis: M.P.O., I.T. and U.Z.; data curation: M.P.O., I.T. and U.Z.; writing—original draft preparation: M.P.O.; writing—review and editing: I.T., U.Z. and D.B.; funding acquisition: M.P.O. and D.B. All authors have read and agreed to the published version of the manuscript.

Funding: This research was funded by the Slovenian Research Agency (research core funding P4-0092 and P1-0255) and the CRP project V4-1622 (The importance of wild pollinators in agricultural crop cultivation and sustainable agricultural management in order to ensure reliable pollination), financed by the Slovenian Research Agency and Ministry of Agriculture, Forestry and Food.

Institutional Review Board Statement: Not applicable.

Informed Consent Statement: Not applicable.

Data Availability Statement: The data presented in this study are available on request from the corresponding author.

Acknowledgments: The authors thank all students for sampling of bumblebees and honeybees and technicians for laboratory work, also thank Slovenian Research Agency and Ministry of Agriculture, Forestry and Food for financial support.

Conflicts of Interest: The authors declare no conflict of interest.

References

1. Potts, S.; Biesmeijer, K.; Bommarco, R.; Breeze, T.; Carvalheiro, L.; Franzen, M.; Gonzalez-Varo, J.P.; Holzschuh, A.; Kleijn, D.; Klein, A.M.; et al. *Status and Trends of European Pollinators. Key Findings of the STEP Project*; Pensoft Publishers: Sofia, Bulgaria, 2015; 72p.
2. Ollerton, J.; Winfree, R.; Tarrant, S. How many flowering plants are pollinated by animals? *Oikos* **2011**, *120*, 321–326. [CrossRef]

3. Garibaldi, L.A.; Steffan-Dewenter, I.; Winfree, R.; Aizen, M.A.; Bommarco, R.; Cunningham, S.A.; Kremen, C.; Carvalheiro, L.G.; Harder, L.D.; Afik, O.; et al. Wild Pollinators Enhance Fruit Set of Crops Regardless of Honey Bee Abundance. *Science* **2013**, *339*, 1608–1611. [CrossRef] [PubMed]

4. Nieto, A.; Roberts, S.P.M.; Kemp, J.; Rasmont, P.; Kuhlmann, M.; Criado, M.G.; Biesmeijer, J.C.; Bogusch, P.; Dathe, H.H.; De la Rúa, P.; et al. *European Red List of Bees*; Publication Office of the European Union: Luxembourg, 2014; pp. 1–84. [CrossRef]

5. Cameron, S.A.; Sadd, B.M. Global Trends in Bumble Bee Health. *Annu. Rev. Entomol.* **2020**, *65*, 209–232. [CrossRef]

6. Glaum, P.; Simao, M.C.; Vaidya, C.; Fitch, G.; Iulinao, B. Big city Bombus: Using natural history and land-use history to find significant envirnmental drivers in bumble-bee declines in urban development. *R. Soc. Open Sci.* **2017**, *4*, 170156. [CrossRef] [PubMed]

7. Woodcock, B.A.; Isaac, N.J.B.; Bullock, J.M.; Roy, D.B.; Garthwaite, D.G.; Crowe, A.; Pywell, R.F. Impacts of neonicotionoid use on long-term population changes in wild bees in England. *Nat. Commun.* **2016**, *7*, 12459. [CrossRef]

8. Whitehorn, P.R.; O'Connor, S.; Wackers, F.L.; Goulson, D. Neonicotinoid Pesticide Reduces Bumble Bee Colony Growth and Queen Production. *Science* **2012**, *336*, 351–352. [CrossRef]

9. Rasmont, P.; Franzen, M.; Lecocq, T.; Harpke, A.; Roberts, S.P.M.; Biesmeijer, J.C.; Castro, L.; Cederberg, B.; Dvorak, L.; Fitzpatrick, U.; et al. Climatic risk and distribution atlas of European bumblebees. *BioRisk* **2015**, *10*, 1–236. [CrossRef]

10. Soroye, P.; Newbold, T.; Kerr, J. Climate change contributes to widespread declines among bumble bees across continents. *Science* **2020**, *367*, 685–688. [CrossRef]

11. Sokol, R.; Michalczyk, M.; Micholap, P. Preliminary studies on the occurrence of honeybee pathogens in the national bumblebee population. *Ann. Parasitol.* **2018**, *64*, 385–390. [CrossRef]

12. Fürst, M.A.; McMahon, D.P.; Osborne, J.L.; Paxton, R.J.; Brown, M.J.F. Disease associations between honeybees and bumblebees as threat to wild pollinators. *Nature* **2014**, *506*, 364–366. [CrossRef]

13. Manley, R.; Boots, M.; Wilfert, L. Emerging viral disease risk to pollinating insects: Ecological, evolutionary and anthropogenic factors. *J. Appl. Ecol.* **2015**, *52*, 331–340. [CrossRef]

14. McMahon, D.P.; Fürst, M.A.; Caspar, J.; Theodorou, P.; Brown, M.J.; Paxton, R.J. A sting in the spit: Widespread cross-infection of multiple RNA viruses across wild and managed bees. *J. Anim. Ecol.* **2015**, *84*, 615–624. [CrossRef] [PubMed]

15. Meeus, I.; Brown, M.J.F.; Graaf, D.C.; Smagghe, G. Effects of Invasive Parasites on Bumble Bee Declines. *Conserv. Biol.* **2011**, *25*, 662–671. [CrossRef]

16. Evison, S.E.F.; Roberts1, K.E.; Laurenson, L.; Pietravalle, S.; Hui, J.; Biesmeijer, J.C.; Smith, J.E.; Budge, G.; Hughes, W.O.H. Pervasiveness of Parasites in Pollinators. *PLoS ONE* **2012**, *7*, e30641. [CrossRef]

17. Goulson, D.; Hughes, W.O.H. Mitigating the anthropogenic spread of bee parasites to protect wild pollinators. *Biol. Conserv.* **2015**, *191*, 10–19. [CrossRef]

18. Graystock, P.; Goulson, D.; Hughes, W.O. Parasites in bloom: Flowers aid dispersal and transmission of pollinator parasites within and between bee species. *Proc. R. Soc. B* **2015**, *282*, 20151371. [CrossRef]

19. Graystock, P.; Ng, W.H.; Parks, K.; Tripodi, A.D.; Muniz, P.A.; Fersch, A.A.; Myers, C.R.; McFrederick, Q.S.; McArt, S.H. Dominant bee species and floral abundance drive parasite temporal dynamics in plant-pollinator communities. *Nat. Ecol. Evol.* **2020**, *4*, 1358–1367. [CrossRef]

20. Koch, H.; Brown, M.J.; Stevenson, P.C. The role of disease in bee foraging ecology. *Curr. Opin. Insect Sci.* **2017**, *21*, 60–67. [CrossRef]

21. Adler, L.S.; Michaud, K.M.; Ellner, S.P.; McArt, S.H.; Stevenson, P.C.; Irwin, R.E. Disease where you dine: Plant species and floral traits associated with pathogen transmission in bumble bees. *Ecology* **2018**, *99*, 2535–2545. [CrossRef] [PubMed]

22. Colla, S.R.; Otterstatter, M.C.; Gegear, R.J.; Thomson, J.D. Plight of the bumble bee: Pathogen spillover from commercial to wild populations. *Biol. Conserv.* **2006**, *129*, 461–467. [CrossRef]

23. Cameron, S.A.; Lim, H.C.; Lozier, J.D.; Duennes, M.A.; Thorp, R. Test of the invasive pathogen hypothesis of bumble bee decline in North America. *Proc. Natl. Acad. Sci. USA* **2016**, *113*, 4386–4391. [CrossRef] [PubMed]

24. Graystock, P.; Blane, E.J.; McFrederick, Q.S.; Goulson, D.; Hughes, W.O.H. Do managed bees drive parasite spread and emergence in wild bees? *Int. J. Parasitol. Parasites Wildl.* **2016**, *5*, 64–75. [CrossRef] [PubMed]

25. Graystock, P.; Goulson, D.; Hughes, W.O.H. The relationship between managed bees and the prevalence of parasites in bumblebees. *PeerJ* **2014**, *2*, e522. [CrossRef] [PubMed]

26. McNeil, D.J.; McCormick, E.; Heimann, A.C.; Kammerer, M.; Douglas, M.R.; Goslee, S.C.; Grozinger, C.M.; Hines, H.M. Bumble bees in landscapes with abundant floral resources have lower pathogen loads. *Sci. Rep.* **2020**, *10*, 22306. [CrossRef] [PubMed]

27. Toplak, I.; Šimenc, L.; Pislak Ocepek, M.; Bevk, D. Determination of genetically identical strains of four honeybee viruses in bumblebee positive samples. *Viruses* **2020**, *12*, 1310. [CrossRef]

28. Gegear, R.J.; Otterstatter, M.C.; Thomson, J.D. Bumble-bee foragers infected by a gut parasite have an impaired ability to utilize floral information. *Proc. R. Soc. B* **2006**, *273*, 1073–1078. [CrossRef]

29. Fauser, A.; Sandrock, C.; Neumann, P.; Sadd, B. Neonicotinoids override a parasite exposure impact on hibernation success of a key bumblebee pollinator. *Ecol. Entomol.* **2017**, *42*, 306–314. [CrossRef]

30. Brown, M.J.F.; Schmid-Hempel, R.; Schmid-Hempel, P. Strong context-dependent virulence in a host-parasite system: Reconciling genetic evidence with theory. *J. Anim. Ecol.* **2003**, *72*, 994–1002. [CrossRef]

31. Schwarz, R.S.; Bauchan, G.R.; Murphy, C.A.; Ravoet, J.; De Graaf, D.C.; Evans, J.D. Characterization of two species of trypanosomatidae from the Honey Bee *Apis mellifera*: *Crithidia mellificae* Langridge and McGhee, and *Lotmaria passim* n. gen., n. sp. *J. Eukaryot. Microbiol.* **2015**, *62*, 567–583. [CrossRef]

32. Stevanovic, J.; Schwarz, R.S.; Vejnovic, B.; Evans, J.D.; Irwin, R.E.; Glavinic, U.; Stanimirovic, Z. Species-specific diagnostics of *Apis mellifera* trypanosomatids: A nine-year survey (2007–2015) for trypanosomatids and microsporidians in Serbian honey bees. *J. Invertebr. Pathol.* **2016**, *139*, 6–11. [CrossRef]

33. Van Der Steen, J.J. Infection and transmission of *Nosema bombi* in *Bombus terrestris* colonies and its effect on hibernation, mating and colony founding. *Apidologie* **2008**, *39*, 273–282. [CrossRef]

34. Otti, O.; Schmid-Hempel, P. *Nosema bombi*: A pollinator parasite with detrimental fitness effects. *J. Invertebr. Pathol.* **2007**, *96*, 118–124. [CrossRef] [PubMed]

35. Otti, O.; Schmid-Hempel, P. A field experiment on the effect of *Nosema bombi* in colonies of the bumblebee *Bombus terrestris*. *Ecol. Entomol.* **2008**, *33*, 577–582. [CrossRef]

36. Graystock, P.; Yates, K.; Darvill, B.; Goulson, D.; Hughes, W.O. Emerging dangers: Deadly effects of an emergent parasite in a new pollinator host. *J. Invertebr. Pathol.* **2013**, *114*, 114–119. [CrossRef] [PubMed]

37. Gisder, S.; Horchler, S.; Pieper, F.; Schüler, V.; Šima, P.; Genersch, E. Rapid Gastrointestinal Passage May Protect *Bombus terrestris* from ecoming a True Host for *Nosema ceranae*. *Appl. Environ. Microbiol.* **2020**, *86*, e00629-20. [CrossRef]

38. Aizen, M.A.; Smith-Ramírez, C.; Morales, C.L.; Vieli, L.; Sáez, A.; Barahona-Segovia, R.M.; Arbetman, M.P.; Montalva, J.; Garibaldi, L.A.; Inouye, D.W.; et al. Coordinated species importation policies are needed to reduce serious invasions globally: The case of alien bumblebees in South America. *J. Appl. Ecol.* **2019**, *56*, 100–106. [CrossRef]

39. Rutrecht, S.T.; Brown, M.J. The life-history impact and implications of multiple parasites for bumble bee queens. *Int. J. Parasitol.* **2008**, *38*, 799–808. [CrossRef] [PubMed]

40. Graystock, P.; Meeus, I.; Smagghe, G.; Goulson, D.; Hughes, W.O. The effects of single and mixed infections of *Apicystis bombi* and deformed wing virus in *Bombus terrestris*. *Parasitology* **2016**, *143*, 358–365. [CrossRef] [PubMed]

41. Peng, W.; Li, J.; Boncristiani, H.; Strange, J.P.; Hamilton, M.; Chen, Y. Host range expansion of honey bee black queen cell virus in the bumble bee, *Bombus huntii*. *Apidologie* **2011**, *42*, 650–658. [CrossRef]

42. Meeus, I.; de Miranda, J.R.; de Graaf, D.C.; Wäckers, F.; Smagghe, G. Effect of oral infection with Kashmir bee virus and Israeli acute paralysis virus on bumblebee (*Bombus terrestris*) reproductive success. *J. Invertebr. Pathol.* **2014**, *121*, 64–69. [CrossRef]

43. Genersch, E.; Yue, C.; Fries, I.; de Miranda, J.R. Detection of Deformed wing virus, a honey bee viral pathogen, in bumble bees (Bombus terrestris and Bombus pascuorum) with wing deformities. *J. Invertebr. Pathol.* **2006**, *91*, 61–63. [CrossRef] [PubMed]

44. Cilia, G.; Zavatta, L.; Ranalli, R.; Nanetti, A.; Bortolotti, L. Replicative Deformed Wing Virus Found in the Head of Adults from Symptomatic Commercial Bumblebee (*Bombus terrestris*) Colonies. *Vet. Sci.* **2021**, *8*, 117. [CrossRef]

45. Pascall, D.J.; Tinsley, M.C.; Obbard, D.J.; Wilfert, L. Host evolutionary history predicts virus prevalence across bumblebee species. *bioRxiv* **2019**, 498717. [CrossRef]

46. Gogala, A. *Čebele Slovenije*, 1st ed.; Založba ZRC: Ljubljana, Slovenia, 2014; pp. 1–180.

47. Toplak, I.; Jamnikar-Ciglenečki, U.; Aronstein, K.; Gregorc, A. Chronic bee paralysis virus and Nosema ceranae experimental co-infection of winter honey bee workers (*Apis mellifera L.*). *Viruses* **2013**, *5*, 2282–2297. [CrossRef] [PubMed]

48. Toplak, I.; Rihtarič, D.; Jamnikar Ciglenečki, U.; Hostnik, P.; Jenčič, V.; Barlič-Maganja, D. Detection of six honeybee viruses in clinically affected colonies of Carniolan gray bee (*Appis mellifera carnica*). *Slov. Vet. Res.* **2012**, *49*, 83–91.

49. Fries, I.; Chauzat, M.P.; Chen, Y.P.; Doublet, V.; Genersch, E.; Gisder, S.; Higes, M.; McMahon, D.P.; Martín-Hernández, R.; Natsopoulou, M.; et al. Standard methods for Nosema research. *J. Apic. Res.* **2013**, *52*, 1–28. [CrossRef]

50. Meeus, I.; de Graaf, D.C.; Jans, K.; Smagghe, G. Multiplex PCR detection of slowly-evolving trypanosomatids and neogregarines in bumblebees using broad-range primers. *J. Appl. Microbiol.* **2010**, *109*, 107–115. [CrossRef]

51. Xu, G.; Palmer-Young, E.; Skyrm, K.; Daly, T.; Sylvia, M.; Averill, A.; Rich, S. Triplex real-time PCR for detection of *Crithidia mellificae* and *Lotmaria passim* in honey bees. *Parasitol. Res.* **2018**, *117*, 623–628. [CrossRef]

Article

The First Detection and Genetic Characterization of Four Different Honeybee Viruses in Wild Bumblebees from Croatia

Ivana Tlak Gajger [1,*], Laura Šimenc [2] and Ivan Toplak [2]

[1] Department for Biology and Pathology of Fish and Bees, Faculty of Veterinary Medicine, University of Zagreb, 10000 Zagreb, Croatia

[2] Virology Unit, Institute of Microbiology and Parasitology, Veterinary Faculty, University of Ljubljana, Gerbičeva 60, 1000 Ljubljana, Slovenia; laura.simenc@vf.uni-lj.si (L.Š.); Ivan.Toplak@vf.uni-lj.si (I.T.)

* Correspondence: ivana.tlak@vef.hr; Tel.: +385-91-2390-041

Abstract: To determine the presence and the prevalence of four different honeybee viruses (acute bee paralysis virus—ABPV, black queen cell virus—BQCV, chronic bee paralysis virus—CBPV, deformed wing virus—DWV) in wild bumblebees, pooled randomly selected bumblebee samples were collected from twenty-seven different locations in the territory of Croatia. All samples were prepared and examined using the RT-PCR methods for quantification of mentioned honeybee viruses. Determined prevalence (%) of identified positive viruses were in the following decreasing order: BQCV > DWV > ABPV, CBPV. Additionally, direct sequencing of samples positive for BQCV (n = 24) and DWV (n = 2) was performed, as well as a test of molecular phylogeny comparison with those available in GenBank. Selected positive field viruses' strains showed 95.7 to 100% (BQCV) and 98.09% (DWV) nucleotide identity with previously detected and deposited honeybee virus strains in the geographic areas in Croatia and neighboring Slovenia. In this article, the first detection of four honeybee viruses with genetic characterization of high diversity strains circulating in wild bumblebees in Croatia is presented.

Keywords: *Bombus* spp.; honeybee viruses; deformed wing virus; black queen cell virus; acute bee paralysis virus; chronic bee paralysis virus; genetic characterization; sequencing; transmission routes

Citation: Tlak Gajger, I.; Šimenc, L.; Toplak, I. The First Detection and Genetic Characterization of Four Different Honeybee Viruses in Wild Bumblebees from Croatia. *Pathogens* **2021**, *10*, 808. https://doi.org/10.3390/pathogens10070808

Academic Editor: Giovanni Cilia

Received: 29 May 2021
Accepted: 24 June 2021
Published: 25 June 2021

Publisher's Note: MDPI stays neutral with regard to jurisdictional claims in published maps and institutional affiliations.

1. Introduction

Requirements for food production and agricultural intensification are resulting in a growing demand for insect pollination services [1]. Bumblebees (*Bombus* spp.) are vital and important insect pollinators to both the agricultural crops and wild plants in natural ecosystems, worldwide [2]. Their commercial rearing has boosted the economic importance of this insect in crop pollination [3]. Free-living bumblebee colonies are integral pollinators within native plant communities throughout temperate ecosystems. Adult bumblebees' robust size, long tongues, and buzz-pollination behavior result in their great pollination effectiveness. Therefore, they are indispensable, particularly for some plant species.

The decline in the number of insects pollinators is determined by various factors, such as climatic niche changes combined with reductions in natural food availability, the lack of the nesting materials, landscape alteration, agricultural intensification, and the spread of pathogens [4,5]. Reports of pollinator numbers declining due to the increased mortality in recent years are alarming on a global level [5–7].

Currently, pollinators are facing increased vulnerability to infectious diseases and other negative environmental stressors, such as harmful pesticides [8]. A reported decline in the abundance and diversity of beneficial insect pollinators can be caused by virus infections. The wild bumblebee population is also under the threat of viral infections [9]. Viruses are spilling over from managed and imported honeybees to free-living insect pollinators, including bumblebees [10,11]. Consequently, there is the possibility of disease occurrence after direct transmission through shared contaminated floral resources or

181

facilitated by changes in host immune status and susceptibility [1]. Moreover, recently, genetically identical strains of deformed wing virus (DWV) were detected in honeybee colonies (*Apis mellifera*) and in *Varroa destructor*, their obligate parasitic mite, making it a vector of this virus disease [10,12].

The importation and deployment of managed honeybee and bumblebee colonies may be a source of pathogen introductions in new geographical areas or alterations in the dynamics of native parasites and causative agents of secondary diseases, e.g., viruses, that ultimately increase disease prevalence in wild bees [1]. Insect pollinator decline has become a worldwide issue [4,5], causing increased concerns over effects on global food production [5], the stability of pollination services [13], and the disruption of the plant–pollinator link [14].

It is known that RNA honeybee viruses, due to their short generation of life and high mutation rates, have already crossed species barriers and have successfully infected a wide range of new insect hosts, such as free-living wild bees—solitary bees and bumblebees, wasps, hoverflies, and ants [9,15].

In recent years, there have been several reports of bumblebees infected with viral honeybee pathogens: DWV, black queen cell virus (BQCV), Israeli acute paralysis virus (IAPV), chronic bee paralysis virus (CBPV), Kashmir bee virus (KBV), and Sacbrood bee virus (SBV) [10,15,16]. In Croatia, the prevalence and regional distribution patterns of seven different honeybee viruses was studied, and simultaneous infection of adult honeybee samples with two to four different viruses was identified [17,18].

The aim of this research was to determine the presence and quantification of different honeybee viruses (ABPV, CBPV, BQCV, and DWV) in wild bumblebee samples originating from 27 geographically different locations. This is the first record of molecular viral examinations, as well as important new phylogenetic comparison information for endemic honeybee virus strains circulating in bumblebees in the territory of Croatia.

2. Results

Collected samples of bumblebees, originating from different locations in Croatia, were found positive with prevalence for ABPV (3.70%), BQCV (88.89%), CBPV (3.70%), and DWV (37.40%) (Figure 1).

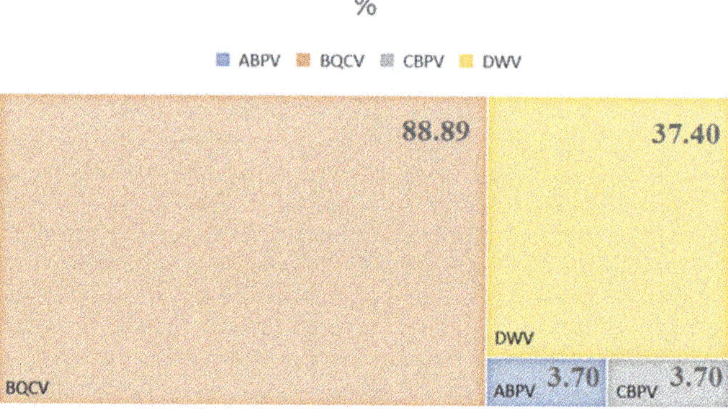

Figure 1. The determined prevalence (%) of the 27 tested bumblebee samples for the presence of four different honeybee viruses (ABPV, BQCV, CBPV, and DWV).

The obtained Ct value for one ABPV-positive sample was 35.04, corresponding to 9.082×10^2 copy number/5 µL. Altogether, BQCV was detected in 24 out of 27 samples, with Ct values from the lowest 17.24 to the highest 29.71 (corresponding to copy number/5 µL from 1.251×10^4 to 1.224×10^8). One bumblebee sample was detected as CBPV-positive with Ct value 41.00. The detected Ct values for ten DWV-positive samples varied from 27.38 to 41.84 with copy number/5 µL from 6.33 to 6.392×10^4 (Table 1).

Table 1. The determined copy number in 27 tested bumblebee samples by real-time RT-PCR assay for the detection of ABPV, BQCV, CBPV, and DWV, collected from 27 different geographical locations in Croatia.

Sample Number	Name of Sample	ABPV Copy Number/5 μL	BQCV Copy Number/5 μL	CBPV Copy Number/5 μL	DWV Copy Number/5 μL	Number of Viruses
1	Bombus-Zac/2018	9.082×10^2	1.251×10^4	0	0	2
2	Bombus-Kri/2018	0	4.415×10^6	0	0	1
3	Bombus-Pas/2018	0	0	0	0	0
4	Bombus-Uglj/2018	0	0	0	1.739×10^2	1
5	Bombus-Bnm/18	0	0	0	1.257×10^4	1
6	Bombus-Mak/2018	0	9.160×10^7	0	1.260×10^4	2
7	Bombus-Mar/2018	0	2.198×10^6	0	0	1
8	Bombus-Kra/2018	0	4.212×10^6	0	0	1
9	Bombus-Var/2018	0	3.511×10^6	0	0	1
10	Bombus-Ses/2018	0	18.09×10^7	0	6.234×10	2
11	Bombus-Zel/2018	0	25.83×10^5	0	0	1
12	Bombus-Kop/2018	0	2.257×10^5	0	0	1
13	Bombus-Zab/2018	0	1.678×10^6	0	0	1
14	Bombus-Dur/2018	0	2.777×10^6	0	0	1
15	Bombus-Ora/2018	0	4.400×10^5	0	0	1
16	Bombus-Bje/2018	0	1.042×10^6	0	0	1
17	Bombus-Dug/2018	0	2.131×10^5	0	0	1
18	Bombus-Jad/2018	0	3.200×10^7	0	3.351×10	2
19	Bombus-Rov/2018	0	1.588×10^5	0	0	1
20	Bombus-Krk/2018	0	8.975×10^3	0	3.08	2
21	Bombus-Dub/2018	0	1.656×10^5	6.33	0	2
22	Bombus-Sib/2018	0	1.110×10^6	0	6.392×10^4	2
23	Bombus-Vet/2018	0	7.72×10^5	0	1.285×10^2	2
24	Bombus-Zag/2018	0	1.309×10^5	0	0	1
25	Bombus-Gos/2018	0	1.348×10^5	0	3.741×10^2	2
26	Bombus-Mur/2018	0	1.360×10^5	0	0	1
27	Bombus-Nez/2018	0	1.224×10^8	0	1.140×10^3	2

For 24 BQCV-positive bumblebee samples, the 653 nucleotide long sequences were successfully determined and compared with those available in GenBank. High genetic diversity among 24 Croatian sequenced BQCV-positive samples in bumblebee were identified, with 95.7 to 100% nucleotide identity to each other. The majority of the identified BQCV strains were closely related with BQCV stains from Slovenia. Strain BQCV Bombus-Gos/2018 (MW488258) has 100% nucleotide identity with bumblebee strain BQCV Bombus BT23/2017 (MH900014) and honeybee isolate 279/2017 (MH899977), both detected in the neighboring country of Slovenia. A group of 17 closely related BQCV-positive samples, located on the same branch of phylogenetic tree, were collected throughout the territory of Croatia, and have from 98.47 to 100% nucleotide identity with isolate 281/2017 (MH899979), detected in Slovenia. The BQCV strain Bombus-Var/2018 (MW488242) has 99.54% nucleotide identities with isolate BQCV 637/2009 (MH899994) detected in Slovenia and 98.82% nucleotide identity with strain Sydney ((MF623171) detected in Australia. Strain Bombus-Vet/2018 (MW488256) has 99.69% nucleotide identity with strain 287-1/2007

(MH899983), while with the strain Bombus-Mak/2018 (MW488239) has 99.08% identity. Detected strain Bombus-Kri/2018 (MW488238) collected in Križevci showed 100% nucleotide identity with Slovenian honeybee strain 1/2008 (MH899946). Strain Bombus-Ses/2018 (MW488243) and Bombus-Nez/2018 (MW488260) have 99.39% nucleotide identities with isolate 1956/2009 (MH899996) identified in Slovenia in 2009 in clinically affected honeybee samples (Figure 2).

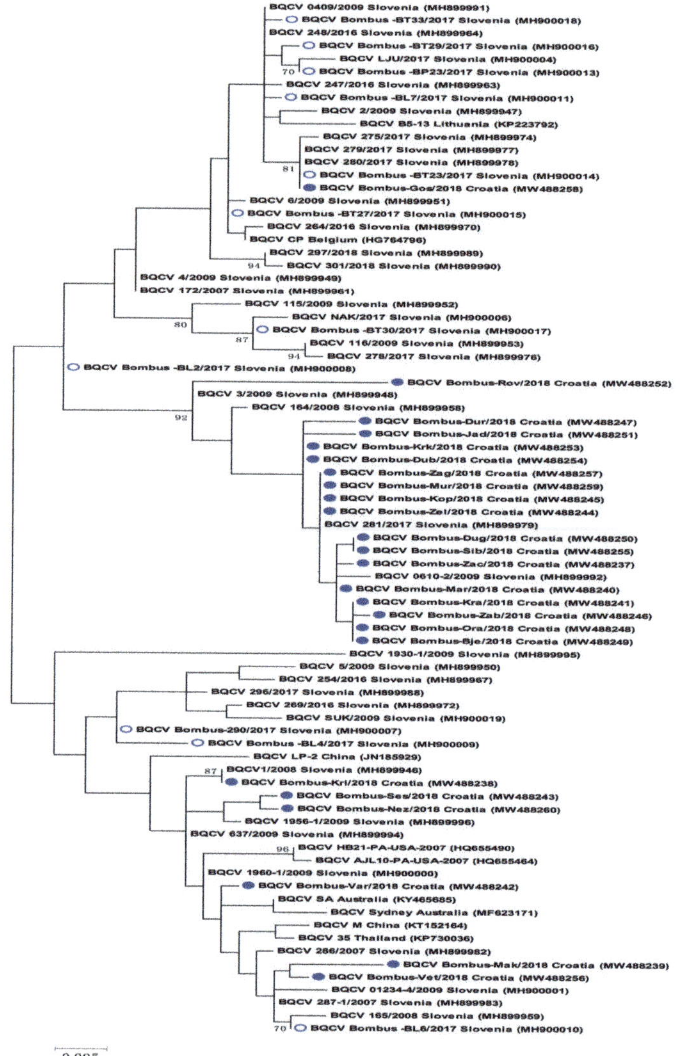

Figure 2. The BQCV tree was constructed from 653 nt long sequences of capsid protein gene (between nt positions 7774 and 8426; numbering according to BQCV isolate Sydney, MF623171) for 24 Croatian and 56 DWV isolates from GenBank. BQCV bumblebee samples from Croatia (•) and BQCV bumblebee positive samples from Genbank (O) are marked on the phylogenetic tree. The maximum likelihood phylogenetic tree with the Tamura 3-parameter substitution model with Gamma distribution. Statistical support for the tree was evaluated by bootstrapping, based on 1000 repetitions. Bootstrap values lower than 70 are not shown.

For 2 samples out of 10 DWV-positive bumblebee samples, the 471 nucleotide long sequences were successfully determined and compared with those available in GenBank. A DWV-positive sample Bombus-Mir/2018 (MW488261) was collected in 2018 in Zagreb and has 98.09% nucleotide identity with the closely related strains YU4 (JF346630) and YU5 (JF346631) in GenBank, collected in 1988 near Turopoljski Lug, which is located about 50 km from Zagreb city. The second DWV-positive sample Bombus-Nik/2018 (MW488262) was collected near Šibenik and has 98.09% nucleotide identity with strain SLO/BM199/2013, detected in *A. mellifera* in 2013 in Slovenia, while with DWV strain MeDWV1 (MW222481) and DWV strain Maryland/2015/422 (MG831202), both from the USA, it shares from 97.45 to 97.66% nucleotide identity (Figure 3).

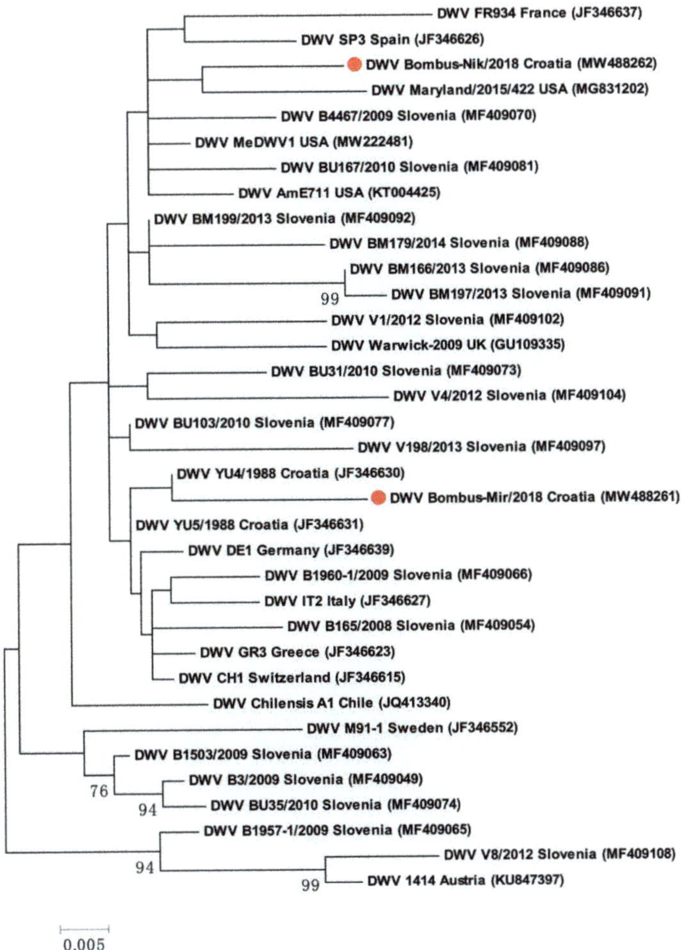

Figure 3. The DWV tree was constructed for two Croatian and 33 DWV isolates from GenBank from 471 nt long sequences of L protein gene (between nt positions 1357 and 1827; numbering according to isolate Austria 1414, KU847397). The two DWV bumblebee samples from Croatia determined in this study are marked (•) on the phylogenetic tree. The maximum likelihood phylogenetic tree with the Tamura 3-parameter substitution model with Gamma distribution. Statistical support for the tree was evaluated by bootstrapping, based on 1000 repetitions. Bootstrap values lower than 70 are not shown.

Genetically closely related strains, previously detected in the same geographic region among honeybees, were also detected in randomly selected bumblebee samples in Croatia. Twenty-six determined sequences from this study were deposited in GenBank with accession numbers MW488237-MW488262.

3. Discussion

This study is the first attempt to constitute an epidemiological baseline regarding the geographical distribution patterns and prevalence of four honeybee viruses in wild bumblebee samples across the territory of Croatia. Results showed that a higher prevalence of BQCV and DWV was determined in tested wild bumblebee samples collected in the continental part of country where the honeybee colonies' density is higher, in comparison with locations at the Adriatic coast and on islands. In particular, high BQCV prevalence was determined at all locations, except in samples originating from the Dalmatian islands of Ugljan and Pašman. Similar results were presented by Alger et al. (2019) and Toplak et al. (2020), which provided support to the observations that honeybee viruses are probably spilling over from managed honeybee colonies to wild bumblebees through visits to the same floral sources [10,11].

Since the current study does not include bumblebee species differentiation, we are not able to discuss possible differences in the species-specific vulnerability of bumblebees to tested honeybee viruses. However, according to Toplak et al. (2020) the different species of bumblebees tested in two consecutive years showed high variability in prevalence for different viruses and species. In addition, their results showed high variability for different species of bumblebees in a short period in the same geographic area [11]. Nevertheless, in our study, the important first data were obtained regarding the prevalence of four honeybee viruses on bumblebees from our region.

This research confirmed that genetically identical or closely related honeybee strains of BQCV and DWV were also identified among tested bumblebees, collected from the same location (geographic area). The high genetic identity with previously determined Slovenian BQCV and DWV strains is not surprising because of the neighboring area and historically long tradition of honeybee pasturing during spring and summer time. Furthermore, those two countries have a good beekeeper connection via trade. The identification of a group of 17 genetically closely related BQCV-positive samples, which were collected throughout territory of Croatia, is probably the result of recent years' transmission events. The opposite observation with the identification of relatively high diversity of BQCV strains among bumblebees is the result of the persistence of this virus for a long period in the population and was like that observed in a previous study in Slovenia [10]. Although honeybee positive samples from Croatia were not included in this study, the genetically very similar strains of BQCV and DWV than identified in bumblebees could be expected in honeybees from the same territory. This was confirmed with the identification of DWV-positive sample Bombus-Mir/2018 (MW488261), which has 98.09% nucleotide identity with the two closely related DWV strains YU4 (JF346630) and YU5 (JF346631), collected more than 30 years ago in Croatia, both DWV-positive samples were collected in *A. mellifera*, from a location near Zagreb city [19].

Interestingly, ABPV and CBPV were detected only in one sample from one location each, with a very low viral load copy number. Due to low positive samples of both ABPV and CBPV, the sequencing and phylogenetic analyses were not possible in this study. In contrast, ten examined bumblebees' samples were DWV-positive (37.40%), which is significantly higher infection prevalence than results presented in previously published studies, where the range was 2.70 to 11% positive bumblebees' samples [9–11,15]. However, general observation for both BQCV- and DWV-positive samples showed that low copy numbers were identified in each pool of five bumblebees' positive sample, suggesting that these viruses are present in bumblebees, but they may have limited impact on bumblebee pathology. Tehel et al. (2020) reported higher viral titers of BQCV, DWV genotype A, and DWV genotype B in bumblebees after experimental inoculation of a pathogen by injection in comparison with oral inoculation [20]. Namely, among more than 30 honeybee-infecting known viruses [21,22], three are characterized by specific clinical symptoms: CBPV, DWV, and SBV [23]. BQCV and ABPV with its belonging complex can show alterations in the morphology of developmental stages of honeybees and adults' behavior. Furthermore, some can be present as inapparent or subclinical infections [24].

It is not yet completely ascertained if field viruses' strains are able to cause clinical manifestation in bumblebees. It is also not clear which environmental stressors are promoting factors for converting an asymptomatic infection into symptomatic and overt. In this study, all collected bumblebees were without visible morphological or behavioral changes, so they were considered clinically healthy. Although the virus presence could also be detected in individual bumblebee tissues or organs, this approach was not applied in our study and was assessed firstly to define the prevalence and diversity of four tested viruses. However, for further research it would be good to use individual bumblebee tissues for the estimation of individual virus tropism for specific tissue and for evaluating the possible effects that those viruses may have on their bumblebee hosts. In addition, according to Manley et al. (2019), honeybee parasitic mite *V. destructor* drives DWV prevalence and titer in honeybees and wild bumblebees [25]. Similarly, experimental injections of DWV under laboratory conditions or natural direct inoculation through *V. destructor* host feeding into insect body haemocoel causes an increase in prevalence and virulence in honeybees [26,27].

In our study, BQCV-positive samples were determined in very high prevalence (24 positive samples/27 total number of samples), which is contrary to the published findings of Dolezal et al. (2016), where same virus was detected extremely rarely in wild bees [28]. However, a recent publication from Slovenia, with the identification of genetically identical strains of ABPV, BQCV, SBV, and Lake Sinai virus (LSV) supports the observation of our study, that identical strains are present in honeybees and bumblebees [10]. Previously published data from Croatian honeybee samples originating from 82 apiaries located in 20 different districts showed a wide spread of honeybee viruses, with 9.75% of CBPV-positive samples, while ABPV, BQCV, and DWV were found in 10.97%, 40.24%, and 95.12% of tested apiaries, respectively [17]. Simultaneous infections with a maximum of two different viruses were detected in 10 (37.03%) of 27 bumblebee samples, and this was lower than the previously observed 64.6% of multiple infections among tested honeybees in Croatia [18].

4. Materials and Methods

4.1. Field Sampling

To determine the presence and the prevalence of four different honeybee viruses (ABPV, BQCV, CBPV, and DWV) in wild bumblebees (*Bombus terrestris*, *Bombus lapidarius*, *Bombus pascuorum*), 27 randomly selected bumblebee samples were collected from a total of 27 sampling locations from the territory of Croatia (Figure 4). Sampling was conducted during July and August 2018. At each site, five clinically healthy bumblebees were taken from flowers, representing one pool sample from each location. Each sample was marked with the number of the sampling location. Collected samples of bumblebees were stored under −70 °C until the start of molecular analyses.

4.2. Molecular Analyses in Laboratory Conditions

For RNA extraction purposes, each pooled sample consisting of five bumblebees' specimens was placed into Ultra-Turrax DT-20 tubes (IKA, Königswinter, Germany) with five mL of RPMI 1640 medium (Gibco, Paisley, UK) and incubated at room temperature for 30 min. Then, prepared samples were homogenized and centrifuged for 15 min at $2500 \times g$. Two milliliters of supernatant were stored from each sample as a suspension for further extraction.

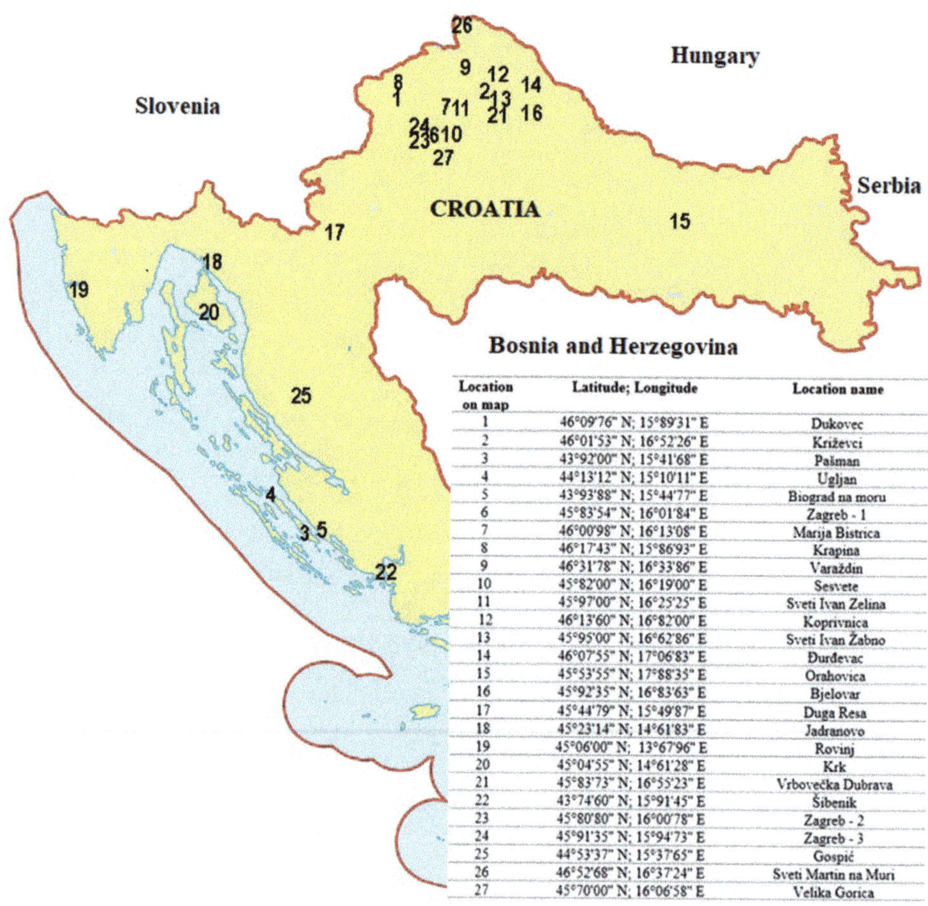

Location on map	Latitude; Longitude	Location name
1	46°09'76" N; 15°89'31" E	Dukovec
2	46°01'53" N; 16°52'26" E	Križevci
3	43°92'00" N; 15°41'68" E	Pašman
4	44°13'12" N; 15°10'11" E	Ugljan
5	43°93'88" N; 15°44'77" E	Biograd na moru
6	45°83'54" N; 16°01'84" E	Zagreb - 1
7	46°00'98" N; 16°13'08" E	Marija Bistrica
8	46°17'43" N; 15°86'93" E	Krapina
9	46°31'78" N; 16°33'86" E	Varaždin
10	45°82'00" N; 16°19'00" E	Sesvete
11	45°97'00" N; 16°25'25" E	Sveti Ivan Zelina
12	46°13'60" N; 16°82'00" E	Koprivnica
13	45°95'00" N; 16°62'86" E	Sveti Ivan Žabno
14	46°07'55" N; 17°06'83" E	Đurđevac
15	45°53'55" N; 17°88'35" E	Orahovica
16	45°92'35" N; 16°83'63" E	Bjelovar
17	45°44'79" N; 15°49'87" E	Duga Resa
18	45°23'14" N; 14°61'83" E	Jadranovo
19	45°06'00" N; 13°67'96" E	Rovinj
20	45°04'55" N; 14°61'28" E	Krk
21	45°83'73" N; 16°55'23" E	Vrbovečka Dubrava
22	43°74'60" N; 15°91'45" E	Šibenik
23	45°80'80" N; 16°00'78" E	Zagreb - 2
24	45°91'35" N; 15°94'73" E	Zagreb - 3
25	44°53'37" N; 15°37'65" E	Gospić
26	46°52'68" N; 16°37'24" E	Sveti Martin na Muri
27	45°70'00" N; 16°06'58" E	Velika Gorica

Figure 4. Sampling location sites (sample number) where a total of 27 samples of healthy bumblebees was collected during July and August 2018. The sample numbers represent different locations on the map of Croatian territory corresponding to the names of those locations, as indicated in accompanying table.

Primers, TaqMan probes, and quantification standards for ABPV [29], BQCV [30], CBPV [31], and DWV [32] were set with reagents according to previously published protocol.

Briefly, the total RNA was extracted from 140 µL of suspension from each sample by the QIAamp viral RNA mini-kit (Qiagen, Hilden, Germany) and recovered from a spin column in 60 µL of elution buffer. Reverse transcription with RT-qPCR assay was made in a single step using QuantiNova Pathogen + IC Kit (Qiagen, Hilden, Germany). The RT-qPCR mix consisted of 5 µL QuantiNova Master Mix, 2 µL 10× Internal (inhibition) Control (IC) Probe Assay, 1 µL IC (1:100), 4.5 µL deionized water, 1 µL forward primer (200 nM), 1 µL reverse primer (200 nM), and 0.5 µL probe (100 nM) and 5 µL of extracted RNA with a total of 20 µL final volume. Thermal cycling was performed on Mx3005P thermocycler (Stratagene, La Jolla, CA, USA) with the following conditions: 20 min 50 °C, 2 min 95 °C, followed by 45 cycles of 15 s 95 °C, 30 s 60 °C, and 30 s 60 °C. In each run, the positive control was included, prepared as a mixed suspension of previously determined positive field samples of four different viruses (ABPV, BQCV, CBPV, and DWV). The negative control was prepared and used in the same way as the positive, while each negative control consisted only of 160 µL of RPMI 1640 medium (Gibco, Paisely, UK) in aliquot.

4.3. Data Processing, Statistical Analysis, and Reporting

The known copy number of the standard for each virus, with 10-fold dilutions from 10^{-3} to 10^{-7}, were prepared and added in each RT-qPCR run. The exact number of RNA viral molecules in individual sample was calculated for positive samples from the standard curve for each of the four honeybee viruses.

The results for each sample were analyzed using MxPro-Mx3005P v4.10 software (Stratagene, La Jolla, CA, USA) and the exact copy number was determined from the standard curve. Results were expressed as number of detected viral copies in 5 µL of extracted RNA.

For sequencing purpose, the BQCV- and DWV-positive samples were amplified by using a specific method of reverse transcription and polymerase chain reaction (RT-PCR), as previously described [33]. Results were evaluated based on the size of RT-PCR products in the agarose gel as positive in the case of the expected product size: for BQCV, 770 nt, and for DWV, 504 nt [33]. In the case of a positive result, the selected RT-PCR products of a single virus were directly sequenced with the Sanger sequencing protocol, using the same primers as used for specific RT-PCR as described previously [10]. Individual sequences were analyzed using the DNASTAR 5.05 (Lasergen, WI, USA) program, and 26 positive samples of two viruses (BQCV $n = 24$ and DWV $n = 2$) were detected in bumblebees together with closely related sequences from GenBank and interpreted according to the results of the nucleotide sequence matching between honeybee and bumblebee samples. Multiple alignments were created using the program MEGA 6.06. Genetic distances were calculated from the alignment based on the Tamura three-parameter model, and phylogenetic trees were generated using the maximum likelihood (ML) statistical method implemented with the Tamura 3-parameter model with Gamma distribution [34]. The test of phylogeny was performed through 1000 bootstrap replicates. Only bootstrap values higher than 70% were presented on phylogenetic trees. The comparative analyses of collected samples at 27 locations and comparison with previously detected viruses in honeybees and with the most closely related sequences available in GenBank were performed.

5. Conclusions

Results of this research confirm that several honeybee viruses (ABPV, CBPV, BQCV, DWV) were found in wild bumblebees in Croatia. It can also be concluded that the presence of the examined viruses was higher in continental parts of the country compared with the Adriatic coast and islands. However, because the virus's presence was not studied in internal bumblebee tissues of separate bumblebee species, in further studies, samples of different bumblebee species will be tested individually to define species-specific prevalence and to evaluate the possible impact that those viruses may have on their bumblebee hosts. Moreover, due to the neighborhood and a good beekeeper connection via trade, virus genetic matches were determined in Croatia and Slovenia.

Author Contributions: Conceptualization, I.T.G. and I.T.; investigation, I.T.G., I.T. and L.Š.; writing—original draft preparation, I.T.G.; writing—review and editing, I.T.G., I.T. and L.Š.; visualization, I.T. and I.T.G. All authors have read and agreed to the published version of the manuscript.

Funding: This research received no external funding.

Institutional Review Board Statement: Not applicable.

Informed Consent Statement: Not applicable.

Data Availability Statement: Data are presented in the manuscript. Twenty-six determined sequences from this study were deposited in GenBank with accession numbers MW488237-MW488262.

Acknowledgments: Authors thank the students of the Faculty of Veterinary Medicine University of Zagreb for their help with the collection and transportation of wild bumblebee samples.

Conflicts of Interest: The authors declare no conflict of interest.

References

1. Graystock, P.; Blane, E.J.; McFrederick, Q.S.; Goulson, D.; Hughes, W.O.H. Do managed bees drive parasite spread and emergence in wild bees? *Int. J. Parasitol. Parasites Wildl.* **2016**, *5*, 64–75. [CrossRef]
2. Osborne, J.L.; Williams, P.H. Bumble bees as pollinators of crops and wild flowers. In *Bumble Bees for Pleasure and Profit*; Matheson, A., Ed.; IBRA: Cardif, UK, 1996; pp. 24–32.
3. Velthius, H.H.V.; van Doorn, A. A century of advances in bumblebee domestication and the economic and environmental aspects of its commercialization for pollination. *Apidologie* **2006**, *37*, 421–451. [CrossRef]
4. Williams, P.H.; Osborne, J.L. Bumblebee vulnerability and conservation world-wide. *Apidologie* **2009**, *40*, 367–387. [CrossRef]
5. Potts, S.G.; Biesmeijer, J.C.; Kremen, C.; Neumann, P.; Schweiger, O.; Kunin, W.E. Global pollinator declines: Trends, impacts and drivers. *Trends Ecol. Evol.* **2010**, *25*, 345–353. [CrossRef]
6. Cameron, S.A.; Lozier, J.D.; Strange, J.P.; Koch, J.B.; Cordes, N.; Solter, L.F.; Griswold, T.L. Patterns of widespread decline in North American bumble bees. *Proc. Natl. Acad. Sci. USA* **2011**, *108*, 662–667. [CrossRef]
7. Cameron, S.A.; Sadd, B.M. Global trends in bumble bee health. *Ann. Rev. Entomol.* **2020**, *65*, 209–232. [CrossRef]
8. McCallum, H.; Dobson, A. Disease, habitat fragmentation and conservation. *Proc. Biol. Sci.* **2002**, *269*, 2041–2049. [CrossRef]
9. Fürst, M.A.; McMahon, D.P.; Osborne, J.L.; Paxton, R.J.; Brown, M.J.F. Disease associations between honeybees and bumlebees as a threat to wild pollinators. *Nature* **2014**, *506*, 364–366. [CrossRef]
10. Toplak, I.; Šimenc, L.; Pislak Ocepek, M.; Bevk, D. Determination of Genetically Identical Strains of Four Honeybee Viruses in Bumblebee Positive Samples. *Viruses* **2020**, *12*, 1310. [CrossRef] [PubMed]
11. Alger, S.A.; Burnham, P.A.; Boncristiani, H.F.; Brody, A.K. RNA virus spillover from managed honeybees (*Apis mellifera*) to wild bumblebees (*Bombus* spp.). *PLoS ONE* **2019**, *14*, e0217822. [CrossRef] [PubMed]
12. Jamnikar-Ciglenecki, U.; Pislak Ocepek, M.; Toplak, I. Genetic diversity of deformed wing virus from *Apis mellifera carnica* (Hymenoptera: Apidae) and Varroa Mite (Mesostigmata: Varroidae). *J. Econ. Entomol.* **2019**, *112*, 11–19. [CrossRef] [PubMed]
13. Biesmeijer, J.C.; Roberts, S.P.M.; Reemer, M.; Ohlemuller, R.; Edwards, M.; Peeters, T.; Schaffers, A.P.; Potts, S.G.; Kleukers, R.; Thomas, C.D.; et al. Parallel Declines in Pollinators and Insect Pollinated Plants in Britain and The Netherlands. *Science* **2006**, *313*, 351–354. [CrossRef]
14. Fontaine, C.; Dajoz, I.; Meriguet, J.; Loreau, M. Functional diversity of plant-pollinator interaction webs enhances the persistence of plant communities. *PLoS Biol.* **2006**, *4*, e1. [CrossRef]
15. McMahon, D.P.; Furst, M.A.; Caspar, J.; Theodorou, P.; Brown, M.J.F.; Paxton, R.J. A sting in the spit: Widespread cross-infection of multiple RNA viruses across wild and managed bees. *J. Anim. Ecol.* **2015**, *84*, 615–624. [CrossRef] [PubMed]
16. Sachman-Ruiz, B.; Narváez-Padilla, V.; Reynaud, E. Commercial *Bombus impatiens* as reservoirs of emerging infectious diseases in central México. *Biol. Invasions* **2015**, *17*, 2043–2053. [CrossRef]
17. Tlak Gajger, I.; Kolodziejek, J.; Bakonyi, T.; Nowotny, N. Prevalence and distribution patterns of seven different honeybee viruses in diseased colonies: A case study from Croatia. *Apidologie* **2014**, *45*, 701–706. [CrossRef]
18. Tlak Gajger, I.; Bičak, J.; Belužić, R. The occurrence of honeybee viruses in apiaries in the Koprivnica-Križevci district in Croatia. *Vet. Arhiv.* **2014**, *84*, 421–428.
19. Forsgren, E.; de Miranda, J.R.; Isaksson, M.; Wei, S.; Fries, I. Deformed wing virus associated with *Tropilaelaps mercedesae* infesting European honey bees (*Apis mellifera*). *Exp. Appl. Acarol.* **2009**, *47*, 87–97. [CrossRef] [PubMed]
20. Tehel, A.; Streicher, T.; Tragust, S.; Paxton, R.J. Experimental infection of bumblebees with honeybee-associated viruses: No direct fitness costa but potential future threats to novel wild bee hosts. *R. Soc. Open Sci.* **2020**, *7*, 200480. [CrossRef]
21. Beaurepaire, A.; Piot, N.; Doublet, V.; Antunez, K.; Campbell, E.; Chantawannakul, P.; Chejanovsky, N.; Gajda, A.; Heerman, M.; Panziera, D.; et al. Diversity and Global Distribution of Viruses of the Western Honey Bee, *Apis Mellifera*. *Insects* **2020**, *11*, 239. [CrossRef]
22. Ullah, A.; Tlak Gajger, I.; Majoros, A.; Dar, S.A.; Khane, S.; Kalimullah; Shah, A.H.; Nasir, S.M.; Khabir, M.N.; Hussain, R.; et al. Viral impacts on honey bee populations: A review. *Saudi J. Biol. Sci.* **2021**, *28*, 523–530. [CrossRef] [PubMed]
23. De Miranda, J.R.; Gauthier, L.; Ribiere, M.; Chen, Y.P. Honey bee viruses and their effect on bee and colony health. In *Honey Bee Colony Health: Challenges and Sustainable Solutions*; Sammataro, D., Yoder, J., Eds.; CRC Press: Boca Raton, FL, USA, 2012; pp. 71–102.
24. Grozinger, C.M.; Flenniken, M.L. Bee viruses: Ecology, pathogenicity, and impacts. *Annu. Rev. Entomol.* **2019**, *64*, 205–226. [CrossRef] [PubMed]
25. Manley, R.; Temperton, B.; Doyle, T.; Gates, D.; Hedges, S.; Boots, M.; Wilfert, L. Knock-on community impacts of a novel vector: Spillover of emerging DWV-B from Varroa-infested honeybees to wild bumblebees. *Ecol. Lett.* **2019**, *22*, 1306–1315. [CrossRef]
26. Martin, S.J.; Highfield, A.C.; Brettell, L.; Villalobos, E.M.; Budge, G.E.; Powell, M.; Nikaido, S.; Schroeder, D.C. Global honey bee viral landscape altered by a parasitic mite. *Science* **2012**, *336*, 1304–1306. [CrossRef] [PubMed]
27. Mondet, F.; De Miranda, J.R.; Kretzschmar, A.; Le Conte, Y.; Mercer, A.R. On the front line: Quantitative virus dynamics in honeybee (*Apis mellifera* L.) colonies along a new expansion front of the parasite *Varroa destructor*. *PLoS Pathog.* **2014**, *10*, e1004323. [CrossRef]
28. Dolezal, A.G.; Hendrix, S.D.; Scavo, N.A.; Carrillo-Tripp, J.; Harris, M.A.; Wheelock, M.J.; O'Neal, M.E.; Toth, A.L. Honeybee viruses and wild bees: Viral prevalence, loads and experimental inoculation. *PLoS ONE* **2016**, *11*, e0166190. [CrossRef] [PubMed]

29. Jamnikar Ciglenecki, U.; Toplak, I. Development of a real-time RT-PCR assay with TaqMan probe for specific detection of acute bee paralysis virus. *J. Virol. Methods* **2012**, *184*, 63–68. [CrossRef]

30. Chantawannakul, P.; Ward, L.; Boonham, N.; Brown, M. A scientific note on the detection of honeybee viruses using real-time PCR (TaqMan) in varroa mites collected from a Thai honeybee (*Apis mellifera*) apiary. *J. Invertebr. Pathol.* **2006**, *91*, 69–73. [CrossRef]

31. Blanchard, P.; Regnault, J.; Schurr, F.; Dubois, E.; Ribière, M. Intra-laboratory validation of chronic bee paralysis virus quantitation using an accredited standardized real-time quantitative RT-PCR method. *J. Virol. Method.* **2012**, *180*, 26–31. [CrossRef] [PubMed]

32. Schurr, F.; Tison, A.; Militano, L.; Cheviron, N.; Sircoulomb, F.; Riviere, M.P.; Ribiere-Chabert, M.; Thiery, R.; Dubois, E. Validation of quantitative real-time RT-PCR assays for the detection of six honeybee viruses. *J. Virol. Methods* **2019**, *270*, 70–78. [CrossRef]

33. Toplak, I.; Rihtarič, D.; Jamnikar Ciglenečki, U.; Hostnik, P.; Jenčič, V.; Barlič-Maganja, D. Detection of six honeybee viruses in clinically affected colonies of Carniolan gray bee (*Apis mellifera carnica*). *Slov. Vet. Res.* **2012**, *49*, 83–91.

34. Tamura, K.; Stecher, G.; Peterson, D.; Filipski, A.; Kumar, S. MEGA6: Molecular Evolutionary Genetics Analysis version 6.0. *Mol. Biol. Evol.* **2013**, *30*, 2725–2729. [CrossRef] [PubMed]

Article

Detection of *Lotmaria passim*, *Crithidia mellificae* and Replicative Forms of Deformed Wing Virus and Kashmir Bee Virus in the Small Hive Beetle (*Aethina tumida*)

Antonio Nanetti [1], James D. Ellis [2], Ilaria Cardaio [1] and Giovanni Cilia [1,*]

[1] CREA Research Centre for Agriculture and Environment, Via di Saliceto 80, 40128 Bologna, Italy; antonio.nanetti@crea.gov.it (A.N.); ilaria.cardaio@crea.gov.it (I.C.)

[2] Entomology and Nematology Department, University of Florida, 1881 Natural Area Dr., P.O. Box 110620, Gainesville, FL 32607-0620, USA; jdellis@ufl.edu

* Correspondence: giovanni.cilia@crea.gov.it; Tel.: +39-051-353-103

Abstract: Knowledge regarding the honey bee pathogens borne by invasive bee pests remains scarce. This investigation aimed to assess the presence in *Aethina tumida* (small hive beetle, SHB) adults of honey bee pathogens belonging to the following groups: (i) bacteria (*Paenibacillus larvae* and *Melissococcus plutonius*), (ii) trypanosomatids (*Lotmaria passim* and *Crithidia mellificae*), and (iii) viruses (black queen cell virus, Kashmir bee virus, deformed wing virus, slow paralysis virus, sacbrood virus, Israeli acute paralysis virus, acute bee paralysis virus, chronic bee paralysis virus). Specimens were collected from free-flying colonies in Gainesville (Florida, USA) in summer 2017. The results of the molecular analysis show the presence of *L. passim*, *C. mellificae*, and replicative forms of deformed wing virus (DWV) and Kashmir bee virus (KBV). Replicative forms of KBV have not previously been reported. These results support the hypothesis of pathogen spillover between managed honey bees and the SHB, and these dynamics require further investigation.

Keywords: honey bee; small hive beetle; invasive pest; trypanosomatids; honey bee virus; deformed wing virus; Kashmir bee virus; replicative virus; strand-specific RT-PCR

Citation: Nanetti, A.; Ellis, J.D.; Cardaio, I.; Cilia, G. Detection of *Lotmaria passim*, *Crithidia mellificae* and Replicative Forms of Deformed Wing Virus and Kashmir Bee Virus in the Small Hive Beetle (*Aethina tumida*). *Pathogens* **2021**, *10*, 372. https://doi.org/10.3390/pathogens10030372

Academic Editor: Lawrence S. Young

Received: 16 February 2021
Accepted: 18 March 2021
Published: 19 March 2021

Publisher's Note: MDPI stays neutral with regard to jurisdictional claims in published maps and institutional affiliations.

1. Introduction

Aethina tumida (Murray 1867), the small hive beetle (SHB), is a coleopteran species belonging to the Nitidulidae family [1]. Native to Sub-Saharan Africa [2], it is a destructive, invasive pest of *Apis mellifera* (western honey bee) colonies [3], and it causes significant damage to brood, pollen, and honey stores [4]. Presently, the SHB is recorded in all continents except Antarctica [3,5–8], having reached North America in 1996; Australia in 2000; and, more recently, countries in Europe, South America, and Asia [9–12]. The SHB is an ecological generalist [4] and creates persistent populations in colonies in areas in which it has been introduced [13].

Honey bees are exposed to pests and pathogens belonging to different groups (viruses, bacteria, fungi, protists, mites, insects, etc.), some of which are responsible for severe health impairment and colony collapse [14–17]. Adult SHBs invade colonies, where they feed, thrive, and reproduce. This allows contact between SHBs and other bee pests and pathogens [18–23].

Lotmaria passim and *Crithidia mellificae* are two trypanosomatid species capable of colonizing the digestive system of honey bees [24,25]. The transmission is deemed to occur by the oral–fecal route [26,27], and the presence of infected faeces within the hive may promote the circulation of the parasite among worker bees [26]. Both pathogens are deemed to impact colony health by altering bee behavior, physiology, immune response, and lifespan [28–31]. Nevertheless, the details of their pathogenic effects are still not fully understood. *Lotmaria passim* has been described only recently [25], and it is presently acknowledged as the most prevalent *A. mellifera* trypanosomatid [32]. Infections have been reported in Asian, European, and South and North American colonies [8], whereas *C. mellificae* infections have been rarely observed [8,33–35].

Deformed wing virus (DWV) is a positive-sense ssRNA virus belonging to the Picornaviridae family within the Iflavirus genus [36,37]. Spread globally [32,36–38], three genetic variants have been acknowledged and identified as types A, B, and C [39,40]. Type A is by far the most widespread [40], and it may generate asymptomatic or symptomatic infections, the latter including deformed or missing wings, shortened abdomens, and premature bee death [36]. Generally, this virus is transmitted through puncture wounds produced by the ectoparasite *Varroa destructor* as it feeds on immature honey bees [41]. However, the infection may be transmitted horizontally by bee-to-bee contact, especially in cases of severe infections [42–46], curbicular pollen, bee products, and floral contamination [47–49].

Kashmir bee virus (KBV) is a positive-sense ssRNA virus of the Dicistroviridae family within the Cripavirus genus [50,51], considered endemic in North America and Australia [52,53] but rarely reported in Europe [54–58]. It is genetically related to acute bee paralysis virus (ABPV) [59], and the two may co-infect the same colony or the same individual bee [59,60]. Low viral titers are generally detected in subclinical colonies; however, viral replication may be triggered by the presence of stressors, including *A. tumida* infestations [46,52,60], with a lethal outcome for different honey bee stages [59,61,62]. Ingestion of contaminated brood food [49,59,63] and *Varroa* feeding behavior [64–66] may elicit the transmission of KBV infections.

Herein, we aimed to assess the presence of the abovementioned pathogens (*L. passim*, *C. mellificae*, KBV, and DWV) in addition to pathogenic bacteria (*Paenibacillus larvae* and *Melissococcus plutonius*) and other bee viruses (ABPV), Israeli acute paralysis virus (IAPV), black queen cell virus (BQCV), sacbrood virus (SBV), chronic bee paralysis virus (CBPV), and slow paralysis virus (SPV, major and minor)) in SHB specimens collected in Florida, USA in 2017. This is an important first step in determining the role SHBs may plan in the movement of pathogens between honey bee colonies.

2. Results

The investigated samples, coming from the same honey bee colony, tested positive for *C. mellificae*, *L. passim*, KBV, and DWV (Table 1). No amplicons were detected for *P. larvae*, *M. plutonious*, ABPV, IAPV, BQCV, SBV, CBPV, SPV major, and SPV minor in SHB individuals and the pool of SHBs.

Table 1. Summary of the *Aethina tumida* (SHB = small hive beetle) samples that tested positive for a given pathogen with the RT-PCR.

Target	Pool (*n* = 30)	SHB 1	SHB 2	SHB 3	SHB 4	SHB 5	SHB 6	SHB 7	SHB 8	SHB 9	SHB 10
Crithidia. mellificae	POS	-	-	POS	-	POS	-	-	POS	-	-
Lotmaria passim	POS	-	POS	-	POS	-	POS	-	-	POS	-
KBV	POS *	POS *	-	-	-	-	-	-	-	-	POS *
DWV	POS *	POS *	POS *	POS *	-	POS *	POS *	-	POS *	POS *	-

POS: positive; POS *: positive samples with replicative virus forms.

One of the SHB individuals was negative for all pathogens, whereas the other nine tested positive for one or two of them. The SHB pool was positive for both trypanosomatid species and the two virus types.

In the SHB individuals, no significant difference was found in the prevalence between *C. mellificae* and *L. passim* positives (bilateral Fisher's exact test: $p = 0.675$). No co-infections with the two were detected.

The frequencies of DWV- and KBV-positive individuals did not significantly differ (bilateral Fisher's exact test: $p = 0.070$). Viral coinfections were found only in one individual

SHB, representing a significantly lower proportion of the positives (bilateral Fisher's exact test: $p = 0.010$).

A strand-specific PCR demonstrated active viral replication of KBV and DWV in PCR-positive samples. Blast analysis on the sequences obtained from positive amplicons confirmed the specificity of the results, with high similarity (99%) to specific virus genome sequences deposited in GenBank. For each virus, the same sequence was recorded in all positive samples. Phylogenetic analysis and pairwise distance analysis indicated the highest homology to DWV type A (Figure 1).

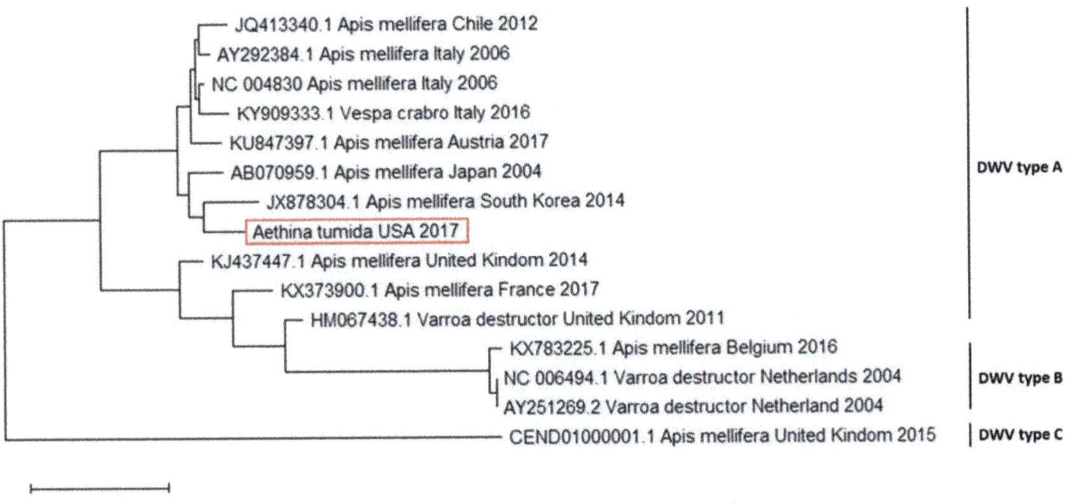

Figure 1. Molecular phylogenetic analysis for RNA-dependent RNA polymerase of deformed wing virus (DWV) using the maximum likelihood method. The evolutionary history was inferred using the maximum likelihood method based on the Tamura–Nei model. The branch lengths of the tree measured the number of substitutions per site. The analysis involved 28 nucleotide sequences. There were 255 positions in the final dataset. Accession number, host, state, and year of available GenBank DWV sequences are shown. DWV sequence accession numbers are reported and associated with year and site of origin and type. The DWV sequence obtained from the tested *Aethina tumida* samples is in a red box.

A similar analysis was conducted for the KBV sequence. A close relationship with sequences found in *A. mellifera* and *V. destructor* from the USA was detected (Figure 2).

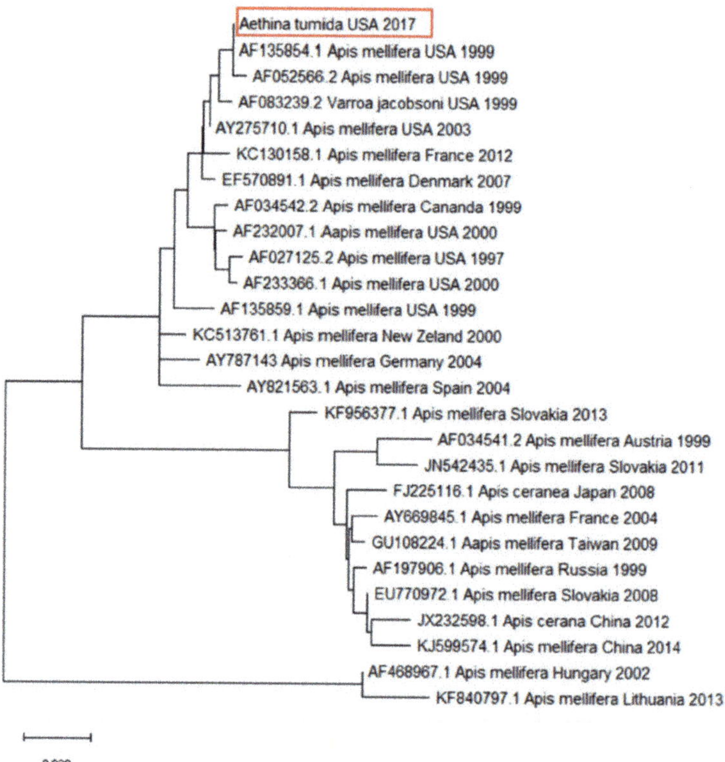

Figure 2. Molecular phylogenetic analysis for RNA-dependent RNA polymerase of Kashmir Bee Virus (KBV) using the maximum likelihood method. The evolutionary history was inferred using the maximum likelihood method based on the Tamura–Nei model. The branch lengths of the tree measured the number of substitutions per site. The analysis involved 35 nucleotide sequences. There were 297 positions in the final dataset. Accession number, host, state, and year of available GenBank KBV sequences are shown. KBV sequence accession numbers are reported and associated with the year and site of origin. The DWV sequence obtained from the tested *Aethina tumida* samples is in a red box.

3. Discussion

To date, only a few instances of individual SHBs bearing bee pathogens have been reported. This is the case for samples from Mexico (positive for *L. passim*, *Apis mellifera* filamentous virus (AmFV), *C. bombi*, *Ascosphera apis*, and *Nosema ceranae* [22]), Florida (positive for *N. ceranae* [20]), and other areas of the USA (positive for DWV, SBV and *P. larvae* [19,20,22,24]).

The present study showed the presence of the honey bee pathogens *L. passim*, *C. mellificae*, DWV, and KBV in SHB adults collected from free-flying colonies in Florida. Furthermore, all SHB samples that were positive for DWV and/or KBV contained replicative viral forms. Although DWV replication in SHB adults is not a new finding [18,23], replication of KBV in SHBs is.

This is not the first time that DWV and KBV have been reported to infect non-Apis hosts. Replicative DWV was found in hornets (Vespa crabro) [67], Asian hornets (V. velutina) [68], and Argentine ants (Linepithema humile) [69]. Replicative KBV has been found in V. velutina [54], Vespula germanica, and Vespula vulgaris [70–72]. However, the current and previous [18,23] detections on coleopterans suggest that DWV and KBV can infect a wide range of potential hosts, thus envisaging a scenario where wild and managed insect species may act as virus reservoirs that fuel reciprocal spillover. Furthermore, the occurrence of replicative DWV and KBV in the same individuals indicates the possibility of viral co-infections in SHBs, as already reported in A. mellifera and other insect species [59,73,74].

The sequence analysis of DWV and KBV resulted in high identity rates to viral sequences identified in *A. mellifera*. The phylogenetic analysis highlighted that the DWV genome detected in the SHB samples belonged to DWV type A, the less virulent genetic variant of this virus [36]. The KBV genome found in the investigated samples bore a close relationship to other KBV outbreaks reported in the USA, thus excluding the involvement of viruses originating from other countries.

The prevalence of *L. passim*-positive individuals detected in this study (40% of all samples) mirrors that reported in a previous survey in which *C. bombi* was also reported in larval SHBs [22]. Additionally, we report for the first time SHB samples positive for *C. mellificae*. Although *C. mellificae* is generally considered less spread than *L. passim* in honey bees [33–35], the prevalence levels of the two trypanosomatids in our individual samples did not significantly differ.

None of the samples was positive for *P. larvae, M. plutonious*, ABPV, IAPV, BQCV, SBV, CBPV, SPV major, or SPV minor. This coincides with the results of previous investigations showing low *P. larvae* [19] and SBV [21] loads in SHB adults. This likely reflects the health of the colonies that were visited by the SHBs prior to sampling.

The finding results highlight the need to clarify pathogen transmission between honey bees and SHB adults better. In the case of DWV, horizontal transmission occurs chiefly by the oral route [18]. In this regard, SHBs are able to trick honey bee adults into feeding them [75,76], possibly acquiring DWV during the exchange of food via trophallaxis. However, the multifaceted host–parasite interaction [77] allows multiple pathways, including oral–oral and fecal–oral transmission. Adult SHBs also may acquire honey bee pathogens by feeding on bee products that are contaminated with multiple microorganism species [41,47,48,78], cannibalizing bee carcasses, or ingesting infected faeces [78–80].

The articulate interactions above and active flying behavior [8] may bring together adult SHBs of different origins that congregate in the same host colony, generating the detected diversity in the pathogen load. On the other hand, the horizontal transmission may occur bi-directionally, as both SHB adults and larvae might defecate inside the hive [77], potentially spreading infected feces that could transmit and perpetuate infective agents within the colony. Infections may also be transmitted vertically. Bee pathogens may be found in SHB larvae [22,23] as consequences of feeding, environmental contamination, and congenital transmission. Nevertheless, the role that SHBs play in the transmission of honey bee pathogens remains unclear.

4. Materials and Methods

4.1. Sample Collection

In summer 2017, one honey bee colony of mixed European origin was selected from an experimental apiary of the University of Florida (Gainesville, FL, USA) based on a conspicuous SHB infestation. No evident signs of other diseases could be detected. Forty SHB adults were randomly sampled alive from the colony combs and hive floor. Once in the laboratory, the collected specimens were randomly separated to compose one pool of thirty adults and ten individual beetle samples.

4.2. Extraction of Total Nucleic Acids

All the SHBs were washed with 95% ethanol to remove possible external microbial contaminants. The ethanol was then allowed to evaporate at room temperature.

A TissueLyser II (Qiagen, Hilden, Germany) was used for 3 min at 25 Hz to crush all SHB samples in separate 2 mL Eppendorf tubes filled to the mark with RNase-free water. The resulting suspensions were then split into two equal aliquots from which nucleic acids were extracted (one for DNA and one for RNA).

DNA and total RNA were extracted with DNeasy Blood & Tissue Kit (Qiagen) and RNeasy Mini Kit (Qiagen) as previously described [20,67]. All samples were eluted in 30 µL DNAase-RNase-free water.

DNA and RNA extracts were stored at −80 °C until analysis. High pure sterile DNA- and RNA-free water was used as a negative control in all analytical steps.

4.3. PCR Assays to Detect Bacteria and Protozoa DNA

The extracted DNA was analyzed by real-time PCR to detect bacteria and trypanosomatids. The primers that were used are reported in Table 2.

For each target gene, a total reaction volume of 15 µL was prepared as previously described [81] using 2x QuantiTect Probe PCR Master Mix (Qiagen), forward and reverse primers (2 µM), forward and reverse probes (500 nM), and 3 µL DNA extract. The real-time PCR assay was performed on a Rotorgene Corbett 6000 (Corbett Research, Sydney, Australia) following the protocols for either gene sequence [54,68]. DNA extracted previously from positive honey bees was used as the positive control for each investigated bacterial and protozoan species.

Table 2. List of primers used to detect bacteria and trypanosomatids in *Aethina tumida*.

Target	Primer Name	Sequence (5′-3′)	Reference
Paenibacillus larvae	AFB-F AFB-R	CTTGTGTTTCTTTCGGGAGACGCCA TCTTAGAGTGCCCACCTCTGCG	[82]
Melissococcus plutonius	MelissoF MelissoR	CAGCTAGTCGGTTTGGTTCC TTGGCTGTAGATAGAATTGACAAT	[83]
Crithida mellificae	Cmel_Cyt_b_F Cmel_Cyt_b_R	TAAATTCACTACCTCAAATTCAATAACATAATCAT ATTTATTGTTGTAATCGGTTTTATTGGATATGT	[84]
Lotmaria passim	Lp2F 459 Lp2R 459	AGGGATATTTAAACCCATCGAA ACCACAAGAGTACGGAATGC	[33]

4.4. PCR Assays to Detect Virus RNA

All RNA extracts were retro-transcribed by M-MLV reverse transcriptase (Invitrogen, Carlsbad, CA, USA) using a blend of oligo-d (T) primers and random hexamers following the manufacturer's instruction. Five microliters of the obtained cDNAs were used as a template for the PCR reactions, performed using HotStarTaqPlus Polymerase Mix (Qiagen). Primers to amplify the viral genomes of the honey bee viruses investigated herein are reported in Table 3. The real-time PCR assay was performed on a Rotorgene Corbett 6000. RNA extracted previously from positive honey bees was used as the positive control for each investigated virus.

Table 3. List of primers used to detect viruses in Aethina *tumida*.

Target	Primer Name	Sequence (5′-3′)	Reference
KBV	KBV 83F KBV 161R	ACCAGGAAGTATTCCCATGGTAAG TGGAGCTATGGTTCCGTTCAG	[85]
DWV	DWV Fw 8450 DWV Rev 8953	TGGCATGCCTTGTTCACCGT CGTGCAGCTCGATAGGATGCCA	[47]
ABPV	APV 95F APV 159R	TCCTATATCGACGACGAAAGACAA GCGCTTTAATTCCATCCAATTGA	[85]
IAPV	IAPV B4S0427_R130M IAPV B4S0427_L17M	RCRTCAGTCGTCTTCCAGGT CGAACTTGGTGACTTGARGG	[86]
BQCV	BQCV 9195F BQCV 8265R	GGTGCGGGAGATGATATGGA GCCGTCTGAGATGCATGAATAC	[85]

Table 3. *Cont.*

Target	Primer Name	Sequence (5′-3′)	Reference
SBV	SBV 311F 79 SBV 380R	AAGTTGGAGGCGCGyAATTG CAAATGTCTTCTTACdAGAGGyAAGGATTG	[85]
CBPV	CPV 304F 79 CPV 371R	TCTGGCTCTGTCTTCGCAAA GATACCGTCGTCACCCTCATG	[85]
SPV major	SPV 8383F 81 SPV 8456R	TGATTGGACTCGGCTTGCTA CAAAATTTGCATAATCCCCAGTT	[59]
SPV minor	SPV Minor F1 SPV Minor R1	ATAGCGCTTTAGTTCAATTGCCAT CTGGAATATGACCATCACGCAT	[38]

KBV: Kashmir bee virus; DWV: Deformed wing virus; ABPV: Acute bee paraylis virus; IAPV: Israeli acute bee paryalis virus; BQCV: Black queen cell virus; SBV: Sac brood virus; CBPV: Chronic bee parylis virus; SPV: slow paralysis virus.

4.5. Strand-Specific RT-PCR

To evaluate the replication of the detected viruses, strand-specific RT-PCRs were performed using specific primers, as previously described [47]. All cDNAs were amplified by PCR for the related viral target. The amplicons were detected on a 2% agarose gel, sequenced (BMR Genomics, Padua, Italy), and analyzed using BLAST [87]. Phylogenetic analysis was performed using the maximum likelihood method based on the Tamura–Nei model using MEGA software [88].

4.6. Statistical Analysis

The prevalence of the individuals that were positive for *C. mellificae* or *L. passim* and of those showing DWV or KBV infections were statistically compared with a bilateral Fisher's exact test under the null hypothesis of equality. The same test was also used to compare the frequency of multiple vs. single viral infections. Due to the small number of samples, the test for independence χ^2 was not used in this case.

5. Conclusions

This investigation suggests that the honey bee trypanosomatids *L. passim* and *C. mellificae* may colonize, and the viruses DWV and KBV successfully infect *A. tumida* adults. Additional studies are needed to determine whether these pathogens generate clinical evidence and signs of infection in SHBs. The horizontal and vertical transmission routes of these pathogens in/between SHBs should also be clarified, as well as the potential, if any, of these pathogens to limit SHB populations in the wild.

Finally, further research is needed to elucidate the epidemiological role that SHBs play in pathogen transmission to honey bees and other insects as a possible dead-end host or vector.

Author Contributions: Conceptualization, G.C.; investigation, G.C.; in-field activity, J.D.E. and I.C.; laboratory activity, G.C.; data curation, G.C. and A.N.; writing—original draft preparation, G.C. and A.N.; writing—review and editing, G.C., J.D.E., I.C. and A.N.; supervision, A.N.; funding acquisition, J.D.E. and A.N. All authors have read and agreed to the published version of the manuscript.

Funding: This research was funded by the Italian Ministry of Agricultural, Food and Forestry Policies, grant number AETHINET ("Monitoraggio e tecniche innovative di diagnosi e di controllo del piccolo coleottero dell'alveare, *Aethina tumida*") (A.N.) and by a cooperative agreement provided by the United States Department of Agriculture, Animal and Plant Health Inspection Service (USDA-ARS), AP17PPQS&T00C079 (J.D.E.).

Institutional Review Board Statement: Not applicable.

Informed Consent Statement: Not applicable.

Acknowledgments: The authors are grateful to Brandi and Branden Stanford of the Entomology and Nematology Department, University of Florida, for their valuable technical support.

Conflicts of Interest: The authors declare no conflict of interest.

References

1. Murray, A. List of Coleoptera received from Old Calabar. *Ann. Mag. Nat. Hist. Lond.* **1867**, *19*, 167–179. [CrossRef]
2. Lundie, A.E. The Small Hive Beetle, Aethina túmida. *Sci. Bull.* **1940**, *220*, 40.
3. Schäfer, M.O.; Cardaio, I.; Cilia, G.; Cornelissen, B.; Crailsheim, K.; Formato, G.; Lawrence, A.K.; Le Conte, Y.; Mutinelli, F.; Nanetti, A.; et al. How to slow the global spread of small hive beetles, *Aethina tumida*. *Biol. Invasions* **2019**, *21*, 1451–1459. [CrossRef]
4. Ellis, J.D.; Hepburn, H.R. An ecological digest of the small hive beetle (*Aethina tumida*), a symbiont in honey bee colonies (*Apis mellifera*). *Insectes Soc.* **2006**, *53*, 8–19. [CrossRef]
5. Al Toufailia, H.; Alves, D.A.; De Bená, D.C.; Bento, J.M.S.; Iwanicki, N.S.A.; Cline, A.R.; Ellis, J.D.; Ratnieks, F.L.W. First record of small hive beetle, *Aethina tumida* Murray, in South America. *J. Apic. Res.* **2017**, *56*, 76–80. [CrossRef]
6. Lee, S.; Hong, K.-J.; Cho, Y.S.; Choi, Y.S.; Yoo, M.-S.; Lee, S. Review of the subgenus Aethina Erichson s. str. (Coleoptera: Nitidulidae: Nitidulinae) in Korea, reporting recent invasion of small hive beetle, *Aethina tumida*. *J. Asia. Pac. Entomol.* **2017**, *20*, 553–558. [CrossRef]
7. Neumann, P.; Pettis, J.S.; Schäfer, M.O. Quo vadis *Aethina tumida*? Biology and control of small hive beetles. *Apidologie* **2016**, *47*, 427–466. [CrossRef]
8. Boncristiani, H.; Ellis, J.D.; Bustamante, T.; Graham, J.; Jack, C.; Kimmel, C.B.; Mortensen, A.; Schmehl, D.R. World Honey Bee Health: The Global Distribution of Western Honey Bee (*Apis mellifera* L.) Pests and Pathogens. *Bee World* **2021**, *98*, 2–6. [CrossRef]
9. Granato, A.; Zecchin, B.; Baratto, C.; Duquesne, V.; Negrisolo, E.; Chauzat, M.-P.; Ribière-Chabert, M.; Cattoli, G.; Mutinelli, F. Introduction of *Aethina tumida* (Coleoptera: Nitidulidae) in the regions of Calabria and Sicily (southern Italy). *Apidologie* **2017**, *48*, 194–203. [CrossRef]
10. da Silva, M.J.V. The First Report Of *Aethina tumida* In The European Union, Portugal, 2004. *Bee World* **2014**, *91*, 90–91. [CrossRef]
11. Palmeri, V.; Scirtò, G.; Malacrinò, A.; Laudani, F.; Campolo, O. A scientific note on a new pest for European honeybees: First report of small hive beetle *Aethina tumida*, (Coleoptera: Nitidulidae) in Italy. *Apidologie* **2015**, *46*, 527–529. [CrossRef]
12. Murilhas, A.M. *Aethina tumida* arrives in Portugal. Will it be eradicated? *EurBee Newsl.* **2004**, *2*, 7–9.
13. Arbogast, R.T.; Torto, B.; Willms, S.; Teal, P.E.A. Trophic habits of *Aethina tumida* (Coleoptera: Nitidulidae): Their adaptive significance and relevance to dispersal. *Environ. Entomol.* **2009**, *38*, 561–568. [CrossRef]
14. McMenamin, A.J.; Genersch, E. Honey bee colony losses and associated viruses. *Curr. Opin. Insect Sci.* **2015**, *8*, 121–129. [CrossRef]
15. Brutscher, L.M.; McMenamin, A.J.; Flenniken, M.L. The Buzz about Honey Bee Viruses. *PLoS Pathog.* **2016**, *12*, e1005757. [CrossRef]
16. Grupe, A.C.; Quandt, C.A. A growing pandemic: A review of Nosema parasites in globally distributed domesticated and native bees. *PLoS Pathog.* **2020**, *16*, e1008580. [CrossRef]
17. Genersch, E.; Evans, J.D.; Fries, I. Honey bee disease overview. *J. Invertebr. Pathol.* **2010**, *103*, S2–S4. [CrossRef]
18. Eyer, M.; Chen, Y.P.; Schäfer, M.O.; Pettis, J.; Neumann, P. Small hive beetle, *Aethina tumida*, as a potential biological vector of honeybee viruses. *Apidologie* **2009**, *40*, 419–428. [CrossRef]
19. Schäfer, M.O.; Ritter, W.; Pettis, J.; Neumann, P. Small hive beetles, *Aethina tumida*, are vectors of *Paenibacillus larvae*. *Apidologie* **2010**, *41*, 14–20. [CrossRef]
20. Cilia, G.; Cardaio, I.; dos Santos, P.E.J.; Ellis, J.D.; Nanetti, A. The first detection of Nosema ceranae (Microsporidia) in the small hive beetle, *Aethina tumida* Murray (Coleoptera: Nitidulidae). *Apidologie* **2018**, *49*, 619–624. [CrossRef]
21. Eyer, M.; Chen, Y.P.; Schäfer, M.O.; Pettis, J.S.; Neumann, P. Honey bee sacbrood virus infects adult small hive beetles, *Aethina tumida* (Coleoptera: Nitidulidae). *J. Apic. Res.* **2009**, *48*, 296–297. [CrossRef]
22. de Landa, G.F.; Porrini, M.P.; Revainera, P.; Porrini, D.P.; Farina, J.; Correa-Benítez, A.; Maggi, M.D.; Eguaras, M.J.; Quintana, S. Pathogens Detection in the Small Hive Beetle (*Aethina tumida* (Coleoptera: Nitidulidae)). *Neotrop. Entomol.* **2020**, 1–5. [CrossRef]
23. Huwiler, M.; Papach, A.; Cristina, E.; Yañez, O.; Williams, G.R.; Neumann, P. Deformed wings of small hive beetle independent of virus infections and mites. *J. Invertebr. Pathol.* **2020**, *172*, 107365. [CrossRef]
24. Arismendi, N.; Caro, S.; Castro, M.P.; Vargas, M.; Riveros, G.; Venegas, T. Impact of Mixed Infections of Gut Parasites Lotmaria passim and Nosema ceranae on the Lifespan and Immune-related Biomarkers in *Apis mellifera*. *Insects* **2020**, *11*, 420. [CrossRef]
25. Schwarz, R.S.; Bauchan, G.R.; Murphy, C.A.; Ravoet, J.; de Graaf, D.C.; Evans, J.D. Characterization of Two Species of Trypanosomatidae from the Honey Bee *Apis mellifera*: *Crithidia mellificae* Langridge and McGhee, and *Lotmaria passim* n. gen., n. sp. *J. Eukaryot. Microbiol.* **2015**, *62*, 567–583. [CrossRef]
26. Ruiz- Gonzalez, M.X.; Brown, M.J.F. Honey bee and bumblebee trypanosomatids: Specificity and potential for transmission. *Ecol. Entomol.* **2006**, *31*, 616–622. [CrossRef]
27. Runckel, C.; DeRisi, J.; Flenniken, M.L. A Draft Genome of the Honey Bee Trypanosomatid Parasite Crithidia mellificae. *PLoS ONE* **2014**, *9*, e95057. [CrossRef] [PubMed]
28. Runckel, C.; Flenniken, M.L.; Engel, J.C.; Ruby, J.G.; Ganem, D.; Andino, R.; DeRisi, J.L. Temporal analysis of the honey bee microbiome reveals four novel viruses and seasonal prevalence of known viruses, Nosema, and Crithidia. *PLoS ONE* **2011**, *6*, e020656. [CrossRef] [PubMed]

29. Ravoet, J.; Maharramov, J.; Meeus, I.; De Smet, L.; Wenseleers, T.; Smagghe, G.; de Graaf, D.C. Comprehensive Bee Pathogen Screening in Belgium Reveals Crithidia mellificae as a New Contributory Factor to Winter Mortality. *PLoS ONE* **2013**, *8*, e72443. [CrossRef] [PubMed]

30. Schwarz, R.S.; Evans, J.D. Single and mixed-species trypanosome and microsporidia infections elicit distinct, ephemeral cellular and humoral immune responses in honey bees. *Dev. Comp. Immunol.* **2013**, *40*, 300–310. [CrossRef]

31. Strobl, V.; Yañez, O.; Straub, L.; Albrecht, M.; Neumann, P. Trypanosomatid parasites infecting managed honeybees and wild solitary bees. *Int. J. Parasitol.* **2019**, *49*, 605–613. [CrossRef]

32. Buendía, M.; Martín-Hernández, R.; Ornosa, C.; Barrios, L.; Bartolomé, C.; Higes, M. Epidemiological study of honeybee pathogens in Europe: The results of Castilla-La Mancha (Spain). *Spanish J. Agric. Res.* **2018**, *16*, e0502. [CrossRef]

33. Arismendi, N.; Bruna, A.; Zapata, N.; Vargas, M. PCR-specific detection of recently described Lotmaria passim (Trypanosomatidae) in Chilean apiaries. *J. Invertebr. Pathol.* **2016**, *134*, 1–5. [CrossRef]

34. Vargas, M.; Arismendi, N.; Riveros, G.; Zapata, N.; Bruna, A.; Vidal, M.; Rodríguez, M.; Gerding, M. Viral and intestinal diseases detected in *Apis mellifera* in central and southern Chile. *Chil. J. Agric. Res.* **2017**, *77*, 243–249. [CrossRef]

35. Castelli, L.; Branchiccela, B.; Invernizzi, C.; Tomasco, I.; Basualdo, M.; Rodriguez, M.; Zunino, P.; Antúnez, K. Detection of Lotmaria passim in Africanized and European honey bees from Uruguay, Argentina and Chile. *J. Invertebr. Pathol.* **2019**, *160*, 95–97. [CrossRef]

36. de Miranda, J.R.; Genersch, E. Deformed wing virus. *J. Invertebr. Pathol.* **2010**, *103*, S48–S61. [CrossRef] [PubMed]

37. Genersch, E.; Aubert, M. Emerging and re-emerging viruses of the honey bee (*Apis mellifera* L.). *Vet. Res.* **2010**, *41*, 54. [CrossRef]

38. Martin, S.J.; Highfield, A.C.; Brettell, L.; Villalobos, E.M.; Budge, G.E.; Powell, M.; Nikaido, S.; Schroeder, D.C. Global honey bee viral landscape altered by a parasitic mite. *Science* **2012**, *336*, 1304–1306. [CrossRef]

39. Mordecai, G.J.; Wilfert, L.; Martin, S.J.; Jones, I.M.; Schroeder, D.C. Diversity in a honey bee pathogen: First report of a third master variant of the Deformed Wing Virus quasispecies. *ISME J.* **2016**, *10*, 1264–1273. [CrossRef]

40. McMahon, D.P.; Natsopoulou, M.E.; Doublet, V.; Fürst, M.; Weging, S.; Brown, M.J.F.; Gogol-Döring, A.; Paxton, R.J. Elevated virulence of an emerging viral genotype as a driver of honeybee loss. *Proc. Biol. Sci.* **2016**, *283*, 20160811. [CrossRef]

41. Yue, C.; Schroder, M.; Gisder, S.; Genersch, E. Vertical-transmission routes for deformed wing virus of honeybees (*Apis mellifera*). *J. Gen. Virol.* **2007**, *88*, 2329–2336. [CrossRef] [PubMed]

42. Ball, B.V.; Allen, M.F. The prevalence of pathogens in honey bee (*Apis mellifera*) colonies infested with the parasitic mite Varroa jacobsoni. *Ann. Appl. Biol.* **1988**, *113*, 237–244. [CrossRef]

43. Nordström, S. Distribution of deformed wing virus within honey bee (*Apis mellifera*) brood cells infested with the ectoparasitic mite Varroa destructor. *Exp. Appl. Acarol.* **2003**, *29*, 293–302. [CrossRef] [PubMed]

44. Shen, M.; Cui, L.; Ostiguy, N.; Cox-Foster, D. Intricate transmission routes and interactions between picorna-like viruses (Kashmir bee virus and sacbrood virus) with the honeybee host and the parasitic varroa mite. *J. Gen. Virol.* **2005**, *86*, 2281–2289. [CrossRef]

45. Lanzi, G.; de Miranda, J.R.; Boniotti, M.B.; Cameron, C.E.; Lavazza, A.; Capucci, L.; Camazine, S.M.; Rossi, C. Molecular and biological characterization of deformed wing virus of honeybees (*Apis mellifera* L.). *J. Virol.* **2006**, *80*, 4998–5009. [CrossRef]

46. Gisder, S.; Aumeier, P.; Genersch, E. Deformed wing virus: Replication and viral load in mites (Varroa destructor). *J. Gen. Virol.* **2009**, *90*, 463–467. [CrossRef]

47. Mazzei, M.; Carrozza, M.L.; Luisi, E.; Forzan, M.; Giusti, M.; Sagona, S.; Tolari, F.; Felicioli, A. Infectivity of DWV Associated to Flower Pollen: Experimental Evidence of a Horizontal Transmission Route. *PLoS ONE* **2014**, *9*, e113448. [CrossRef]

48. Mockel, N.; Gisder, S.; Genersch, E. Horizontal transmission of deformed wing virus: Pathological consequences in adult bees (*Apis mellifera*) depend on the transmission route. *J. Gen. Virol.* **2011**, *92*, 370–377. [CrossRef]

49. Chen, Y.; Evans, J.; Feldlaufer, M. Horizontal and vertical transmission of viruses in the honey bee, *Apis mellifera*. *J. Invertebr. Pathol.* **2006**, *92*, 152–159. [CrossRef]

50. de Miranda, J.R.; Cordoni, G.; Budge, G. The Acute bee paralysis virus–Kashmir bee virus–Israeli acute paralysis virus complex. *J. Invertebr. Pathol.* **2010**, *103*, S30–S47. [CrossRef]

51. Valles, S.M.; Strong, C.A.; Oi, D.H.; Porter, S.D.; Pereira, R.M.; Vander Meer, R.K.; Hashimoto, Y.; Hooper-Bùi, L.M.; Sánchez-Arroyo, H.; Davis, T.; et al. Phenology, distribution, and host specificity of Solenopsis invicta virus-1. *J. Invertebr. Pathol.* **2007**, *96*, 18–27. [CrossRef] [PubMed]

52. Berenyi, O.; Bakonyi, T.; Derakhshifar, I.; Koglberger, H.; Nowotny, N. Occurrence of Six Honeybee Viruses in Diseased Austrian Apiaries. *Appl. Environ. Microbiol.* **2006**, *72*, 2414–2420. [CrossRef] [PubMed]

53. Tentcheva, D.; Gauthier, L.; Zappulla, N.; Dainat, B.; Cousserans, F.; Colin, M.E.; Bergoin, M. Prevalence and seasonal variations of six bee viruses in *Apis mellifera* L. and Varroa destructor mite populations in France. *Appl. Environ. Microbiol.* **2004**, *70*, 7185–7191. [CrossRef] [PubMed]

54. Mazzei, M.; Cilia, G.; Forzan, M.; Lavazza, A.; Mutinelli, F.; Felicioli, A. Detection of replicative Kashmir Bee Virus and Black Queen Cell Virus in Asian hornet Vespa velutina (Lepelieter 1836) in Italy. *Sci. Rep.* **2019**, *9*, 1–9. [CrossRef]

55. Cersini, A.; Bellucci, V.; Lucci, S.; Mutinelli, F.; Granato, A.; Porrini, C.; Felicioli, A.; Formato, G. First isolation of Kashmir bee virus (KBV) in Italy. *J. Apic. Res.* **2013**, *52*, 54–55. [CrossRef]

56. Porrini, C.; Mutinelli, F.; Bortolotti, L.; Granato, A.; Laurenson, L.; Roberts, K.; Gallina, A.; Silvester, N.; Medrzycki, P.; Renzi, T.; et al. The Status of Honey Bee Health in Italy: Results from the Nationwide Bee Monitoring Network. *PLoS ONE* **2016**, *11*, e0155411. [CrossRef]

57. Siede, R.; Büchler, R. First detection of Kashmir bee virus in Hesse, Germany. *Berl. Munch. Tierarztl. Wochenschr.* **2004**, *117*, 12–15.

58. Ward, L.; Waite, R.; Boonham, N.; Fisher, T.; Pescod, K.; Thompson, H.; Chantawannakul, P.; Brown, M. First detection of Kashmir bee virus in the UK using real-time PCR. *Apidologie* **2007**, *38*, 181–190. [CrossRef]

59. de Miranda, J.R.; Dainat, B.; Locke, B.; Cordoni, G.; Berthoud, H.; Gauthier, L.; Neumann, P.; Budge, G.E.; Ball, B.V.; Stoltz, D.B. Genetic characterization of slow bee paralysis virus of the honeybee (*Apis mellifera* L.). *J. Gen. Virol.* **2010**, *91*, 2524–2530. [CrossRef]

60. Evans, J.D. Genetic Evidence for Coinfection of Honey Bees by Acute Bee Paralysis and Kashmir Bee Viruses. *J. Invertebr. Pathol.* **2001**, *78*, 189–193. [CrossRef]

61. Pettis, J.; Van Engelsdorp, D.; Cox-Foster, D. Colony collapse disorder working group pathogen sub-group progress report. *Am. Bee J.* **2007**, *103*, 595–597.

62. Todd, J.H.; De Miranda, J.R.; Ball, B.V. Incidence and molecular characterization of viruses found in dying New Zealand honey bee (*Apis mellifera*) colonies infested with *Varroa destructor*. *Apidologie* **2007**, *38*, 354–367. [CrossRef]

63. Chen, Y.P.; Siede, R. Honey Bee Viruses. *Adv. Virus Res.* **2007**, *70*, 33–80.

64. Shen, M.; Yang, X.; Cox-Foster, D.; Cui, L. The role of varroa mites in infections of Kashmir bee virus (KBV) and deformed wing virus (DWV) in honey bees. *Virology* **2005**, *342*, 141–149. [CrossRef]

65. Hung, A.C.F. PCR detection of Kashmir bee virus in honey bee excreta. *J. Apic. Res.* **2000**, *39*, 103–106. [CrossRef]

66. Hung, A.C.F.; Shimanuki, H. A scientific note on the detection of Kashmir bee virus in individual honeybees and Varroa jacobsoni mites. *Apidologie* **1999**, *30*, 353–354. [CrossRef]

67. Forzan, M.; Sagona, S.; Mazzei, M.; Felicioli, A. Detection of deformed wing virus in Vespa crabro. *Bull. Insectology* **2017**, *70*, 261–265.

68. Mazzei, M.; Forzan, M.; Cilia, G.; Sagona, S.; Bortolotti, L.; Felicioli, A. First detection of replicative deformed wing virus (DWV) in Vespa velutina nigrithorax. *Bull. Insectology* **2018**, *71*, 211–216.

69. Sébastien, A.; Lester, P.J.; Hall, R.J.; Wang, J.; Moore, N.E.; Gruber, M.A.M. Invasive ants carry novel viruses in their new range and form reservoirs for a honeybee pathogen. *Biol. Lett.* **2015**, *11*, 20150610. [CrossRef]

70. Anderson, D.L. Kashmir bee virus: A relatively harmless virus of honey bee colonies. *Am. Bee J.* **1991**, *131*, 767–771.

71. Quinn, O.; Gruber, M.A.M.; Brown, R.L.; Baty, J.W.; Bulgarella, M.; Lester, P.J. A metatranscriptomic analysis of diseased social wasps (Vespula vulgaris) for pathogens, with an experimental infection of larvae and nests. *PLoS ONE* **2018**, *13*, e0209589. [CrossRef] [PubMed]

72. Singh, R.; Levitt, A.L.; Rajotte, E.G.; Holmes, E.C.; Ostiguy, N.; vanEngelsdorp, D.; Lipkin, W.I.; dePamphilis, C.W.; Toth, A.L.; Cox-Foster, D.L. RNA Viruses in Hymenopteran Pollinators: Evidence of Inter-Taxa Virus Transmission via Pollen and Potential Impact on Non-Apis Hymenopteran Species. *PLoS ONE* **2010**, *5*, e14357. [CrossRef] [PubMed]

73. Evans, J.D.; Schwarz, R.S. Bees brought to their knees: Microbes affecting honey bee health. *Trends Microbiol.* **2011**, *19*, 614–620. [CrossRef] [PubMed]

74. Hung, A.C.; Adams, J.R.; Shimanuki, H. Bee parasitic mite syndrome. (II). The role of Varroa mite and viruses. *Am. Bee J.* **1995**, *135*, 702–704.

75. Ellis, J.D.; Pirk, C.W.W.; Hepburn, H.R.; Kastberger, G.; Elzen, P.J. Small hive beetles survive in honeybee prisons by behavioural mimicry. *Naturwissenschaften* **2002**, *89*, 326–328. [CrossRef]

76. Ellis, J.D. Reviewing the confinement of small hive beetles (*Aethina tumida*) by western honey bees (*Apis mellifera*). *Bee World* **2005**, *86*, 56–62. [CrossRef]

77. Cuthbertson, A.G.S.; Wakefield, M.E.; Powell, M.E.; Marris, G.; Anderson, H.; Budge, G.E.; Mathers, J.J.; Blackburn, L.F.; Brown, M.A. The small hive beetle *Aethina tumida*: A review of its biology and control measures. *Curr. Zool.* **2013**, *59*, 644–653. [CrossRef]

78. Chen, Y.P.; Pettis, J.S.; Collins, A.; Feldlaufer, M.F. Prevalence and Transmission of Honeybee Viruses. *Appl. Environ. Microbiol.* **2006**, *72*, 606–611. [CrossRef]

79. Gusachenko, O.N.; Woodford, L.; Balbirnie-Cumming, K.; Ryabov, E.V.; Evans, D.J. Evidence for and against deformed wing virus spillover from honey bees to bumble bees: A reverse genetic analysis. *Sci. Rep.* **2020**, *10*, 16847. [CrossRef]

80. Felicioli, A.; Forzan, M.; Sagona, S.; D'Agostino, P.; Baido, D.; Fronte, B.; Mazzei, M. Effect of oral administration of 1,3-1,6 β-glucans in DWV naturally infected newly emerged bees (*Apis mellifera* L.). *Vet. Sci.* **2020**, *7*, 52. [CrossRef]

81. Cilia, G.; Garrido, C.; Bonetto, M.; Tesoriero, D.; Nanetti, A. Effect of Api-Bioxal® and ApiHerb® Treatments against Nosema ceranae Infection in *Apis mellifera* Investigated by Two qPCR Methods. *Vet. Sci.* **2020**, *7*, 125. [CrossRef]

82. Dobbelaere, W.; de Graaf, D.C.; Peeters, J.E. Development of a fast and reliable diagnostic method for American foulbrood disease (*Paenibacillus larvae* subsp. *larvae*) using a 16S rRNA gene based PCR. *Apidologie* **2001**, *32*, 363–370. [CrossRef]

83. Roetschi, A.; Berthoud, H.; Kuhn, R.; Imdorf, A. Infection rate based on quantitative real-time PCR of Melissococcus plutonius, the causal agent of European foulbrood, in honeybee colonies before and after apiary sanitation. *Apidologie* **2008**, *39*, 362–371. [CrossRef]

84. Xu, G.; Palmer-Young, E.; Skyrm, K.; Daly, T.; Sylvia, M.; Averill, A.; Rich, S. Triplex real-time PCR for detection of Crithidia mellificae and Lotmaria passim in honey bees. *Parasitol. Res.* **2018**, *117*, 623–628. [CrossRef]

85. Chantawannakul, P.; Ward, L.; Boonham, N.; Brown, M. A scientific note on the detection of honeybee viruses using real-time PCR (TaqMan) in Varroa mites collected from a Thai honeybee (*Apis mellifera*) apiary. *J. Invertebr. Pathol.* **2006**, *91*, 69–73. [CrossRef]

86. Kajobe, R.; Marris, G.; Budge, G.; Laurenson, L.; Cordoni, G.; Jones, B.; Wilkins, S.; Cuthbertson, A.G.S.; Brown, M.A. First molecular detection of a viral pathogen in Ugandan honey bees. *J. Invertebr. Pathol.* **2010**, *104*, 153–156. [CrossRef]

87. Altschul, S.F.; Gish, W.; Miller, W.; Myers, E.W.; Lipman, D.J. Basic local alignment search tool. *J. Mol. Biol.* **1990**, *215*, 403–410. [CrossRef]
88. Kumar, S.; Stecher, G.; Li, M.; Knyaz, C.; Tamura, K. MEGA X: Molecular Evolutionary Genetics Analysis across Computing Platforms. *Mol. Biol. Evol.* **2018**, *35*, 1547–1549. [CrossRef]

MDPI
St. Alban-Anlage 66
4052 Basel
Switzerland
Tel. +41 61 683 77 34
Fax +41 61 302 89 18
www.mdpi.com

Pathogens Editorial Office
E-mail: pathogens@mdpi.com
www.mdpi.com/journal/pathogens